W9-BNN-698

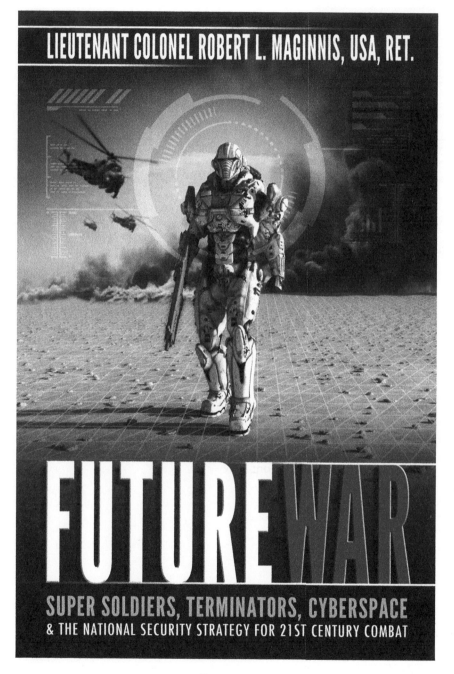

LIEUTENANT COLONEL ROBERT L. MAGINNIS, USA, RET.

FUTURE WAR

SUPER SOLDIERS, TERMINATORS, CYBERSPACE
& THE NATIONAL SECURITY STRATEGY FOR 21ST CENTURY COMBAT

Defender

Crane, MO

Future War: Super-soldie̶   e National
Security Strategy for 21st-
by Robert L. Maginnis

Defender
Crane, MO 65633
©2016 Defender Publishing
All rights reserved. Published 2016.
Printed in the United States of America.

ISBN: 978-0-9964095-7-5
A CIP catalog record of this book is available from the Library of Congress.
Cover illustration and design by Jeffrey Mardis.

Scripture quotations are from the following versions, which are abbreviated as noted.
Berean Study Bible: BSB
English Standard Version: ESV
International Standard Version: ISV
King James Version: KJV
New American Standard Version: NASV
New International Version: NIV
New King James Version: NKJV
New Living Translation: NLT

Acknowledgments

I gratefully acknowledge...

I owe special mention and thanks to my wife Jan for putting up with my absence to research and write this important volume. It is a sacrifice for both of us and I'm appreciative to the Lord that he gave me a Proverbs 31 wife.

I'm also indebted to Don Mercer, a dear brother in Christ who gave up many hours to edit my work, encourage me, and provide many very wise recommendations. He is a true blessing!

Further, I appreciate Tom Horn for trusting me to write on such a tough issue. He understands the times in which we live and he is a true man of God.

I prayed throughout the process of writing this book asking for God's guidance, and I know that if this work helps even one person it is because He gave me the inspiration and wisdom to boldly proclaim the truth. Praise the Lord!

Robert Lee Maginnis
Woodbridge, Virginia

Dedication

This volume is dedicated to America's Christian founders who stayed true to their faith in spite of great opposition. May similar men and women be raised up today to bring America back to God through faith in Jesus Christ. Amen!

Contents

Contents

Foreword

HOPE FOR AMERICA is dangling on a thread as we face dual wars from within our culture and an exploding set of assaults from a frightening array of outside enemies. No wonder more Americans than ever believe our once-great country is spinning out of control, racing at Mach speed to the prophetic end times.

Future War provides a glimpse into our future world with the help of our intelligence community's portrayal of a frightening new reality full of "black swans" (unpredictable events), world and domestic mega-trends, sobering security threats, and new realities built around a very strange set of emerging global coalitions of state, non-state, and other power brokers.

Future War exposes what only the sci-fi community would dare to imagine: "cyborgized" super-soldiers, cyber threats beyond our wildest imagination, sobering outer-space developments, and the most horrendous weapons of mass destruction in the hands of more actors than ever. It demonstrates how each of these sci-fi-like developments plays a role in the coming prophetic end times.

Future War concludes with a glimmer of hope for our troubled and lost world. It provides a detailed plan of action to guide America's future governments and Congress. Those plans have a biblical worldview foundation that tackles five of the most daunting macro challenges facing America today.

Every concerned American and especially Christians must study *Future War* to understand our frightening future, the sobering threats each of us will face, and how especially Bible-believing Christians must get involved in responding to their obligation to engage the world (Romans 13:1–7).

Introduction

Spinning-out-of-Control America

AMERICANS SENSE their world is spinning out of control because hope is dangling on a thread, physical and economic dangers abound, and our feckless leaders are pushing us at warp speed in the wrong direction over a cliff. No wonder there is so much talk about the biblical end times!

America has been on the edge of the precipice at least two other times in its history. Our forefathers faced significant trepidation in 1861 as the nation risked ending the American experiment by launching into a civil war. Secondly, our great-grandparents faced potential national disaster in the mid-1930s through the early 1940s as they simultaneously suffered the devastating effects of the Great Depression and faced global armed threats from the rampaging imperial Japanese in Asia and the German Nazi campaigns in Eastern Europe. At that time, hope understandably dangled on a very thin thread, and for many, the American spirit was drained almost to empty. Our forefathers were desperate for sound leadership and renewed hope.

Eventually and thankfully, strong leaders did rise to the occasion and American hope was restored—but not without steep costs. President

Abraham Lincoln rose to the cry of his countrymen to save the republic from its worst-ever internal threat that took four years to resolve and claimed 2 percent of the population, an estimated 620,000 lives, and simultaneously emptied America's coffers.[1] Years later, President Franklin Roosevelt rallied the nation to defeat economic disaster as well as the military tyrants in both Europe and Asia. Thankfully, our response to the Second World War pulled America out of the economic pit at a very steep cost of 407,000 lives and $341 billion in 1945 dollars—or more than $3.5 trillion in current dollars.[2]

Are the present crises really different than those faced by our forefathers who weathered the Civil War and the Great Depression-Second World War period? Yes, and this is why: Today, our enemies are both among us and across the world. We are battling for the very heart of America as well as for our literal survival in an incredibly dangerous new world order.

America's cultural war is for the heart of America. At stake is nothing less than the undoing of the results of the American Revolution, which won our liberty from Great Britain's tyranny and our religious freedom, said Dennis Prager, a politically conservative, syndicated radio talk-show host who moderated a seminar on "How to Save America" at the Conservative Political Action Conference in 2016.[3] "The American Revolution's ideal of [a] small and limited government and a great citizenry are being undermined by the left," said Prager. The political right represents what the American Revolution gave us, which Prager defines as the American "trinity": liberty, "In God We Trust," and *e pluribus unum.*

The left is undermining all three aspects of the American "trinity," and time is short if we have any hope to restore our country.

First, the left prefers material equality to liberty. Seventy percent of millennials (those born in the early 1980s to the early 2000s), according to Pew Research, "do not believe in freedom of speech if the speech will hurt somebody's feelings," Prager explained. After fifty years of left-wing cultural indoctrination, Americans no longer believe in freedom of speech. Rather, the vast majority of Americans today "believe in the right not to be offended."[4]

The second aspect of the American value system—the "trinity"—is "In God We Trust." "This country was founded on the belief that freedom is only possible and small government is only possible if people are accountable to a morally judging God," Prager, a Jew, said. The origins of America were "God centered because either God gives us rights or people give us rights and there is all the difference in the world."[5]

The third part of the American "trinity" is our national motto, *e pluribus unum*, which means "from many one." That is being undermined by the left's doctrine of multiculturalism. We are no longer all Americans—the so-called melting pot of humans from all peoples, races, languages, religions, and ancestries. No, today we are no longer just Americans but African-Americans, Hispanic-Americans, or Native-Americans, as Prager interprets the left's new culture. That was not the way the country was founded, but that's what the left has made of contemporary America.[6]

"That's the battle in a nutshell," Prager said, and that's the cancer now destroying America from within.[7]

It's bad enough that we face a significant cultural war for America's soul, but our spinning-out-of-control feeling is also attributed to a host of other domestic and international challenges.

Economically, the world is riding a roller coaster of confidence with very volatile equity markets such as in China, the beginning of what might be the European Union's implosion, massive American federal debt that exceeds $19 trillion, weak currencies, and high unemployment and underemployment. Juxtapose these pocketbook uncertainties with a panoply of gut-wrenching challenges: intrusive and corrupt government fueled by misguided and bloated bureaucracies; massive global migration; deadly epidemics like Ebola; genocide among Christians and other minorities in places like the Middle East; almost constant state-on-state conflicts on multiple continents; non-state violent actors driven by radical Islamic ideology creating fear at every level on every continent; the rapidly escalating proliferation of weapons of mass destruction (radiological, nuclear, chemical, and biological) in the hands of rogue regimes like Iran; and the terrifying cyber threats to our critical infrastructure that could shut down the electricity in our homes and empty our bank accounts.

Compounding American trepidation and a sense that the nation is spinning out of control is the collapse of traditional life. Traditional marriage and family are rarer today than ever before; children suffer the results of massive fatherlessness. Society judges both pre-born and aged lives as totally disposable if not an unnecessary burden. Meanwhile, the pop culture pushes radical views about relationships like "homosexual marriage" and defines sex by personal preference, not by one's genitalia. The cultural "elite" insist our young children must embrace transgender dressing rooms, and radical ideologies are foisted involuntarily on our children in their public schools.

American mass media adds to the sense of a lack of control by sensationalizing what years ago were abnormal relationships, but are now the new norm—such as "my two dads or my two moms"; further, American "entertainment" media is as raunchy as Europe's anything-goes programming. The nation's pop music is incredibly misogynistic, filled with vulgarities and virtually every debauchery known to man. As a result, our society has become incredibly coarsened, disgusting, self-absorbed, and dehumanized.

The 2016 U.S. presidential campaign evidenced a polarized America like never before; accusatory rhetoric was at an all-time high; and there seemed to be a lack of worthy candidates with relevant and trustworthy leadership experience. The U.S. Congress enjoys the lowest level of the public's confidence in history, and there is shrinking regard for the presidency, according to polls.[8] One can't help but wonder whether the divide can be bridged ever again.

High technology contributes to the feeling of a lack of control as well. Our lives are no doubt made easier by high-tech machines, but as a result, we have become very hurried and lonely. We are becoming more insular, almost empty, human beings—thanks to machines and our frenetic lifestyles. Meanwhile, transhumanism, a value-free and high-tech scientific movement that is embraced by progressives, is going mainstream and harnessing genomic technology promising to make us smarter, live longer, and be physically stronger. And now, thanks to modern science, our comic book superheroes are becoming reality as the Pentagon and our adversaries like China are on the verge of creating super-soldiers vis-a-vis the fictional Iron Man.

Couple these modern challenges and new-age realities with the inevitable disturbing and seemly more frequent natural disasters—earthquakes, tsunamis, hurricanes, tornadoes, floods, and fires—that create a soreness in the pit of our souls.

Thanks to modern media, our every waking hour is filled with moment-by-moment reports of a potpourri of crises; high-tech, social value shifts; and political shenanigans that contribute to our anxiety and to the belief that in fact our dizzy world is spinning out of control. That explains in part (and justifiably so) why some people, especially conservative Christians steeped in eschatology, conclude we are rushing headlong into the biblical end times.

Is there hope that this spinning will stop and globalized chaos will recede? Yes, but it will take strong visionary leaders and a comprehensive strategy. America once again needs wise leaders who share a hopeful moral vision for our future country, leaders who are willing to make hard decisions like Lincoln and Roosevelt did in their troubled times. Failing the emergence of strong moral leadership and the formulation of a comprehensive, twenty-first century, all-American strategy of hope, this country is doomed to just muddle through and diminish in all the ways it was once great. Or, as many who follow biblical eschatology believe, all these crises might just be building to the end times.

Future War addresses the issue of the unhinged America in three sections, including a restoration plan for the twenty-first century America.

The first section demonstrates that most citizens perceive the United States is spinning out of control, and a major reason for that widespread view appears to be the nation's cultural implosion—declining liberty, no trust in God, and multiculturalism—that is fanned by a leadership failure on critical domestic issues as well as overseas, where our security interests are challenged at every turn.

This section anticipates the United States' future will get much worse than it is today, which leads us to examine the characteristics and influencers of our future alarming world: global megatrends, potential game-changers that might influence those trends, the impact of the inevitable catastrophic events—black swans—and likely alternative worlds that may emerge from the anticipated chaos.

The first section concludes by identifying specific challenges that must be immediately resolved. Then we examine the type of leader best prepared to tackle those challenges in a spinning-out-of-control America.

Section II acknowledges that the foundation of our spinning-out-of-control condition is the fragmenting of our institutions and our morality, which explains why many biblical-worldview Christians believe we are rushing toward the prophetic end times. Then we anticipate the logical progression of this future chaotic state to examine a dangerous mix of frightening actors and emerging sci-fi-like technologies that create a sobering mixture of global threats to our collective security: a new, nonpolar geopolitical security environment; a morphing, increasingly dangerous global terrorist threat; a transhumanist technological revolution that threatens to remake man into Frankenstein-like monsters, including a terrifying cyborg warrior class for robotic-like war-making; the reality of weaponizing outer space; the cyberworld's threats to control every aspect of our lives; and the growing near-term threat of catastrophic attacks by any number of ideologically charged enemies armed with increasingly cruel weapons of mass destruction.

Each of these threats appears linked, albeit subtly, to specific prophetic signs of the end times that are explored at the conclusion of each of the section two chapters.

Section III is about America's future strategic plans to address the new realities, given the aforementioned megatrends and the emerging threats. The collision of these sobering volatile ingredients makes us wonder whether the biblical end times are imminent. However, we offer a detailed strategic plan for America should the Lord tarry, which, frankly, can't be too far into the future unless we soon experience a global miracle of peace.

This section assumes Christians will participate in their government as never before and do so with a fully engaged biblical worldview and an eye for prophecy—believing the time is short. It examines what a biblical worldview means and how that view is carried out in terms of biblical values and principles within every part of our government and culture. Then those values and principles are put into action through a comprehensive Christian vision for America, with very specific strategies for five major challenge areas: national security, national intelligence, foreign affairs, economy, and culture.

Section III's chapters begin with a profile of the nation's security strategies by outlining a six-part approach to revamping the Pentagon's critical task of protecting America, albeit this time from a biblical world-view perspective. Similarly, it then examines the nation's national intelligence community, which is hard to understand mostly because it is so secretive. However, the need for secrecy doesn't give it license to run roughshod over American civil liberties—and yes, we have every right to expect it to operate within the law while gathering the intelligence needed to keep us safe.

Our wayward foreign policy establishment is the next topic, and we begin with an explanation as to how it works (or doesn't) and the things that drive our foreign policy decisions. Then we suggest why our dealings with other nations seem so dysfunctional. That chapter advances an American foreign policy from a biblical worldview that addresses tough topics like immigration and Christian genocide in the Middle East.

Pocketbook issues have every American's interests today, especially in light of the nation's near economic collapse in 2008 followed by the tepid recovery under President Obama. Chapter 14 examines a biblical view of economics and then applies that perspective to a plan to return our free-enterprise system to a healthy state that begins with protecting the foundational family unit and outlines a plan to address our colossal deficit and our out-of-control government spending.

Finally, we examine our Beavis and Butt-Head (a crass adult animated sitcom) culture that is spiraling into the slimy pit of immorality, and then we explore a five-part vision and plan to restore our culture to a moral place until the Lord's return.

Throughout *Future War,* we keep returning to the Scriptures for insights about the future. In Matthew 8:28 (NKJV), the Bible relates a story of Jesus' confrontation with "two demon-possessed men, coming out of the tombs, *exceedingly fierce,* so that no one could pass that way" (emphasis added). Other accounts describe the men as incredibly strong; no chains could hold them and they were possessed by many demons.

In 2 Timothy 2:24–3:9, Paul warns his disciples "that in the last days *perilous times* shall come" (emphasis added). "Perilous times" and "exceedingly fierce" are indicated by the same Hebrew words, and these two uses are the only ones in the Bible.

Reconciling these two English versions of the same word, it is an easy shift to a single translation: "raging insanity." In short, God is describing a world of chaos and evil as we approach the end times, a world spinning out of control fueled by raging insanity. That's where we are today.

Future War is a sobering call to action for Christian Americans, who must take off their blinders and engage our government in perhaps the last chance we have to salvage our once-great nation. We must do so with a clear biblical worldview powered by Christian values and principles. And even then, it may just be too late, given the avalanche of signs that are crashing at an accelerating speed toward the prophetic end times.

SECTION I

Our Current Lost State of Affairs

Chapter 1

Spinning out of Control and Wrong Direction

IN OCTOBER 2014, President Barack Obama spoke to a fundraising event at the White Street Restaurant in New York City. He assured fellow Democrats, "There's a sense possibly that the world is spinning so fast and nobody is able to control it." Obama guaranteed his supporters that he has things under control.[9]

The president promised the New York City audience of his steady hand on world affairs. "On every single issue of importance, when there are challenges and there are opportunities around the world, it's not Moscow they call; it's not Beijing. They call us," Mr. Obama said. The president's pledge aside, most Americans believe the world is in fact spinning out of control, and Mr. Obama shares some of that responsibility for the widely held view that America is on the wrong path.

Why do Americans perceive that our country is on the wrong path? To a large extent, it's because there is an absence of American strategic leadership in spite of Mr. Obama's lofty rhetoric about his foreign policy "successes." In fact, his foreign policies, coupled with his domestic "successes" (liberal social policies, high corporate taxes, Obamacare, politicized federal bureaucracy like the Internal Revenue Service, and

11

whittling away at our Second Amendment rights), are major factors contributing to widespread angst among Americans.

As bad as Mr. Obama's foreign and domestic policies may be, he is only the latest contributor to the modern chaos that most Americans sense.

Widespread Spinning and Wrong Direction Perception

Americans consistently over the past couple of decades have told pollsters by large majorities that the world is spinning out of control—or, in other words, everything seems to be heading in the wrong direction. Recently, for example, six national surveys taken between February 11 and March 20, 2016, found that an average of 64.8 percent of Americans believe the country is heading in the wrong direction.[10]

Such widespread expression of "wrong direction" is relatively long in the tooth. Pollsters started asking about the level of American dissatisfaction with the country's direction in the early 1990s. The *New York Times* found, for example, in 2002 that 35 percent of Americans believed "things have pretty seriously gotten off on the wrong track." Responses to that question have consistently risen and remained at super majorities over the subsequent years to the present.

In 2008, four in five Americans (81 percent) shared the "wrong-track" view and, according to the *New York Times*, their negative view at the time was attributed to a spike in bad news coming from the front lines of the Iraq war coupled with dire news that the economy was slipping into recession.[11] Other national polls since that time found much the same angst.

In September 2014, the George Washington University Battleground Poll found that 70 percent of voters thought things in the U.S. had gotten off on the wrong track. The number-one reason for the "wrong-direction" answer given by voters was that they "have issues with Obama/think the country has a lack of leadership."[12] President Obama's party lost the U.S. Senate in 2014, which is a reflection of that widespread "wrong-direction" sentiment.

Not surprisingly, these "wrong-direction" feelings appear to be correlated with Americans' fears associated with everyday life issues flooding the news.

A 2015 survey conducted by Chapman University asked American adults about their level of fear regarding eighty-eight different issues across a variety of topics ranging from crime, the government, disasters, personal anxieties, and technology to many others.

Chapman found the highest levels of fear expressed were about manmade disasters such as terrorist attacks, followed by worries about technology, including corporate and government tracing of personal data, and concerns about government, such as Washington corruption and Obamacare.[13] Those fears, according to a Chapman spokesman, appear to influence our actions. One-third (32.6 percent) of those surveyed who registered above-average fear of government said they voted for a particular candidate due to their fears.[14]

There appears to be a strong association between American fear as identified by Chapman and the average citizen's sense that the country is heading in the wrong direction. Certainly, polling shows that as concern for terrorist attacks climb, so does the sense that the country is on the wrong path and, inevitably, the person in the Oval Office catches much of the blame.

A December 2015 NBC/*Wall Street Journal* poll found that the late 2015 terrorist attacks in Paris and San Bernardino, California, have vaulted terrorism to the American public's top concern and simultaneously drove President Obama's job rating to the lowest in more than a year, with seven in ten Americans saying the country is headed in the wrong direction—the highest score since August 2014.[15]

Arguably, we could lay the blame for the national "wrong-way" perceptions at the feet of modern media: cable, broadcast, Internet, and social (YouTube, Twitter, Facebook). These modes of delivering information dump large doses of distressing (personal, international, and domestic) news on our collective laps at rapid fire, 24/7 with no let-up, ushering into our daily existence giant doses of insomnia.

Blaming the media for all the out-of-control stress is wrong, though. Reporting and the 24/7 technologies that drive information into our brains are not responsible for the earthquakes, wars, and genocides; after all, public media provides a service, albeit sometimes biased, that we have the option to switch off. Few of us do in fact turn it off because it's so addictive. Our appetite for "news," whether about our friends, families, or foreign wars, seems insatiable. But there is a deeper issue that forms the background for much of the bad news that we consume every day that contributes to the present sense of being out of control.

The Spinning Started with Soviets' Downfall

The seemingly out-of-control rush of bad news can be blamed in part on the unraveling of international order, which began with the collapse of the Soviet Union in 1991. But just as important as the collapse of the Soviet Union was the American failure to seize that opportunity to set the world on a better course. Ultimately, it was America's leadership failure that significantly contributed to present-day chaos.

The former Soviet Union was a definable enemy during the nearly five-decade-long Cold War that gave the West an "evil empire" upon which to focus all its energy. Now, decades after the end of the Cold War, Americans find they no longer face one giant evil threat but numerous lesser enemies and associated crises. To our collective worry, the result of replacing that giant enemy with a potpourri of lesser "evils" is a much more unstable world that feeds our fears that drive perceptions the world is in fact spinning out of control.

The proliferation of "numerous enemies" was preventable, but over the ensuing decades, American leaders lacked the insight and wisdom to take advantage of the (notwithstanding brief) unipolar world they inherited once the Soviet Union crumbled. Rather than act responsibly and with vision, American leaders failed to seize the moment and to grasp the opportunity.

America acted like other Western nations and cashed in on the demise of the Soviet Union by disengaging from much of the world militarily and diplomatically. What emerged over time from that neglect and lapsed discernment were more real threats that would eventually draw the U.S. into new conflicts. That's when American leadership started to really crumble, especially in the last twenty years.

What should America, the sole remaining superpower, have done differently in the wake of the collapsed Soviet Union?

The end of the Cold War created a unique opportunity to make profound changes and prevent the vacuum left by Russia's withdrawal from superpower competition being filled by others. Instead, America saw the collapse of the Soviet Union as an Isaiah 2:3–4 moment: "They shall beat swords into plowshares, and their spears into pruning hooks: nation shall not lift up sword against nation, neither shall they learn war anymore." American leaders should have recognized the immediate need for action to arrest the resurgent and anti-West Islamic powers that today have the Middle East in chaos. History will prove this was a grave miscalculation.

The evaporation of the superpower rivalry that held multiethnic nations together gave way to the reemergence of ethno-nationalism in many developing world nations. But America turned away from those countries where instability rapidly grew in places like the Republic of Georgia and in old tension pockets like Armenia versus Azerbaijan.

The Cold War victory also ushered in market-based democratic governments over centralized, government-controlled economies in many countries. But these new capitalists quickly realized the old order had to go before capitalism could flourish. America took that transition on faith and did far too little across the globe to mentor.

Then we were caught flat-footed, naively unprepared for the rapidly emerging new world order. We were thrust into that new order after the September 11, 2001, attacks as large majorities of Americans accepted the need to take strong military action to deter future attacks. President George W. Bush rode that popular sentiment to launch military operations into Afghanistan with the intent of destroying al Qaeda and to deter similar future attacks. The outcome was an accelerated entrance into the new reality.

Mr. Bush evidently thought the public's mandate to attack al Qaeda in Central Asia gave him license to expand his war on terror elsewhere in the Middle East. History will prove that was a serious and costly misjudgment and unfortunately not the first.

Admittedly, President Bush had a hard sell to persuade skeptical Americans that attacking Iraq was in the nation's best interests. But in time, he convinced Congress using faulty intelligence to grant him war authority. He even trotted out his much-respected Secretary of State Colin Powell to convince a skeptical United Nations Security Council to issue a resolution authorizing military action to remove the Iraqi dictator Saddam Hussein.

Some of us knew at the time that Bush's Iraq war was a fool's errand. Prior to that war, I worked with the U.S.-based Iraqi exile community who had remained in close contact with key Iraqi military leaders in Baghdad. Based on their personal experience and information from trusted agents inside Iraq, my friends dismissed the Bush administration's allegations that Hussein retained weapons of mass destruction. On multiple occasions, my friends and I met with Bush administration officials to include members from the intelligence agencies to present information that contradicted the administration's go-to-war messages. Unfortunately, our insistence to go slow was dismissed as out of hand. I had top officials laugh or literally turn their backs to me and walk away when I insisted they were leading us to war based on wrong information.

The Iraq invasion went ahead and, as predicted, no weapons of mass destruction were found, and sectarian civil war erupted thanks in part to then Ambassador Paul Bremer's misguided decision to disband the Iraqi army rather than putting it to work to help stabilize the country. That decision also gave a green light to neighbor Iran to jump into the cat-bird seat in Baghdad, where it remains today as the master puppeteer manipulating that mostly Shia-based government to favor Tehran. America paid a steep price for that unnecessary war: 4,497 of America's finest and a cool $1+ trillion.[16]

President Obama inherited the Iraq war, which he declared finished in December 2011, another foolish presidential mistake. He should have worked hard to leave a robust stability force behind to secure our costly investment. Predictably, thanks to Tehran's mischievous puppeteering and Baghdad's corrupt sectarian government, Obama three years later grudgingly found it necessary to send thousands of fresh troops back to help Iraq's flailing government. So, our military was sent to retrain many of the same soldiers who evidently forgot all the training we provided them the previous decade at the cost of tens of billions of taxpayer dollars as well as to help push back against the surging Islamic State of Iraq and Syria (ISIS), which in 2014 gobbled up a large swath of northern Iraq and now terrorizes much of the world.

In 2009, on another war front, Obama surged seventeen thousand reinforcing troops atop thirty-eight thousand already in Afghanistan to "stabilize a deteriorating situation…which has not received the strategic attention, direction and resources it urgently requires."[17] This was another tragic mistake by a clueless president, an arrangement that has now cost 2,382 American lives and squandered hundreds of billions of dollars on a war in a hopeless country stuck in the fourteenth century.[18] Unfortunately, our sacrifice in Afghanistan will continue thanks to Obama's mid-2016 decision to increase our role from training to include air strikes and accompanying the Afghans on combat missions, a decision that kicks the tough decision beyond his commitment to quit that war by the end of 2017.[19]

In his own words, President Obama also "led from behind" a miniwar favored by then-Secretary of State Hillary Clinton by backing NATO allies to topple Libya's Muammar al-Gaddafi, a man who had fully cooperated with the U.S. by disbanding his weapons of mass destruction and helping with valuable intelligence about radical Islamists. Today that country is partly in the hands of ISIS, teeters on the brink of civil war, and fosters instability across all of Northern Africa.

Obama then piled atop his other geopolitical Mideast failures the carcass of another failed regime. He ignored his own "red lines," such as Syrian President Bashar al-Assad will not use chemical weapons against his people, and he then squandered millions of dollars on the half-baked plan to train so-called moderate Syrian rebels as America's proxy ground force to fight ISIS. General Lloyd Austin, the commander of U.S. Central Command, testified at the time to Congress that only "four or five" of the first fifty-four U.S.-trained moderate Syrian fighters remain in the fight against ISIS.[20] Shortly after that statement, Obama disbanded the effort.

When there is a perceived just cause to employ the military in harm's way, American citizens will rally to the call. But, these repeated failures by presidents from both major political parties understandably make Americans skeptical about efforts to join foreign wars.

It isn't surprising that a 2014 *Pew Research/USA Today* poll found a shifting of American opinions regarding the United States' role in solving global problems writ large. That shift corresponded with Russia's 2014 in-your-face challenge to the pro-Western government in Ukraine and, as mentioned above, the rapid rise of ISIS and subsequent threat to the homeland. Today, a plurality (39 percent) of Americans say the U.S. is doing too much to solve world problems, but that's a drop from the 51 percent in November 2013.

The tension between America doing too much and not doing enough to solve global problems appears to be related to our fear and the sense that the world is spinning out of control. After all, once ISIS came to the American public's attention—and especially as ISIS-associated terrorism killed Americans here at home like the December 2, 2015, attack in San Bernardino, California, and the June 12, 2016 attack in an Orlando, Florida, nightclub—Americans began to wake up to the reality that the Islamist group is a major threat and accepted that more deterrence was necessary. But at that point, American leadership had long ago failed.

Not All the President's Fault

It is easy to lay all the blame for today's insecurity on President Obama and the Democrats. The fact is the international environment that emerged from the end of the Cold War was terribly unstable. Specifically, the collapse of the former Soviet Union, the rise of China and Western retrenchment were byproducts of the post-Cold War era that gave us the leadership failures outlined before. Unfortunately, those leadership failures further destabilized the broader global geopolitical situation.

After the Cold War, Moscow and Beijing came to understand that the West didn't win as much as communism lost. Rather, the West—in particular, the U.S.—squandered the opportunity to remake the world. Why? The West tried to apply the model it used to help crumble the Iron Curtain—democratic politics and capitalist economics—but it failed when applied elsewhere in places like Iraq, as demonstrated earlier. In fact, the West's efforts backfired, and from those ashes emerged radical Islamists such as the caliphate-declaring ISIS.

The new international order not only made the world less stable but also contributed to the globalization of commerce and information, arguably good outcomes. But at the same time, the globalization trend put advanced military technologies into more hands, thus narrowing the capability gaps between Western powers and former enemies like Russia and emergent non-state players like al Qaeda and ISIS, which predictably eroded the West's strategic deterrence.

Obama's policies in particular do share some of the blame for this erosion. He shepherded a very weak economic recovery post-2008, imposed high corporate tax rates, and weakened the military through draconian (sequestration) cuts in spite of escalating threats abroad and at home.

Obama's neglect of American interests and capabilities has also been destructive to American security in general. Specifically, the significant downsizing of our military might in Europe and reduced engagement with new foreign partners in Eastern Europe encouraged Russia's remilitarization and eventually Moscow's incursion into Crimea and eastern Ukraine.

Similarly, Washington's failures in Asia led to an escalation in the Cold War-like relationship with China resulting from the construction of man-made islands in international waters and their military occupation. Chinese expansionism, new territorial claims and defiant, aggressive military confrontation are a serious threat to American interests.

Furthermore, Obama's so-called "strategic patience" policy with North Korea encouraged that rogue regime to accelerate its nuclear program, such as the January 2016 test of a "hydrogen bomb," launching a "satellite" in February 2016, and in December 2015 successfully testing a submarine-launched ballistic missile. As of this writing, North Korea has threatened to attack Manhattan with a nuclear missile.

These post-Cold War security threats created serious global stability problems.

Richard Haass, president of the Council on Foreign Relations, explained the post-Cold War new international environment in *Foreign Affairs:* "Our present times are unsettled, chaotic and dangerously diffuse because familiar methods of diplomacy, statecraft and war changed in the beginning of the twentieth century to a large degree because the time of 'unipolar' power is over and replaced by 'multipolar power.'"[21]

"Unipolar power" came about with the end of the Cold War and the demise of the former Soviet Union, as outlined above. That collapse gave the United States the one-time opportunity to make the world in its image, but as subsequent events indicate, it failed in that mission; as a result, a multipolar power mixture of state and non-state actors seized pieces of the modern geopolitical pie.

Today the potpourri of new state powers more often than not violates past international agreements at will and escape accountability, such as Russia's violation of the 1987 Intermediate-range Nuclear Forces (INF) Treaty. In 2014, the U.S. State Department officially accused Moscow of violating the treaty, but there has been no accountability for Russia's flagrant violations.[22]

Even our past best allies now show a cold shoulder or outright disdain for America's failing leadership. Paris turned to the European

Union, not NATO, for security support after the November 2015 terror attacks. Our longtime ally Egypt, which received $1.3 billion in military aid from the U.S. in 2016 and $150 million in economic assistance,[23] signed a $5.6 billion jet-fighter deal with France in 2015,[24] and Baghdad has become especially cozy with Moscow, as evidenced by its contract to buy 30 Mi-28 helicopters and an unspecified number of MiG-29 jet fighters.[25]

Further evidence that America has lost its strategic leadership role is the general shift in international structures. Beijing and Moscow created the Shanghai Cooperation Organization, and the International Monetary Fund welcomed the Chinese yuan to the world community's reserve currencies. Then China developed an alternative to the SWIFT international funds transfer system, the Asian Infrastructure Investment Bank.

The world's pillar of security has crumbled, thanks to miserably poor and incredibly naïve American leadership. After giving the world seventy years of peace, our abandonment of the lead thanks to Obama has come at a high cost for us and the balance of the world.

Why all the changes? The transition from bipolar (U.S. and Union of Soviet Socialist Republics) to unipolar (America alone), and now to a multipolar or nonpolar world, happened thanks to the absence of strategic leadership, the U.S. in particular. After all, only the U.S. could provide the needed strategic leadership given its means, reach, and credibility. Previously, the U.S. demonstrated strategic leadership by helping to establish major security organizations like the North Atlantic Treaty Organization (NATO), and at virtually every turn over the past half century, the U.S. has led in the building of alliances that delivered more global order. That's no longer the case.

Will America reclaim the mantle of strategic leadership or follow the course set by Obama to "lead from behind" and let the world continue to spin further out of control?

It is understandable that Americans don't like being the world's policeman, because that job can come with a high cost in blood and treasure. But more than our collective physical well-being is at stake. True strategic leadership is "about seizing opportunities and creating conditions conducive to the attainment of U.S. national security objectives."

Like it or not, only the U.S. can fill the current global leadership vacuums because it is in a unique position. Unlike any other nation, America enjoys a global span of influence. Failing to provide that strategic leadership as we've seen in the recent past has had serious consequences.

American Enterprise Institute scholar Robert Kagan explained the implication of such a failure for the Middle East and beyond:

> America's unwillingness to play that role has reverberations and implications well beyond the Middle East. What the U.S. now does or doesn't do in Syria will affect the future stability of Europe, the strength of trans-Atlantic relations and therefore the well-being of the liberal world order.[26]

Failing to provide that leadership has costs for us, our allies and the innocents. It's hard to say how that failure will be manifested in the future other than to suggest global stability will continue to erode as the many state and non-state actors grow in power to seed conflicts that inevitably will impact American interests.

Daniel Goure with the Lexington Institute makes a keen observation about the world's cry for strategic leadership: "When one surveys the present international environments three facts are clear: Europe will not lead, China cannot lead and Russia must not lead. That leaves, only the United States." Will the U.S. reclaim the mantle of global strategic leadership?[27]

Conclusion

During the post-Cold War period, instead of leading, America took a backseat, allowing others to emerge and gain influence and power, and generally create a chaotic situation that fuels the general sense of widespread disarray.

Most Americans perceive that the world is spinning out of control, and that in particular here at home everything seems to be heading in the wrong direction. There is good evidence to back up that conclusion, and the singular root for that feeling appears to be America's failure to provide strategic leadership in the wake of the Cold War.

It is also clear that only the U.S. is positioned to turn the global anarchy into a better, more stable forward direction. But it is also dangerously clear that President Obama is incapable of reversing his failed policies. In fact, his recent actions, not the least of which is his increased favoritism to radical Islam abroad and at home, will lead America into even more danger with potentially cataclysmic consequences.

We can be hopeful, but it is prudent to take a sobering look at the global trends that indicate that, if not quickly reversed, the widespread perception of spinning could get much worse. In fact, if America doesn't step into the strategic leadership gap very soon, a host of sobering global trends outlined in the next chapter will validate those who believe these are indeed the end times.

While our hope for alleviating the sense that America is spinning out of control should not be placed in a single elected national leader or the winning party in the 2016 election and beyond, it is incumbent on every American to carefully choose those who have the best chance of restoring a Christian worldview to our leadership and direction. Unfortunately, we witnessed some candidates in the 2016 campaign that clearly cannot fulfill that criteria and would only accelerate the spinning.

Chapter 2

Future Trends in an Out-of-Control World

THE PREVIOUS CHAPTER demonstrated that most Americans perceive that the world is spinning out of control, and a major reason for that widespread view appears to be America's failure to provide strategic leadership in the wake of the Cold War.

Even though history is not necessarily a prologue for the future, still, it is important we study not only our past failures but also the environment we anticipate in the future. Understanding possible future environments will help us avoid repeating past failures and simultaneously raise up the right type of leaders and build strategic plans that provide for the best possible outcomes for America.

This chapter anticipates the characteristics and influencers of our future world with the help of forward-thinking efforts such as the analysis published by the National Intelligence Council (NIC), the *Global Trends 2030: Alternative Worlds*. The NIC states: "We are at a critical juncture in human history, which could lead to widely contrasting futures. It is our contention that the future is not set in stone, but is malleable." Specifically, the NIC considers current global megatrends, potential game-changers that might alter or influence those trends, some potential future black swans, and suggests alternative worlds that might emerge from these ingredients. The chapter concludes with a suggestion that the forecasted outcomes by the NIC and others represent some, perhaps not all, of the prophetic signs of the end times.[28]

Global and Domestic Trends

The following global and domestic trends provide a framework for thinking about the future. They are likely to play as significant factors for our thinking and strategic planning for the time to come.

The NIC identifies four megatrends worth our consideration: individual empowerment, diffusion of power, demographic patterns, and food, water, and energy nexus.

Individual empowerment: The NIC forecasts that individual empowerment "will accelerate owing to poverty reduction, growth of the global middle class, greater educational attainment, widespread use of new communications and manufacturing technologies, and healthcare advances." NIC states the expansion of the middle class is a tectonic shift whereas for the first time in history a majority of the world's people will emerge as the most important social and economic sector. Individual empowerment that results from middle-class growth is significant, because it is "both a cause and effect of most other trends," such as economic growth of developing countries and use of communications and manufacturing technologies. The downside of this trend is individual access to disruptive technologies—sophisticated convention and bioterror weapons as well as cyber instruments.

Domestically, the middle class is not expected to expand like much of the balance of the world. In 2015, 120.8 million adults were in middle-income households, compared with 121.3 million in lower- and upper-income households combined. The Pew Research Center indicates that the share in the upper-income tier grew more than did the middle-income tier. Fully 49 percent of U.S. income went to upper-income households in 2014, up from 29 percent in 1970. Meanwhile, the share accruing to middle-income households was 43 percent in 2014, down substantially from 62 percent in 1970.[29]

Diffusion of power: Chapter 1 laid out the case that we no longer live in a post-Cold War, unipolar world led by the U.S.; rather, there are numerous centers of power—and that trend is likely to continue. In the future, power will shift to many networks and coalitions; thus, power is diffused across more players.

The diffusion of power will be especially noteworthy in the collective Asian countries that will likely surpass North America and Europe combined in terms of aggregate power—gross domestic product, population, military might, and technology. In late 2014, the BBC reported that China, based on IMF data, had surpassed the U.S. as the world's largest economy.[30]

The major change to anticipate is the economic role of the developing world. The NIC indicates that future economic health by 2030 will be dictated mostly by the developing world, not the West. Further and surprisingly, the "nature of power is shifting to networks that influence state and global actions." Specifically, according to the NIC, no country will be hegemonic—ruling or dominant. Rather, networks composed of state and non-state actors will form to influence global policies, and those networks will constrain policymakers "because multiple players will be able to block policymakers' actions at numerous points."[31]

Demographic patterns: As the world's population expands from the current 7.1 billion to 8.3 billion in 2030, the demographics will shift in ways that are especially challenging to the United States. Specifically, the West and its Western allies are aging faster than the balance of the world, which has production and resource implications. That means there will be a shrinking number of youthful states, which likely will exclude much of the West and will explain why Western leaders will try to attract young people from those states to help fuel Western economies.

Border-control issues will become very challenging due to significant migration for many European countries. We saw a hint of those problems in 2015 and 2016 with the tsunami of migrants moving into Europe from North Africa and the Middle East. It is noteworthy that in 2015, German Chancellor Angela Merkel invoked Germany's long-range need for youthful immigrants to justify her open-borders policy that backfired because many of the immigrants were young poorly skilled men who soaked up social welfare, failed to assimilate, behaved violently, and are not expected to contribute to the economy for many years. In fact, the large numbers of military-age males have turned out to be largely Islamic thugs and even a few terrorists that will likely never assimilate.

The demographic shift is significant in the U.S. as well with broad implications for the future, which contributes to the left's multicultural desires for America.

America's immigrant population is growing. Over the past half-century, immigrants and their descendants have accounted for just over half the nation's population growth, and as a result, have reshaped America's racial and ethnic composition. The Pew Research Center projects that if current demographic trends continue, future immigrants and their descendants will become a bigger source of population growth. For example, between 2015 and 2065, Pew projects immigrants will account for 88 percent of the U.S. population increase, or 103 million people, as the nation grows to 441 million.[32]

It appears that Obama's campaign promise to "fundamentally change America" is coming true in part thanks to immigration, which his administration encourages. A Center for Immigration Studies (CIS) indicates immigrant "offspring are a difficult group because their split identity causes many to turn to gang formation or, as some rightly fear, to jihad activities among Muslim youth. Today at least 61 million immigrants and their children live in the U.S., that number grew from 13.5 million in 1970, and an increase of 18.4 million since 2000." The CIS study indicates that the "number of immigrants and their young children grew six times faster than the nation's total population from 1970 to 2015—353 percent vs. 59 percent."[33]

Food, water, and energy nexus: The expansion of the world's population creates increased demands on resources. The NIC anticipates demands for food, water, and energy will grow as much as 50 percent, and, evidently, the NIC also bets the climate-change prognosticators are correct. As a result, shepherding those resources will become more difficult as a result of intensifying weather patterns, "with wet areas getting wetter and dry and arid areas becoming more so."

Resource scarcity is not necessarily the long-term trend, however. There will be a scramble for resources where some nations may do well, but others may lack.

There is good news for the United States, however. Its growing use of hydraulic "fracking" technologies to extract natural gas and oil significantly expanded its reserves from thirty to one hundred years.

Besides the four megatrends identified by the NIC above that potentially affect much of the world, there are social and spiritual trends here in the U.S. that bear watching. These trends impact the American family, their relationship with their government, and the nation's moral climate—and they may suggest the end times are rapidly approaching.

The American family is morphing into something very different than it was only a few decades ago.

The "Leave It to Beaver" homes are trending to extinction: The "Leave It to Beaver" world is rapidly being replaced by a dramatic rise in kids living with a single parent. In 2014, just 14 percent of children younger than 18 lived with a stay-at-home mother and a working father who were in their first marriage. By comparison, in 1960, just half of children were living in this type of arrangement, and by 1980, the share had dropped to 26 percent. The biggest change has been the increase in children living with single parents—up to 26 percent from 9 percent in 1960, and, most dramatically, the majority (54 percent) of black children are living with single parents.[34]

Americans are recycling more marriages: In 2013, fully four in ten new marriages included at least one partner who had been previously married, according to Pew Research Center analysis. Today, almost forty-two million adult Americans have married more than once, up from twenty-two million in 1980, and the number of remarried adults has tripled since 1960. At least two phenomena contribute to this demographic trend: a rising divorce rate and the aging of the population, which makes more Americans available for remarriage.[35]

More American dads are at home with the kids: The number of fathers who do not work outside the home was up to two million in 2012. High unemployment in the past decade contributed to that increase, but the biggest contributor to the growth in the number of "stay-at-home fathers" is the rising number of fathers who are at home primarily to care for their families while their wives are the bread-winners. That number doubled since 1989, when 1.1 million stayed at home. While most stay-at-home parents are mothers, fathers represent a growing share of all at-home parents—16 percent in 2012, up from 10 percent in 1989.[36]

Americans are not replacing themselves: The U.S. birth rate dipped in 2011 to the lowest ever recorded. The overall U.S. birth rate, which is the annual number of births per thousand women in the prime child-bearing ages of 15 to 44, declined 8 percent from 2007 to 2010. The birth rate for U.S.-born women decreased 6 percent during these years, but the birth rate for foreign-born women plunged 14 percent and it fell 23 percent among Mexican immigrant women. Today's rate, 63.2 per thousand women of childbearing age, is the lowest since at least 1920. The U.S. birth rate peaked most recently in the baby boom years, reaching 122.7 in 1957, nearly double today's rate.[37] The U.S. total fertility rate dropped to 1.87 in 2013[38]—well below the replacement level of 2.1, the rate to maintain the population's status quo.[39]

Trust in U.S. government declines: America's founders framed our Constitution to maximize the citizen's liberties and limit government. Over the past few decades, government grew well beyond what our founders ever intended. That has resulted in an almost adversarial relationship between government and the citizenry, which explains a serious trend of less and less trust in government.

Historically, the number of Americans who said they trusted their government has trended down for decades. In the 1960s, a majority of Americans said they trusted their government "always or most of the time." That trust started to decline with the Watergate scandal in 1974, when only 36 percent expressed good faith in government, but it rose again to 50 percent after the September 11, 2001, attacks on America.[40]

However, recent national polling found that few Americans (13 percent) agree that the U.S. government "can be trusted to do what is right always or most of the time." Conversely, nearly all Americans (75 percent) believe their government works properly just "some of the time."

State of moral decline: America will continue to decline morally. Some will rightly argue that as Americans granted their government more say over our lives, those same "leaders" used their power to remove religion from everyday society, and as a result, morality fell off a cliff.

Today, most (72 percent) Americans, according to a 2015 Gallup poll, believe that the state of morality in America is getting worse. The perception of a moral decline reflects the role of morality in the socio-cultural and political scene, especially regarding issues like same-sex marriage, hip-hop music with explicit lyrics, and growing comfort with having a child out of wedlock.[41]

Certainly from a biblical perspective, America is on a course to moral bankruptcy. Just consider the scriptural injunctions regarding our open defiance to God's Word: abortion (Exodus 20:13), homosexuality (Romans 1:26–27), apathy (Jeremiah 5:31), sexual addiction (1 Thessalonians 4:5), churches becoming spiritually neutralized (Galatians 3:3), and many more.

Prophecy watchers point out that the decline in morality and the rise of lawlessness and chaos are well documented as end-times indicators. There is no reason to expect a reversal of this trend at this time.

America trending against the spiritual: Americans are abandoning the Christian church in record numbers, which is also a reflection of a general moral decline.

Today, Americans are evenly divided on the importance of attending church. Half (49 percent) say it is "somewhat" or "very" important, the other (51 percent) say it is "not too" or "not at all" important. Age is highly correlated with the perceived importance of church. Specifically, one-third of millennials (those 30 and under) take an anti-church stand, while elders (those over 68) are most likely (40 percent) to view church attendance as "very" important, compared to one-quarter (24 percent) who deem it "not at all" important.[42]

Little wonder church attendance is down across America, given the declining widespread view about the importance of church. The Barna Group found overall church attendance has dipped from 43 percent as recently as 2004 to 36 percent today. The nature of churchgoing has also trended away from regular attendance (people present at services three or more times each month or more frequently). Now, many show up for church only once every four to six weeks and still consider themselves regular churchgoers.[43]

Millennials, those most likely to abandon church, cite as their reasons: church irrelevance, hypocrisy, and the moral failures of its leaders. Further, two out of ten unchurched millennials believe God is missing in church.[44]

A few years ago, evangelist Billy Graham wrote that his wife, Ruth, read a draft of a book he was writing at the time. "When she finished a section describing the terrible downward spiral of our nation's moral standards and the idolatry of worshiping false gods such as technology and sex, she startled me by exclaiming, 'If God doesn't punish America, he'll have to apologize to Sodom and Gomorrah.'"

Reverend Graham explained, "She was probably thinking of a passage in Ezekiel where God tells why he brought those cities to ruin. 'Now this was the sin of…Sodom: she and her daughters were arrogant, overfed and unconcerned; they did not help the poor and needy. They were haughty and did detestable things before me. Therefore I did away with them as you have seen'" (Ezekiel 16:49–50, NIV).[45]

Anticipated Shifts between Now and 2030

In addition to the global and domestic trends just discussed, the NIC anticipates seven global shifts between now and 2030 that must be weighed before considering the alternative worlds: the growth of the global middle class, wider access to lethal and disruptive technologies, a shift in economic power to the east and south, unprecedented aging, urbanization, food and water pressures, and U.S. energy independence. The coincidence of the trends and shifts will have a synergistic impact on the emerging world in the immediate decades.

- Growth of the global middle class: The middle class is defined by economic power that is expected to rapidly expand in absolute numbers to a growing percentage of the population. This will influence resource demands, migration, demands on government, and, as a result, national stability. A Reuters report indicates that the middle class in real terms will more than double in size, from 2 billion today to 4.9 billion by 2030. Meanwhile, the middle classes in Europe and America will shrink from 50 percent of the total to just 22 percent. Asia is the big winner, with an expected 64 percent of the globe's middle class by 2030.[46]

- Wider access to lethal and disruptive technologies: Modern technology is a mixed blessing; while helping improve the quality of life, it can also be harnessed to rob many of their lives. The wide availability of high-tech killing and otherwise destructive weapons—precision-strike capabilities, cyber instruments, biological weapons—empowers individuals and small groups with the means to perpetrate massive violence.

- Definitive shift of economic power to the east and south: The Western-aligned nations' projected share of global income is expected to fall significantly to under half by 2030. Meanwhile, China's financial assets are projected to double by 2020 compared to the where it was in 2012.

- Unprecedented and widespread aging: The West is rapidly entering the post-mature age category, and countries in Europe, South Korea, and Taiwan will be there by 2030. The graying trend may prompt immigration from the poorer to the richer countries as workforce shortages become more acute. The NIC comments on this possible outcome. Specifically, it notes that immigration may pick up in Europe, but many of those who come may not assimilate and may rapidly grow, thus eroding social cohesion and promoting reactionary politics. We are already seeing this across much of Europe, and especially in Germany, where in 2015 a tsunami of immigrants sparked a political backlash against Chancellor Angela Merkel's open-door policy.

- Urbanization: The world's population is moving to urban areas, which will climb to nearly 60 percent, or 4.9 billion by 2030 and 70 percent by 2050. Africa is expected to experience the highest urbanization growth, and urban centers will deliver the greatest economic growth (80 percent). To put these figures into perspective, in 1900, only 13 percent of the population lived in urban areas, that number rose to 30 percent in 1950 and to 50 percent in 2010. By 2025, the world will have 640 cities with populations over one million, and that's up from just 114 in 1960.[47]

- Food and water pressures: Demand for food (35 percent) and water (40 percent) are expected to dramatically rise by 2030, with almost half of the world facing severe water stress, especially in Africa and the Middle East.

- U.S. energy independence: By 2030, the U.S. will have sufficient natural gas to meet its needs for many decades, and oil production will significantly reduce our net trade balance, thus fueling economic expansion. Further, America's energy prosperity will dramatically impact the Organization of Petroleum Exporting Countries (OPEC) through an oil-price collapse, which will negatively impact those oil-export economies as well as Russia, a country that heavily depends on oil revenues.

The Eighth Shift: Nontraditional Spiritual Lives

The religious life of nations is rarely given significant attention by the NIC. Yet it can be the determining factor in how a nation perceives and interacts with its people and the world, and ultimately determines a nation's culture. Look at the age-old tensions in the Middle East between various Islamic factions as well as Muslims versus Christians and/or Jews.

It appears that Western culture is turning off to traditional Christianity and a personal relationship with Jesus Christ and alternatively seeking their "spirituality" from science, technology, experiences, and pluralism. This isn't surprising given the biblical prophecy that in the end times, man will become spiritual heretics (2 Timothy 3:1–5). It is, however, a wake-up call for Christians to recognize what is happening and get serious about their faith.

David Bryce, a research fellow at the University of Pennsylvania in the Positive Psychology Center, speculates about the future of spirituality. His analysis suggests a widespread biblical apostasy whereby mankind will choose spiritual practices based on evidence about their outcomes such as stress reduction. He suggests technology will infuse our spirituality with "meditation training" improved "with non-invasive brain stimulation technology." Future psychiatrists might prescribe a "spiritual experience to help ease the dying process, treat victims suffering from PTSD [post-traumatic stress disorder], or provide a meaningful rite of passage for young adults." Finally, Bryce suggests that spirituality will become more pluralistic or characterized as an anything-goes view. "The spirituality of the future will be more socially inclusive and will foster more respect for the differences between us."[48]

Potential Game-Changers that Might Alter or Influence the Trends

The NIC scopes six key "game-changers" that inevitably interact with one another and with the aforementioned megatrends to further transform the world by 2030. These "game-changers" are outlined below, accompanied by my analysis.

Crisis-prone global economy: The NIC questions whether "global volatility and imbalances among players with different economic interests result in collapse." Or, alternatively, it asks rhetorically whether the development of multiple growth centers will lead to resiliency and if the economic decline of the U.S. will lead to greater multipolarity.

The NIC does not anticipate near-term that the U.S. will return to the pre-2008 growth rates, and it forecasts that non-financial debt will continue to increase and future recessions will tend to be deeper and require longer recoveries. Further, even though the NIC does not rule out another 2008-like global economic crisis, it cites the type of crises we may experience, such as the default-related crisis between the nation of Greece and the European Union, and further as we saw in June 2016, the EU may be imploding as a financial entity as evidenced by United Kingdom's referendum-based decision to leave the EU.

Further, a 1930s Great Depression-type event is unlikely because, as the NIC suggests, America's current workforce is in a better position than it was nearly a century ago, but it warns that labor productivity must increase to remain competitive—and that's where technology could help.

It suggests that the east and south offer the best hope for near-term economic growth. Specifically, the developing world already provides half of current economic growth and 40 percent of global investment. Not surprisingly for China watchers, that country's contribution to economic growth is one and a half times the size of the U.S. contribution today.

All is not bright for developing nations over the next couple of decades, however. There will be challenges to include slowed economic growth, which will drive down income growth, creating inequities between rural and urban sectors, increasing constraints on resources such as water, and demanding greater investment in technology to keep those economies humming.

Governance gap: The NIC anticipates that as power becomes more diffuse, more players will therefore need to participate in solving major transnational challenges, and, due to human nature, decision-making will become more complex. This will result in a trend toward fragmentation, which could push the world in different directions.

The governance gap—autocracy versus democracy—will continue

due to rapid political and social changes. China's growing middle class is expected to trigger democratization, and social networking will enable citizens to better challenge government. Simultaneously, the same technology will allow governments to better monitor their citizens.

Potential for increased conflict: The NIC anticipates fewer **major** armed conflicts over the next couple of decades due to maturing age structures and many disincentives, especially among great powers. However, the aggregate number of conflicts may well increase, chiefly intrastate conflicts that "contain a politically dissonant, youthful ethnic minority," such as Kurds in Turkey and Iranian-backed Shia in Lebanon. Other contributors to conflict include the competition for natural resources as well as the disproportionate level of young men to the overall population, which could contribute to more unrest—especially where meaningful employment is scarce, the radicalizing influence of Islam, and the failure of Muslim immigrants to assimilate in Western countries.

Another factor that could contribute to conflict is changes to the international system, such as America's unwillingness to act as the world's policeman. Further, there are three other factors that, if they collide, could result in conflict: the national interests of key players like China and Russia; the increase in competition for resources such as in the arctic regions; and the growing availability of war-making technologies.

The current Islamist terrorism threat may well diminish over the next couple of decades, according to the NIC, but terrorism as a tactic will continue for some states because "of a strong sense of insecurity." Emerging at the same time will be individual actors with access to lethal technologies like cyber that will sell their services to the highest bidder, including terrorists.

Wider scope of regional instability: The NIC anticipates that regional instability might spill over to create more global insecurity. Specifically, the Middle East and South Asia are likely to witness further instability. As the world reduces its dependence on the Mideast's abundant oil and gas, the region will likely sink into economic stagnation as a result of failure to meet the challenge to diversify its economies.

One wild card for the Mideast is whether the Islamic Republic (Iran) continues on a course to develop nuclear weapons and what that implies for regional competitors like Saudi Arabia. There are numerous credible reports that Riyadh has a deal with ally Pakistan to provide nuclear weapons as Tehran more than likely continues its program in spite of assurances made to the West in 2015.

South Asia—mostly Pakistan and Afghanistan—are anticipated to continue low growth, which will pose stiff instability challenges among those large youthful populations. India will face some of the same challenges, but it is better positioned for growth. Due to age-old tensions, the parties are unlikely to develop a stabilizing regional framework.

China is the key to Asian stability. Growing Chinese power, nationalism, and questions about U.S. commitment to remain engaged in the region will feed insecurity.

Europe is walking a tightrope regarding its future as a security provider. Although it played roles in the Mideast, across the continent, security investment is a mere shadow of what it was during the Cold War. Likely, the 2030 Europe will focus on crises at home, many of which come from the immigrants from North Africa and the Middle East who refuse to assimilate. However, Europe's security situation is dependent on Russia and whether it elects integration with the West, which now appears unlikely, or further adventurism, as in Ukraine.

Impact of new technologies: Technology breakthroughs will help productivity and solve significant challenges while presenting new ones. Information technology is a significant factor because process power and data storage are becoming very inexpensive, and networks and the cloud provide anytime global access to fuel the free markets. This will maximize individual productivity and quality of life as well as minimize resource consumption. Challenges to the exploding information technology include cybercrime and growth in Orwellian-state surveillance.

Manufacturing will continue to benefit from the technology explosion. Robotics will continue to change work patterns and productivity and diminish outsourcing and the length of supply chains. Of course, the negative impact of more technology is the potential to reduce human labor requirements.

Optimistically breakthrough technologies will help meet global demand for more food, water, and energy. Specifically, advances in genetically modified crops, precision agriculture, water irrigation, and other trends will continue to improve our collective lives, although a developing backlash trend against genetically modified organism crops could play a counter role.

Health technologies will extend longevity, which corresponds to the anticipated skyrocketing middle-class populations.

Finally, technology will create new weapons systems that are more robotic and deadly. These technologies will help create super soldiers—ones that are stronger, smarter, and require less sleep than America's best contemporary warriors.

Role of the United States: The preceding chapter made the case that the U.S. failed to provide the required strategic leadership during the post-Cold War period, which resulted in the current multipolar world with its growing instability. The NIC's sixth game-changer is perhaps the hardest to predict: Will the U.S. be able to work with new partners to reinvent the international system to shape the future global order?

Little doubt the U.S. will remain "first among equals," at least over the next couple of decades due to its significant range of powers. However, the NIC rightly concludes that "the era of American ascendancy in international politics that began in 1945—is fast winding down." Although there is still the opportunity to reverse that trend, time for radical redirection is short.

America is joined in the "winding-down" trend in terms of economic prowess, but also in international political influence. The NIC foresees that the U.S. will less often project power outside of an alliance or coalition of key partners. This is indicative of a rapid change that is reflected in national security strategies and even in the Pentagon's go-to-war doctrine.

Even though the U.S. is credited with inventing the Internet and led in piloting global communication technologies, that edge is losing steam while others, to include non-state actors, harness the capabilities.

The U.S. global influence in 2030 will be determined by its interim roles in crises. Remember, the U.S. became a global power starting with its tepid involvement in the First World War. Only after the attack on Pearl Harbor on December 7, 1941, was then-President Franklin Roosevelt forced to commit the nation to global war, which eventually led to America's ascendency to global power status where it remained until the present. That status is at risk today, and the outcome depends on American future economic, diplomatic, and security policies and actions.

A pregnant question for now is: Will any nation or coalition emerge to replace the U.S. to erect a new international order? At this point, none of the emerging powers (China, Russia, Brazil, and India) appear ready or able to climb on top of the international community to set a future course. That outcome could change should a new cohesive bloc of nations and power partners emerge that shares a unitary alternative vision. Failing American leadership and the emergence of a new power bloc, the world in 2030 is likely to face widespread anarchy. This potential is addressed in section II of this volume.

Potential Future Black Swans

The old world once believed all swans were white, that is, until black swans were discovered in Australia. That account illustrates a severe limitation to our life experience because sometimes things happen—black swans—that are rare, deliver extreme impact, and were retrospectively predictable. The terrorist attack of September 11, 2001 was a black swan; had the risk been reasonably conceivable on September 10, it would not have happened. "Isn't it strange to see an event happening precisely because it was not supposed to happen?" wrote Nassim Nicholas Taleb in the *New York Times*.[49]

The NIC identifies eight possible future black swans that could deliver extreme impact but no one really believes they are possible. However, as the collective world looks back if any of these do in fact occur, then there will be claims that they were predictable.

- **Severe pandemic:** "No one can predict which pathogen will be the next to start spreading to humans, or when or where such a development will occur. An easily transmissible novel respiratory pathogen that kills or incapacitates more than one percent of its victims is among the most disruptive events possible. Such an outbreak could result in millions of people suffering and dying in every corner of the world in less than six months." At this writing, most readers recall the Ebola breakout in West Africa in 2014 and the publicity associated with the American care providers who fell ill before being evacuated to the U.S. for treatment. That Ebola outbreak was mostly contained to Liberia, Guinea, and Sierra Leone, where it claimed more than 11,300 lives and infected more than 28,500 people across three countries.[50]

- **Much more rapid climate change:** "Dramatic and unforeseen changes already are occurring at a faster rate than expected. Most scientists are not confident of being able to predict such events. Rapid changes in precipitation patterns—such as monsoons in India and the rest of Asia—could sharply disrupt that region's ability to feed its population." We must be careful about jumping to conclusions about the cause of climate change, however. There are myths that man-made climate change supporters propagate such as the false narrative that 97 percent of all scientists agree with them. That number is based on murky terminology and sourced to a 2009 article by a student/teacher couple at the University of Illinois. They conducted an online poll that concluded most (nine in ten) scientists agreed that temperatures were rising and mankind was the driving force.[51]

- **Euro/European Union collapse:** The NIC estimated that Greece's exit from the Eurozone could cause eight times the collateral damage as the Lehman Brothers bankruptcy, provoking a broader crisis regarding the EU's future. So far that hasn't happened, but it still could. In 2015, Greece received its third bailout in five years. That bailout requires Athens to implement austerity measures and economic reforms, which includes raising the retirement age, cutting pensions, liberalizing the energy market, expanding property tax, and selling state assets.[52]

- **A democratic or collapsed China:** "China is slated to pass the threshold of U.S. $15,000 per capita purchasing power parity in the next five years or so—a level that is often considered a possible trigger for democratization. Chinese 'soft' power could be dramatically boosted, setting off a wave of democratic movements. Alternatively, many experts believe a democratic China could also become more nationalistic. An economically collapsed China would trigger political unrest and shock the global economy," NIC states.

- **A reformed Iran:** "A more liberal regime could come under growing public pressure to end the international sanctions and negotiate an end to Iran's isolation. An Iran that dropped its nuclear weapons aspirations and became focused on economic modernization would bolster the chances for a more stable Middle East." This is mostly hopeful and arguably naïve thinking on the part of the NIC. Unless there is a revolution that removes the Shia clerical power structure, reform, much less the emergence of a liberal regime, is highly unlikely although not impossible.

- **Nuclear war or weapons of mass destruction/cyberattack:** "Nuclear powers such as Russia and Pakistan and potential aspirants such as Iran and North Korea see nuclear weapons as compensation for other political and security weaknesses, heightening the risk of their use. The chance of non-state actors conducting a cyber-attack—or using WMD—also is increasing." It is noteworthy both North Korea and Iran surprised the world with their nuclear and ballistic missile progress in the past, and we should expect further and more dramatic surprises in the future. There are plausible scenarios in which either regime might deploy a nuclear weapon, such as in the case of North Korea, if the regime believed it was about to be overthrown or if Iran's ayatollahs decided it is time to bring the necessary chaos in order to hasten their savior's arrival.

- **Solar geomagnetic storms:** "Solar geomagnetic storms could knock out satellites, the electric grid, and many sensitive electronic devices. The recurrence intervals of crippling solar geomagnetic storms, which are less than a century, now pose a substantial threat because of the world's dependence on electricity."
- **U.S. disengagement:** "A collapse or sudden retreat of U.S. power probably would result in an extended period of global anarchy; no leading power would be likely to replace the United States as guarantor of the international order."

Certainly the missing black swan is the return of Jesus Christ, which many Bible-believing Christians anticipate could happen soon, given the many prophetic signs evident today. That event will ultimately make all the trends, game-changers, and the above unexpected events totally irrelevant. However, in the interim, we must deal with the potential unpredictable catastrophic events identified above and the challenges they present to our rapidly changing chaotic world.

Alternative Worlds that Might Emerge from These Ingredients

Chapter 1 indicated that 1991 was the beginning of a transition period: the post-Cold War, unipolar world dominated by the United States. It became by a quirk of history the American era to reshape the world or to abandon the strategic leadership role. Unfortunately, the U.S. demonstrated a mixed record of leadership that resulted in today's multipolar world that is trending toward instability.

One could give America a failing grade for its record of stewardship or, at best, wait another couple of decades and let historians decide the grade. Now, in 2016, the path forward for the U.S. is not clear cut. The future could either become the best of times or the worst, but we can reasonably forecast that the rate of change experienced in the recent past will accelerate in the future.

The NIC used the aforementioned analyses to create four possible scenarios that may become the future reality, given the aforementioned megatrends and game-changers influenced by the tectonic shifts and the ever-unpredictable black swans. Examine each of these possible scenarios before considering which, if any, in fact best fits biblical prophecy about the end times.

Stalled Engines

The NIC foresees that the most plausible worst-case scenario by 2030 is an interstate conflict in Asia rather than a full-scale conflagration on the order of World War II. Rather, the world's power centers would more likely join a conflict intending to stop it with the objective of limiting political and economic damage.

This scenario best comes to fruition if the U.S. and Europe elect to abandon global leadership roles. What might prompt Western retrenchment? The NIC speculates that global economic growth will stagnate and, as a result, limited resources will be used for primarily domestic programs rather than foreign policing.

Fusion

This is the best-case scenario in which a conflict in Asia draws in the U.S., Europe, and China to impose a ceasefire. These parties collaborate, leading to positive change such as worldwide cooperation to deal with global challenges, which then leads to political reform and the reformation of multinational institutions.

This scenario depends on strong strategic leadership focused on international challenges rather than domestic requirements. As a result of improved bilateral relationships, all economies grow faster and the GDP improves. Domestically, American per capital incomes rise with the help of technological innovation.

Genie Out of the Bottle

The NIC states that this is a world scenario of extremes. Inequalities lead to increased political and social tensions. The U.S. remains the top power due to its emerging energy independence, but it abandons its previous role as the "global policeman." Meanwhile, the twentieth century's primary energy producers suffer from declining revenues (read OPEC, Russia, and Iran), and because they failed to diversify their economies, social unrest escalates. China is split between the thriving coastal cities and other regions, a cocktail that increases social discontent.

Sputtering economies will fuel the possibility of conflict as some countries fail and the wealthier nations refuse to provide sufficient aid to prop up those developing lackluster economies.

Non-State World

This scenario is unique because states take a backseat to a host of non-state actors—nongovernmental organizations and multinational businesses as well as megacities. This group is expected to lead in tackling global challenges and enjoy the support of the growing middle classes. These assorted groups form "hybrid" coalitions with states to tackle big challenges.

Traditional and polar-opposition types of governments from authoritarian to democratic are challenged severely to operate in such a complex, numerous-actor arena. The NIC indicates the more agile countries led by elites would prosper in such an environment, while larger countries lacking political cohesion would lag. Meanwhile, the non-state groups operate freely across traditional national boundaries in the new hyper-globalized world, where "expertise, influence, and agility count for more than 'weight" or 'position.'"

This patchwork world may solve some problems through networking and cooperating, but they are expected to meet major power opposition on some fronts. Security threats will be among the most difficult for the new patchwork world, especially given the access to lethal and disruptive technologies.

These four scenarios all seem plausible given the parameters outlined. However, there are other possible alternatives given the interaction of the trends, the game-changers, black swans, and the unseen and impossible-to-anticipate but expected influence of prophesied signs.

Alternative World's Best Fit with Biblical Prophecy

The NIC's four alternative-world scenarios present indicators of what some Christians believe are biblical signs of the end times. The scenarios do suggest some prophesied signs, but the NIC and other secular analyses fail to consider the moral and spiritual issues of our day and where they are leading.

Consider that in the times of Noah and Lot, the people were warned about God's coming judgment and destruction, but few listened. Instead, they treated Noah and Lot as "extremists" and, as a result, they were unprepared for what happened. Scripture tells us the same will happen at the return of Jesus Christ, because the people of that future time will not heed the warnings (Luke 17:26–30).

It could be that the NIC is correct about trends and game-changers, but it may be that it misses some of the black swans or trends (developments) that in fact are biblical signs of the end times. Many such harbingers are evident today, however. Consider a few here that seem to resonate with the NIC trends. Others will be addressed elsewhere in this volume.

Knowledge Will Increase

> Even to the time of the end: many shall run to and fro, and knowledge shall be increased. (Daniel 12:4, KJV)

Technology is developing today at a breathtaking pace in terms of scientific discoveries. For example, the words "running to and fro" may be a reference to modern means of transportation and especially the frenetic lifestyle many of us live. It could also be about something deeper as it applies to biblical truths. As one commentator said, "God's people have been running to and fro in His word freely and knowledge about great Bible truths have been revealed."[53]

Nature Will Roar

> And there will be great earthquakes, and in various places plagues and famines; and there will be terrors and great signs from heaven. (Luke 21:11, KJV)

Although this is not considered in the trends outlined above, it is a scriptural sign that can't be ignored: a black swan of sorts, like the tsunamis that struck Malaysia and Northern Japan or the earthquake that devastated much of Nepal in 2015. Matthew 24:8 declares the signs will be like "birth pangs" which means, as every mother who has given birth knows, the pains will increase in frequency and intensity the closer we get to the Lord's return.

Over the past twenty years, the incidence of natural disasters has increased. Specifically:

- The number of natural disasters in the U.S. increased 100 percent between 1980 and 2014.[54]
- Financial losses due to lightening, tornados, floods, and severe storms more than doubled from 1980 to 2014.

- Tornado reports have increased from approximately 200 in 1950 to nearly 1,700 in 2011.[55]
- Half of the ten most expensive natural disasters in the U.S. occurred in this century.[56]

Moral Decline and Humanism Erupts:

But know this, that in the last days perilous times will come: For men will be lovers of themselves, lovers of money, boasters, proud, blasphemers, disobedient to parents, unthankful, unholy, unloving, unforgiving, slanderers, without self-control, brutal, despisers of good, traitors, headstrong, haughty, lovers of pleasure rather than lovers of God. (2 Timothy 3:1–4, NKJV)

This sign is undeniable, given the "news" that spews from every media source. Mankind is consumed with personal betterment and abandons everything for materialism—hedonism is his god.

Hedonism leads to nihilism and then to abandonment in despair. No wonder society is plagued by abortion, homosexuality, domestic violence, blasphemy, pornography, alcoholism, and drug abuse. Today, people are doing what is right in their own eyes (Isaiah 5:20; Judges 21:25).

In Noah's time, "the earth was filled with violence" (Genesis 6:13, KJV), and in Lot's day, the people were given over to "sexual immorality and perversion" (Jude 1:7, NIV). Similarly, "when you see these things happening, you know that the kingdom of God is near (Luke 21:31, NIV).

Certainly, today is marked by violence and sexual immorality, and homosexuality has become a "norm" as it was in Lot's time. Since 2001, fourteen countries, including the U.S., legalized homosexual "marriage." Beyond homosexuality, today's society is full of sexual immorality, as evidenced by rampant sexually transmitted diseases and ever-increasing licentiousness in our public media.

Heresy Invades the Church

> The time will come when they [professing Christians] will not
> endure sound doctrine, but wanting to have their ears tickled,
> they will accumulate for themselves teachers in accordance to
> their own desires; and will turn away their ears from the truth,
> and will turn aside to myths. (2 Timothy 4:3–4, NASV)

The Christian world is full of false christs, apostasy, and persecution
that promote the occult and spiritualism. The entertainment media
is alive with examples of seducing spirits and fraudulent pop-culture
doctrines.

Increased Christian Persecution

> Then shall they deliver you up to be afflicted, and shall kill you:
> and ye shall be hated of all nations for my name's sake. And then
> shall many be offended, and shall betray one another, and shall
> hate one another. (Matthew 24:9–10, KJV)

Today, Christians are being threatened due to their faith unlike any other
faith group and rivaling anytime in history. The Christian genocide in
the Mideast is especially noteworthy, but the continued persecution of
Christians across the globe suggests it will get worse.

War More Frequent

> And ye shall hear of wars…. For nation shall rise against nation
> and kingdom against kingdom. (Matthew 24:6–7, KJV)

Wars and conflicts seem to be on the increase although they are not
on the scale of world wars. There were approximately fifty-two ongoing armed conflicts and wars raging in 2015 all around the world, and,
depending on the scenario above, the incidence of armed conflict could

grow in the near future.[57] Keep in mind, war is a sign. Jesus said, "So likewise ye, when ye shall see all these things, know that it is near, even at the doors" (Matthew 24:33, KJV).

Technology Controls Man

> And that no man might buy or sell, save he that had the mark, or the name of the beast, or the number of his name. (Revelation 13:17, KJV)

How might the beast stop people from buying and selling if they refuse the mark? Obviously, today's worldwide financial system operates electronically and can exclude anybody from purchasing virtually anywhere.

We now have the technology to implant microchips inside the body, and when the chip is read, it could be used to approve the purchase—or a decoder can see the number printed with invisible markings on the hand or forehead. These are present realities, and there are those pushing for everyone to receive the mark.

Implanting microchips in humans is taking place as well. A Swedish company, Bionyfiken, piloted a program that allows employees to unlock doors, access their printing accounts, and pay for lunch—all with the simple wave of their hand, as scanners read the chip implanted within. Hannes Sjoblad, Bionyfiken's chief "disruption officer," predicts a future in which "big corporates and big government come to us and say everyone should get chipped."[58]

Perhaps this is the mark of the beast foretold in Revelation 13:16–17 (KJV):

> He causes all, both small and great, rich and poor, free and slave, to receive a mark on their right hand or on their foreheads, and that no one may buy or sell except one who has the mark or the name of the beast, or the number of his name.

Conclusion

Former President William Clinton said, "In a world with no systems, with chaos, everything becomes a guerilla struggle, and this predictability is not there. And it becomes almost impossible to save lives, educate kids, develop economies, whatever."[59] There is a case to be made that given the trends, game-changers, and potential for devastating black swans, the world is becoming more unpredictable, or as biblical prophecy indicates, many of these gauges are in fact signs that the end times are approaching at a fast pace.

The wild card in our human understanding of the above information is God's plan. Will God give America and the world clear-eyed strategic leaders with morally sound plans to turn the tide of our chaotic world? Or have we gone beyond the tipping point, and the coming decade or two will in fact usher in the prophetic end times?

Chapter 3

Bring World Back from the Cliff or End Times

SO FAR WE have established that most Americans believe the world is spinning out of control, and that outcome appears linked in part to America's failing strategic leadership given the unique opportunity she inherited to reshape the world in the wake of the collapse of the former Soviet Union. Further, chapter 2 provided a look into the possible future by examining U.S. government analysis concerning how the world might transform in the next couple of decades thanks to global and domestic megatrends, significant emerging game-changers, and the impossible-to-predict black swans. That unpredictable future is already beginning to evidence itself via biblical prophecy signs that will usher in the end times.

This chapter prepares the reader for the rest of the volume by identifying problems that need to be resolved given the forecasted emerging world and the type of leaders best prepared to tackle the future challenges in a spinning-out-of-control world environment.

Major Problems Facing America That Need Fixing

Our problems are both man-made and products of nature. Ideologi-cal conservatives will insist that the next president must begin his or her administration by reversing many "problems" created by President Obama's executive actions, policies, regulations, and appointments. Some of Obama's politically created problems can in fact be reversed by a stroke of the executive pen, while others will take the full cooperation of Congress and time. Yet others may not be rescinded easily because of momentum and costs. Whatever the case may be, the list of prob-lems facing America today is very long and involves many aspects of society, our government, and the geopolitical world. Failure to address them threatens to further diminish America, but they could contribute to God's end-times plan. Notable prophecy scholars in fact postulate that a weak America is necessary for nations like Russia, Iran (Persia), and several Islamic nations to form a coalition and attack Israel (Ezekiel 38–39) with little fear of consequences.

Many of our more serious problems pre-existed Mr. Obama, but he certainly exacerbated these complications, because he was a divisive figure who pitted Americans against one another: black against white, straight against gay, young against old, poor against wealthy, and men against women. Remember, Obama launched into many controversial issues, such as when he disparaged the Cambridge police for acting "stu-pidly," and when he pre-judged racially charged cases like the shooting of Trayvon Martin and the questionable police shooting in Ferguson, Missouri, before the defendant got his day in court.[60]

Mr. Obama's progressive supporters nonetheless celebrate his "accom-plishments" to remake America, such as the Affordable Healthcare Act—or, as it is popularly known, Obamacare—which failed to live up to its billing to provide coverage for the uninsured and because of which many people were kicked out of their previous insurance plans in spite of presi-dential promises otherwise, and for most others whose premiums radically increased. Obama abused our wallets in the name of equity, such as pushing for the Dodd-Frank Wall Street and banking regulations that failed to reform anything, and his almost trillion-dollar so-called economic stimulus was supposed to create hundreds of thousands of shovel-ready jobs, but it too fell flat other than to enrich some progressive supporters like those in the alternative (read "green") energy business.

Obama's radical social actions were especially distressing. The Human Rights Campaign, the world's most powerful and well-funded lesbian-gay-bisexual-transgender (LGBT) lobby organization, compiled lists celebrating the magnitude of Obama's social record. That record includes three LGBT-inclusive legislative acts signed into law; fifteen openly LGBT federal judicial appointments; seven openly LGBT federal ambassadorial appointments; and 111 federal policy/regulation changes to promote the LGBT agenda, such as lifting the military's forever ban on open homosexual service and the promotion of so-called homosexual marriage within the federal government.[61]

Obama's opponents are quick to review the president's litany of "accomplishments" that evidence corruption, dishonesty, and incompetence. They include the following, which are representative of perhaps more than 100 one hundred that the next administration must mine and then vacate as appropriate.

- Obama's policies resulted in 11.4 million Americans leaving the workforce.[62]
- "Fewer Americans are at work today than in April 2000, even though the population since then has grown by 31 million," said Mortimer Zuckerman, editor and chief *U.S. News & World Report*.[63]
- One-third (35.4 percent or 110,489,000) of Americans are either on welfare or receiving unemployment benefits.[64]
- Nearly 11 million Americans receive disability benefits, an all-time record, according to the Social Security Administration.[65]
- America lost its AAA credit rating from Standard & Poor's because of Obama's overspending.[66]

- Obama's "Cash for Clunkers" program, which was intended to scrap old cars for new but cost $1.4 million for every job created and did little to reduce carbon emissions.[67]
- After British Petroleum had a huge oil spill in the Gulf of Mexico, Obama bungled the clean-up process and slowed oil production from other companies that had done nothing wrong, which led to higher oil prices.
- Obama has helped drive up the cost of gas by consistently blocking the Keystone pipeline.
- When he was running for office in 2008, Obama claimed that, "Under my plan, no family making less than $250,000 a year will see any form of tax increase. Not your income tax, not your payroll tax, not your capital gains taxes, not any of your taxes." He lied.

Obama's foreign policy record draws considerable fire. He withdrew prematurely from Iraq in 2011, then totally ignored that country, even as the Islamic State of Iraq morphed into the Islamic State of Iraq and Syria (ISIS) or Daesh and then seized much of Iraq and Syria. Once ISIS became an undeniable threat to the region, Obama admitted to having no strategy, but floundered miserably as many people fell victim to ISIS atrocities. Then, by the fall of 2015, Obama claimed to have found a plan, but at the time, then-U.S. Central Command General Lloyd Austin admitted one part of that strategy, training Syrian proxies to fight ISIS, failed miserably—only five trained rebels joined the anti-ISIS fight. Meanwhile, Obama's so-called sixty-nation air "coalition of the willing" enjoyed limited success in part due to his overly restrictive rules of engagement.

The major problem created by Obama's failed foreign policy is that he made America irrelevant. Although his speeches and press conferences claimed enlightened intentions, his actions delivered irrelevancy. He failed at the Arab Spring by irresponsibly tossing leaders like Egypt's Hosni Mubarak under the bus, and our best regional ally, Israel, suffered consistent snubbing from the Obama White House.

Obama's signature Afghanistan war—his campaign promised "the right war" as opposed to Iraq, "the wrong war"—received his seventeen thousand fresh troop surge in 2009, but that effort failed as well. In 2015 President Obama decided to keep our forces in that country past his presidency.

Meanwhile, Mr. Obama's 2012 security policy promised a "pivot to Asia," which was intended to put more emphasis on China's misbehavior, but it died from neglect. Beijing took advantage of Obama's failed "pivot" to claim all the South China Sea as sovereign territory, and then the communist giant built islets in the middle of international marine transit lanes, arming the islets with sophisticated air defense missiles and direct fire weapons thus threatening our shipping traffic as well as every regional neighbor's claim over the energy reserves under the sea, the prize that likely prompted Beijing's aggressive move.

Lest we forget, Obama campaigned for president on the promise to make America more respected in the world. Now, in the opinion of much of the international community, America under Obama's leadership has become a useless ally with little or no credibility and an ineffectual opponent. Our enemies do not fear us, they jeer us.

Out of fairness to Mr. Obama, many of the domestic and international problems America faces today have others' fingerprints on them as well. History will sort out the blame, but the future presidents and new Congresses will inherit a raft of challenges such as those outlined above as well as some very serious, longer-range issues discussed below.

Consider some of the most serious longer-range challenges waiting for future leaders.

There is a significant wealth disparity among Americans. From 2009 to 2012, some 95 percent of new income has gone to the top 1 percent of the population—that's evidence of income mobility, which means the rich get richer and the poor get poorer. This disparity discourages those struggling to make ends meet while those at the top of the wealth chain use their influence to maintain their disproportionate wealth.

America has a serious education problem. Ask the average American, and he or she is likely to naively insist that we have the world's best education system. However, based on the 2012 Programme for International Student Assessment, a test administered every three years, U.S. schools rank seventeenth out of thirty-four countries among fifteen-year-old students in terms of math, and American students score near the average in science and reading, and rank twenty-first in science and seventeen in reading.[68] The education problem goes to the root of our future competitiveness.

Healthcare is a serious challenge. American healthcare ought to be much better, given the skyrocketing costs. However, the World Health Organization's (WHO) ranking of healthcare systems put the U.S. at the thirty-seventh spot, and a 2013 study by the Institute of Medicine ranked the U.S. as the lowest among seventeen developed nations. WHO's ranking of 191 member states used "performance indicators to assess the overall level and distribution of health in the populations, and the responsiveness and financing of health care services."[69] The Institute of Medicine's study found that "Americans on average die sooner and experience higher rates of disease and injury than the populations of 16 other high-income countries."[70]

Financial safety net is crumbling. Our financial safety net for America's aging population is quickly becoming unsustainable as well. Social Security was created to protect the aged, but at the current rate of expenditure, that system will become insolvent by 2030.[71]

Government is too big. Perhaps the most obvious problem hasn't until now been mentioned: a bloated federal government. It seems as if our federal government keeps growing in size and influence, and demands ever more of our money to fuel its solvent ways. This is perhaps the major challenge our future presidents and Congress must address, and one that must be conquered if we are ever to turn around the fiscal decline of America.

Our collective fear given these and other significant problems is that gradually we are losing our freedom to big government and other interests. We must never forget what president Ronald Reagan warned about our responsibility today for our children's generation.

> Freedom is never more than one generation away from extinction. We didn't pass it to our children in the bloodstream. It must be fought for, protected, and handed on for them to do the same, or one day we will spend our sunset years telling our children and our children's children what it was once like in the United States where men were free.[72]

What Type of Leaders Are Best Qualified to Shepherd the Significantly Challenged America Forward?

Our future leaders need to pay close attention to history and especially the men who founded America. Their example offers lessons that at the very minimum inform today's leaders and help our citizens consider the type of leaders the nation needs in order to attack the significant problems outlined earlier in a world that seems to be spinning out of control.

Future leaders and especially our president must understand that human nature cannot be changed, but its tendencies must be mitigated—and that's the role of our Constitution. That's why the Constitution establishes checks, balances, and rights—the fabric of America. And that's why America became a bottom-up system of self-government, not a monarchy, oligarchy, or other authoritarian form of rule.

Admittedly, as Winston Churchill famously said, "Democracy is the worst form of government...except for all the others."[73] That's a warning to all Americans to protect what we have and select leaders carefully who embrace our time-tried form of government.

It seems that every generation rediscovers the American founders and interprets them for its own age. Our founders were real heroes who overcame significant challenges such as the Revolution in 1776 and framing and then winning support for the U.S. Constitution, the lon-gest-enduring document of its type in the world.

Today, Americans understandably bemoan the lack of greatness in the present crop of leaders who "don't measure up" to the likes of found-ers George Washington and Thomas Jefferson. We compare today's leaders with the founders to justify why the principles we live by are "legitimate" in the ultimate sense.

Our founders possessed a unique moral vision that must be embraced by twenty-first century leaders if we are ever to right the country's course. That moral vision centered on a special kind of freedom and liberty.

George Washington called it "republican liberty," which has noth-ing to do with the contemporary Grand Old Party. Rather, Washing-ton explained the American experiment was an "ordered liberty," which meant a historic test of whether human beings could enjoy equality of rights and opportunities while governing themselves without a king or a military dictator.[74] Robert Kraynak, a Colgate University professor, wrote that our founders governed by a moral vision based on four ideas: a republican form of government, constitutionalism, natural law, and cultural traditions.[75]

This moral vision provides a litmus for future leaders, because they were critical to the American experiment then and arguably now.

First, America enjoys a republican form of government that derives its authority from the people. Leaders must reject the temptation to ignore the people's will and begin to act like kings. Mr. Obama—and, to a lesser extent, president George W. Bush—abused their executive authority well beyond what the founders ever intended.[76]

Second, the founders embraced the Constitution, which put lim-its on the power of the government and established a shared system of power. One of the beauties of our Constitution-based government is the checks and balances of the three branches of government (Articles I, II and III). Future leaders must protect that constitutional checks-and-balances system.[77]

Third, the founders embraced the intangible higher law called natural law as part of their moral vision for America. Liberty was rooted in the moral order of the universe, evidenced in our declaration of independence, which asserts that our liberty and rights come from "the laws of nature and of nature's God."[78]

Professor Kraynak explains that "the natural law doctrine of the Declaration is the moral, philosophical, and theological underpinning of the Constitution and the republican form of government." Our freedom is grounded in "God-given natural rights and would make no sense if the universe were meaningless or indifferent, a view embraced by existentialists who embrace a moral relativism."[79]

Finally, our founders' moral vision is built on cultural traditions, the idea of moral order to social practices such as customs, habits, and manners derived from the heritage of Western civilization. The Judeo-Christian ethic and its biblical beliefs are what made America a shining "city on a hill" that evidences moral virtues for the exercise of responsible liberty by citizens and their leaders.

The moral vision was intended to inspire citizens and exercise their God-given natural rights along with the virtues of courage, moderation, justice, and prudence at the local and personal level.

What is the relevance of the founders' moral vision today? Does the founders' moral vision still provide guidance?

Three hotly debated issues are worthy of consideration by future leaders based on the founders' moral-vision formula: size and scope of government, American exceptionalism, and real freedom. How a leader addresses these issues speaks volumes about his or her governing philosophy.[80]

First, the size and scope of government are a major problem identified in the previous section. Clearly, most of our founders embraced "republican liberty" and therefore favored limited government, because they believed people had to assume responsibility for their own lives. They no doubt would reject the modern "cradle-to-grave" big government that progressives have forced on America, the so-called entitlement power.

Second, is America an exceptional country that ought to use its influence to lead others to freedom and self-government? George Washington was an "isolationist," but he did believe that America had a positive message for others to emulate. Mr. Obama denied America's exceptionalism, but Mr. Bush embraced it—likely the founders would have threaded the needle between the two views of America's role in today's world.

Third, future leaders must have a vision for freedom. Is American-style freedom really moral relativism, in which anything goes that doesn't violate law? Or, as the founders would likely say, the American style of freedom balances the ideal of a free society with the cultural traditions that make it possible and desirable.

The founders, according to Professor Kraynak, were "enlightened traditionalists"—they understood the special preconditions of ordered liberty, which distinguished liberty from license because it was based on certain God-given truths…they took for granted that the cultural preconditions of ordered liberty would always be there—in churches and stable family life, in local communities and social manners, in competitive economic life, in public education…in today's moral climate…they would be deeply disturbed by the degeneration of the culture and would be on the conservative side of the culture wars, since responsible liberty is not moral relativism, libertarianism, or mere self-expression.[81]

Today we are moving toward a destructive mixture of centralized state power and permissive freedom, an irrational combination of the all-powerful state and rootless individualism that undermines ordered liberty at both ends by rejecting limited government and vital cultural traditions.

This generation must rediscover the American founders and restore the balance they considered necessary for the proper exercise of republican liberty.[82]

Alexander Hamilton embraced the moral vision requirement for American leaders, but he also argued for our president to be given vast powers. Hamilton argues in Federalist Paper Number 70 for "energy in the executive," and said "it was a leading character in the definition of good government." He wrote that such energy "is essential to the protection of the community against foreign attacks and other dangers, including the enterprises and assaults of ambition, or faction and anarchy."[83]

Are there certain traits our future leader ought to demonstrate in addition to embracing the founders' moral vision?

John Gartner wrote an insightful article in *Psychology Today* about the traits of an effective president. The piece cites an unnamed historian who surmised that "no past president could function effectively in today's environment."[84] Gartner goes on to explain that contemporary top American leaders need to be "slightly deranged to run for the office today."[85]

There are key psychological traits needed by a future successful president that Gartner identifies with the help of presidential historians, biographers, and political scientists. The most important of these traits is judgment, and we must eschew the opposite characteristics in our top leader, such as impulsivity, arrogance, and a tendency to move and think too fast. These traits work against what judgment requires: measured, sober, thoughtful, and patient study[86].

Gartner applied his analysis to three recent presidents to build a list of desirable traits. President Bill Clinton evidenced a number of very desirable traits for a would-be president: extroversion, care for people, and an insatiable curiosity about everything.[87]

Clinton's extroversion drove him in every aspect of life—evident in his hyper-sexuality, eating, and talking, but also in his "driving need for contact with people, people, and more people," wrote Joe Klein in *The Natural: The Misunderstood Presidency of Bill Clinton*.[88] That's why Clinton was a natural campaigner. He could go for long periods without sleep and famously wore out the soles on three pairs of shoes in a single campaign.

Mr. Clinton is a very intelligent man. Hillary Clinton said of her husband, "He is insatiably curious about everything." That trait made him the ultimate policy wonk. Klein illustrates, "He read every line of the 2,000-page budget, and sought open—and, it seemed, endless—

debate on every item." Biographer Bob Woodward wrote, "The staff did their best to move Clinton along, but the president resisted, hungering always for more detail."[89]

President George W. Bush evidenced dominance, confidence, vision, aggression, and judgment. One saw Bush's traits during the election debate when he attacked Democratic opponent Vice President Al Gore. He attacked Gore as a liar, evidence of Bush's dominance. Gartner says Bush was known as a strong leader, which is often why Americans are attracted to military commanders like George Washington, Dwight Eisenhower, or even David Petraeus.[90]

Few can forget the image of strong leadership demonstrated by Bush as he stood on the rubble of the post-9/11 World Trade Center with a bullhorn in his hand shouting encouraging words to those desperately trying to rescue victims from the rubble.

Bush was a visionary much like President Roosevelt. Jay Winik wrote in *1944: FDR and the Year that Changed History* that Roosevelt was a leader who could "see over the horizon." Bush similarly saw the "axis of evil" that became his mantra to justify attacking Iraq in 2003.[91]

Visionaries often demonstrate less-than-admirable contrary traits. Roosevelt, for example, was "all ego, all vanity," wrote Winik. Bush seemed to have too much aggressive energy with a tendency toward impulsive actions. Gartner surmised that Bush has a short attention span, likes to make snap judgments, and displays a "bullying impatience with doubters and even friendly questions." These foibles arguably contributed to Bush's failures.[92]

President Obama is more of a philosopher king-type president as opposed to Bush's alpha male. Obama is an introvert like former President Jimmy Carter. He is highly thoughtful and works slowly, deliberately. He told a reporter, "We probably spent much more time trying to get the policy right than trying to get the politics right."[93]

Obama is also a visionary, but of a different sort than Bush. "Being led by Barack Obama is like being trumpeted into battle by Miles Davis. He makes you want to sit down and discern," wrote David Brooks with the *New York Times*.[94]

Gartner concludes his trait analysis by explaining that the most reliable indicator of true toughness is toughness under pressure, something we should want in every president. Some will label it as "resilience" or the quality of being "indomitable."

Conclusion

The twenty-first century problems outlined above exist in a rapidly changing and troubling world, which demands leaders with our founders' type of moral vision and a host of time-proven character traits to guide America to renewed greatness or usher it into the prophesied end times.

It is incumbent upon Americans to chart a path ahead that tries to right the many problems facing our country while mitigating the negative impacts of the emerging megatrends and game-changers outlined in this section.

What America needs now is a twenty-first century, all-American strategy that addresses the challenges and problems identified above and are attacked by leaders equipped with the right set of time-proven character traits.

America needs men like William Wilberforce, an English politician and Christian champion of the movement to abolish the slave trade. Thanks to Wilberforce's tenacity to overcome many obstacles, his campaign led to the Slavery Abolition Act of 1833, which abrogated slavery in most of the British Empire. We need that Wilberforce-like, dogged determination in our champions facing the challenges in the twenty-first century.

New World Order
Threats and Risks

We are experiencing a global moral/spiritual decline that makes it seem as if the world is heading at high speed to a cataclysmic ending—the prophetic end times. Section I provided a glimpse of our future world and set the stage for building a twenty-first century strategy, but that process can't begin until we thoroughly understand the actors, technologies, and environments that present emergent threats and risks that must be tamed or controlled. These challenges are linked, albeit subtly, to specific prophetic signs of the end times.

Chapter 4 profiles the threats and risks associated with the new multipolar world from a sovereign-state perspective. No longer do we live in the bipolar Cold War era or the unipolar period dominated exclusively by the United States. The current and especially the future decades are marked by a multipolar world stage where numerous sovereign nations will compete for scarce resources and dominance, which presents significant threats and risks.

Chapter 5 picks up from chapter 4 to address the non-state actor threats and risks, especially terror groups that will grow in number, capabilities, and influence to threaten even more instability across the globe. They will radicalize more of the world's growing unconnected population through the sophisticated use of social media and arm them with an array of killing means.

Chapter 6 addresses the role technology is playing on the world stage, especially in terms of human capability and capacity. The transhumanist movement—altering man's mental and physical personhood—introduces intriguing advances in terms of intelligence, longevity, and strength, but it also represents some very serious and frightening threats and risks.

Modern technology also gives mankind some very significant capabilities beyond creating transhumanist Frankensteins. Specifically, in chapter 7 we will address the emerging significant changes in the use of outer space followed by the emergent cyberworld in chapter 8, and then incredibly daunting threats and risks associated with modern weapons of mass destruction in chapter 9.

These insights are shocking and exciting, and together they illustrate a set of threats and risks for America's future leaders to counter when assembling their twenty-first century, all-American strategy. Failure to address any one of these pieces could be catastrophic, and each appears to be associated with a prophetic end-times sign.

Chapter 4

Nonpolar New World Order

THE TWENTY-FIRST CENTURY will be characterized by a form of international order known as nonpolarity. Richard Harass, the president of the Council on Foreign Relations, defines nonpolarity as "a world dominated not by one or two or even several states but rather by dozens of actors possessing and exercising various kinds of power."[95]

This chapter will examine what nonpolarity means for the new world order and in particular the U.S. to include an examination of the likely consequences, as well as profile an example of some of the world-stage actors. Also, the chapter examines whether the contemporary nonpolar international environment is a prophetic sign of the end times.

Return of the Multipolar World

The twentieth century began as a multipolar world, but then two world wars and a bipolar international system emerged featuring the former Soviet Union and the United States. Once the Soviet Union collapsed in 1991, the U.S. became the world's sole superpower, or the unipolar leader, which came with the opportunity—some say the responsibility—to remake the world in its image.

America failed to rise to the occasion to make a better world in the wake of the collapsed Cold War in spite of its sole proprietorship of unipolar status. Too quickly, the nation lost its way as the world's sole superpower, and a new nonpolar or multipolar world order reemerged with numerous centers of influence where no one player dominates. But this time, the traditional sovereign nation-state power centers were joined by many new centers.

Today's multipolar world has six obvious sovereign nation-state power centers: the United States, China, the European Union, Russia, India, and Japan. Together, these six powers control half of the world's population and three-quarters of the world's gross domestic product. These nation-state centers are joined by other power centers: regional powers like Brazil in South America; intergovernmental organizations like the United Nations; global corporations like Exxon Mobile; media outlets like *Fox News* and *CNN*; terrorist organizations like the Islamic State of Iraq and Syria; as well as non-government organizations like Samaritan's Purse and World Vision.

The U.S. remains the most influential of these centers with a large military and a giant economy. But, after many years on the ascent, today America appears to be on the decline in influence while other centers are rising in terms of wealth. Our dollar is weakened compared to other world currencies and our financial leverage is declining across the world. Diplomatically we must depend on others to leverage regimes like Iran regarding its nuclear program.

In spite of America's decline, no great power has emerged to take the unipolar seat, although China now has the world's leading economy mostly because of America's fiscal stumbling. Meanwhile, the emergence of globalization spreads the wealth among more actors—but not without risks.

The major risk associated with globalization is that it encourages nonpolarity by granting leverage to other than nation-state governments due to the flow of commerce. Nation-states must now share power with other, non-state centers; thus, influence is defused, shared, and not as easily coordinated to address world crises, which contributes to more instability (read "the likelihood of war").

Multipolar International Systems Trend to Be Unstable

A review of modern history is instructive regarding past nonpolar international periods. Those periods tended to be unstable and war-prone, a likely pretext for the new nonpolar world order.

Seventeenth-century Europe's multipolar order crumbled with the advent of the Thirty Years War (1618–48). That conflict was triggered by the inevitable disagreements among men: religion, land, and politics. The Peace of Westphalia ended that conflict and established the concept of a sovereign state, which led to our international system of states.[96]

French Emperor Napoleon Bonaparte interrupted the system of states in the early nineteenth century with a series of conquering campaigns. Once Napoleon fell, the European powers held the Congress of Vienna to reestablish the previous order that resulted in the creation of the Concert of Europe, a balance of power that existed until 1914 and the outbreak of World War I.[97]

The twentieth century was marked by further changes to the multipolar international system. World War I destroyed the Concert of Europe's multipolarity success and replaced it with a new system of alliances. During and immediately following World War I, President Wilson and other European leaders worked to create the League of Nations, a multilateral organization headquartered in Geneva, Switzerland, whose purpose was to be a forum for resolving international disputes. Although the U.S. cooperated with the League, it never officially joined. The League lasted until Adolf Hitler's Nazi Germany invaded Poland in 1939, triggering the start of the Second World War.[98]

After World War II, the world tried yet another multipolar

experiment called the United Nations, but soon the concept of multipolarity was shoved aside by the emergence of a new bipolar world order: two nuclear giants, the United States and the Soviet Union. Virtually every country aligned with one or the other in the Cold War's bipolar new international world system.

Obviously, multipolarity hasn't fared well throughout history, because it tends to wax and wane, creating an unstable and unpredictable world with shifting alliances and rising players that change the balance of power to create yet another fresh effort at world order.

This lesson applies today with our new nonpolar new world order. However, things today aren't really that different from the past, even with the world's interconnectedness.

Keep in mind that the twentieth century was economically interconnected much like today, but that interconnectedness died with the advent of the First World War. Further, the multipolar League of Nations world order and multinational United Nations effort at world order failed and continues to fail to provide global peace and security. Ultimately, global powers like the U.S. have stepped into chaotic situations to lead because multipolar, multinational efforts need singularity in leadership and purpose, which comes only from a superpower leader.

Dr. Henry Kissinger, who served as national security adviser and secretary of state under Presidents Nixon and Ford, is skeptical about the world's new multipolar future. Kissinger wrote in the *Wall Street Journal*, "The concept of order that has underpinned the modern era [of multipolarity] is in crisis." He places much of that blame on the West and its squandered unipolar opportunity.[99]

Kissinger explained that "the search for world order has long been defined almost exclusively by the concepts of Western societies." Specifically, world order for America is identified with the spread of liberty and democracy, which was expected to "achieve just and lasting peace" where applied.[100]

Therefore, America's goal for international order came through its efforts to spread democracy. Further, accompanying democracy, America advocated free markets as the necessary medicine to "uplift individuals, enrich societies and substitute economic interdependence for traditional international rivalries."[101]

Unfortunately, as Kissinger explained, most of the world never really shared America's idealism, and now the world stands at a turning point. Specifically, "the international order thus faces a paradox: Its prosperity is dependent on the success of [economic] globalization, but the process produces a political reaction that often works counter to its aspirations."[102]

He illustrates the point with the situation in the Middle East. That region may have prospered from world trade—mostly energy sales—but today, in spite of America's war in Iraq to replace a dictator with a democratic government, that government and others in the region are in chaos.

The sage of foreign affairs also believes that the new multipolar world order is failing for lack of an effective mechanism to deal with issues of consequence. This is a poke at the United Nations and other intergovernmental forums that host discussions on "pending tactical issues" but fail to address relevant issues.

Rather, Kissinger argues, "The contemporary quest for world order will require a coherent strategy to establish a concept of order within the various regions and to relate these regional orders to one another." He warns that should one country use its military to "establish a concept of order," it could "produce a crisis for the rest of the world."[103]

Multipolar System Is Net Negative for America

The new international political order will deliver a host of negative consequences for the United States and others, primarily because there are more power centers and increased threats.

Having more power centers makes reaching consensus about crises more difficult. And just as reaching consensus becomes more difficult, the world seems to face more and diverse threats than ever: rogue regimes, transnational terrorist groups, commercial power centers leveraging critical commodities, and global financial groups manipulating currencies. The expanded list of multipolar actors and their influence over world to their parochial favor creates an incredibly complex international arena that, frankly, will more often than not seem to most observers to contribute to the sense that the world is spinning out of control.

Consider some of the consequences of the mix of many actors and more threats/crises.

First, globalization encourages a painful trade consequence. Globalized trade works best when all nations refrain from economic warfare, but today's many authoritarian regimes tend to violate those rules. They get away with such abuse because there is no international enforcement mechanism against the likes of Internet crime and the infrastructure that fuels globalized trade. China, a major American trade partner, also abuses its access to American markets by undermining the promise of free and unsubsidized trade. That hurts globalized trade, as does a growing lack of confidence in international finance.

Second, the nonpolar world has no single power center that is really in charge of managing flare-ups, which means the inevitable confrontations potentially metastasize into something far worse: more wars. For example, the rise of China undermines the Asian balance of power and lights up its old rivalry with Japan. Already, China's expanding missile arsenal and renewed confrontation over some disputed islands in the East China Sea are considered a threat to Japan because, in part, America is giving more indicators that it is retrenching in its global commitments to Tokyo.

Nuclear power India is a rising power center and it faces a credible menace from nuclear-armed and hostile neighbor Pakistan. A war between Pakistan and India could result in an Armageddon-like conflagration for all of Asia.[104]

Russia is growing very aggressive, as evidenced by its 2014 annexation of Crimea, support to the ongoing insurgency in eastern Ukraine, involvement in the Syrian civil war, and extensive saber-rattling about NATO's incursions in East Europe. Meanwhile, Moscow significantly increased military expenditures over the last decade and uses its petroleum and natural gas sales to leverage relations with western European customers and former satellites.

Third, the emerging small power centers/states and non-state actors undermine future stability the most. These actors have military capabilities that could endanger global security, such as North Korea's growing nuclear warhead and ballistic missile arsenal. An attack from North Korea on Japan or South Korea will trigger an immediate reaction from the U.S. that could lead to a nuclear domino effect in eastern Asia.[105]

Regional power centers like Iran armed with nuclear capabilities could represent a significant cause for concern for global security, especially as the regime repeatedly threatens to attack Israel, an action that could quickly involve the rest of the world. Further, Iran's emerging nuclear weapons arsenal will inevitably persuade other countries in the region like Saudi Arabia and Egypt to develop new security measures and alliances with external powers and seek their own nuclear weapons.

Finally, the new multipolar world encourages globalist efforts to exercise far more control over the world's finances. Consider a statement by globalist Paul Volcker, the former chairman of the U.S. Federal Reserve board, who anticipates that the global community is poised to leap on the world's brewing financial crisis with a fresh set of market tools.

> Fortunately, there is also good reason to believe that the means are now available to turn the tide. Financial authorities, in the United States and elsewhere, are now in a position to take needed and convincing action to stabilize markets and to restore trust… the point is the needed tools to restore and maintain functioning markets are there.[106]

Volcker's statement suggests he believes that, given the right financial crisis, world governments can work together to do what they want, such as infringe upon basic freedoms. The U.S. already experienced a distasteful example of politicians using a financial crisis to seize power, and few Americans want to see a similar episode imposed by globalists.

President Obama and Congress, in the wake of the American financial meltdowns of 2007 to 2009, took advantage of that crisis to take many groundbreaking and unprecedented actions that involuntarily conscripted the American taxpayer to take on enormous risks and crippling debt that now exceeds $19 trillion.

Such a scenario on the world stage is unfortunately more likely given the current level of imbalance among world financial markets. But first, the scale of a future financial crisis must be world threatening before globalists like Volcker and currency speculator George Soros can reach a backdoor consensus—but they will then enact draconian changes. You can take that to the bank!

We came close to such a scenario following the Asian financial instabilities in the late 1990s. With markets in crisis and Russia's markets collapsing, the World Bank and the International Monetary Fund called for enormous bailout and emergency financing programs to "fix the global financial architecture."[107] The United Nations report on the subject is especially chilling. It called for the creation of a world financial authority, the establishment of an international credit insurance corporation, and significant exchange-rate system reform to include the adoption of a single world currency, issued by a world monetary authority.

Multipolar Players

It's always helpful before any game to consider the players and their track records. Their past performance is an indicator as to how they might perform in the future. That's also true about power centers in the new nonpolar world. Consider what three of the nonpolar world's leading power players—a nation-state, intergovernmental organization, and corporation—might do in the future.

Sovereign Nation-State Players

China is the most powerful emerging nation state today, not only economically but also militarily. China's socialist market economy ranks second in the world in terms of gross domestic product, but first in terms of purchasing power. Until 2015, China had the world's fastest-growing economy, averaging 10 percent annually. It is also the world's largest trading nation, a member of the World Trade Organization, and a party to a number of free-trade agreements.[108]

China's financial future isn't clear, but there are optimists who encourage Beijing to speed up its economic engines. Specifically, billionaire globalist George Soros is practically giddy about China's role in the nonpolar new world order. Soros called for the communist regime to "own" the new order and step up to the plate. "I think you really need to bring China into the creation of a new world order, financial world order," Soros told the *Financial Times*. "I think you need a new world order, that China has to be part of the process of creating it and they have to buy in, they have to own it in the same way as the United States owns…the current order."[109]

It's evident that China's president, Xi Jinping, is trying to transform that country economically while simultaneously pushing to extend its diplomatic, military, and trade reach. Specifically, Xi is trying to shift the country's economy from an export-driven one to one based on consumer demand. That's evidence that he no longer believes that China's future growth will come from cheap labor-producing cheap exports. This view may be a reaction to China's slumping growth, which is expected to slow from 6.8 percent for 2015 to 6.3 percent for 2016.

China intends to possess a globally capable and high-tech military the rival of the American armed forces. For three decades, Beijing made double-digit annual increases to its military budget, putting it in many areas almost on par with the U.S. military in terms of size and sophistication.

In early 2016, the Chinese defense ministry announced the construction of its first indigenous aircraft carrier and the second ship for the People's Liberation Army (PLA) Navy after the Liaoning (originally a Ukrainian vessel) entered service in 2012.

Former Central Intelligence Agency acting director John McLaughlin profiled the emerging Chinese military threat. He offered that China's ultimate character—whether it will be an adversary, competitor, or partner with the U.S.—remains a mystery. But in terms of its military, there is no doubt about its significant growth.[110]

China's military modernization investments have centered mainly on creating a more capable blue-water navy, a smaller streamlined army, less vulnerable nuclear forces, a significant cyberforce, and an expeditionary command-and-control capability—which means China can project its military action virtually anywhere on the globe. Chinese ships and soldiers are currently deployed around Africa, in the Middle East, and throughout Latin America.

This massive, sophisticated force is much larger and more capable than is necessary to be the regional hegemon—China's long-held public explanation for building up its military. For example, China fields a submarine fleet larger than America's in terms of sheer numbers of attack vehicles. The PLA's recent defense white papers indicate the goal of an over-sized submarine capability is to gain regional power while protecting the mainland, especially the areas stretching far from China's coastline to what it calls the two island chains, an area that extends from Japan to Indonesia.

China in the South and East China Seas defies the international consensus on maritime and aviation freedoms. It is building artificial islands atop coral reefs five hundred miles from the Chinese mainland that are armed with air-defense systems and jet fighters, and Beijing claims 90 percent of the South China Sea in defiance of international law that limits territorial waters to only twelve miles from its mainland shore.

On March 10, 2016, Director of National Intelligence James Clapper said China's military buildup on disputed islands in the South China Sea is nearing completion. "Based on the pace and scope of construction at these outposts, China will be able to deploy a range of offensive and defensive military capabilities and support increased PLA [Navy] and [Chinese coast guard] presence beginning in 2016," Clapper said.[111]

Senate Armed Services Committee Chairman Senator John McCain (R-AZ) asked Mr. Clapper for the assessment of the impact of Chinese island-building on Beijing's ability to deploy armed forces to the region. McCain responded to Clapper's statement, "Whatever Chinese officials may be saying publicly, Beijing is using coercion and the threat of force to unilaterally change the status quo and challenge the rules-based order in the Asia-Pacific." The senator continued, "As China militarizes the South China Sea, the credibility of America's commitments to the security of our allies and partners in the region hangs in the balance."[112]

China's aggressive move to dominate the areas far from its shores is a telling development. Keep in mind, 50 percent of the world's tanker trade passes through the South China Sea and, according to the World Bank, the oceanic energy reserves beneath those waters could total seven billion barrels of oil and nine hundred trillion cubic feet of natural gas. This is sufficient incentive for the Asian giant to break others' "china."

Intergovernmental Organization Players

The new nonpolar world order includes a host of intergovernmental organizations (IGOs), nongovernmental organizations that operate internationally to mediate political bargaining and act as a catalyst for coalition formation. Members of IGOs are primarily sovereign states, and the list of IGOs should be familiar: World Trade Organization, Council of Europe, Organization for Economic Co-operation and Development, and of course, the best known IGO of all, the United Nations.

The United Nations is the world's largest and arguably most influential IGO that exercises great influence and no doubt is a key player in the new world order. Its critics have called the U.N. "the most dangerous threat to American sovereignty," as it has a history of taking aim at Western culture and acting to create a "one-world order."

There is some merit to that criticism, because the U.N. long ago lost sight of its founding purpose to become a mechanism for the global progressive movement.

The United Nations officially came into existence in October 1945 in San Francisco at a meeting of representatives from fifty countries.[113] Article 1 of the U.N.'s charter outlines the organization's four-part purpose:[114]

1. To maintain international peace and security, and to that end: to take effective collective measures for the prevention and removal of threats to the peace, and for the suppression of acts of aggression or other breaches of the peace, and to bring about by peaceful means, and in conformity with the principles of justice and international law, adjustment or settlement of international disputes or situations which might lead to a breach of the peace;

2. To develop friendly relations among nations based on respect for the principle of equal rights and self-determination of peoples, and to take other appropriate measures to strengthen universal peace;

3. To achieve international cooperation in solving international problems of an economic, social, cultural, or humanitarian character, and in promoting and encouraging respect for human rights and for fundamental freedoms for all without distinction as to race, sex, language, or religion; and

4. To be a center for harmonizing the actions of nations in the attainment of these common ends.

Over the U.N.'s seventy-year history, it has grown from fifty member countries that formed the first quorum in 1945 to 184 member nations today.

One widely heard criticism is the U.N. has embraced moral relativism. Dore Gold, a former Israeli ambassador to the U.N., wrote a book, *Tower of Babble: How the United Nations Has Fueled Global Chaos*, in which he argues the U.N. lost sight of its original purpose. Originally

U.N. membership was limited to democratic and free countries that declared war on the Axis powers of World War II, according to Gold. That limitation was soon scuttled by U.N. membership, and today much more than half of the member countries are closer to dictatorship than democratic rule. [115]

The criticism of the U.N. also involves specific social issues like abortion. It is frequently accused of providing support for programs that promote forced abortions and coercive sterilizations. For example, the U.N. Population Fund is the organization's abortion culprit, which was caught aiding a Peru population control program in the 1990s.

Another frequent U.N. criticism is that the organization seeks to rob free nations of their sovereignty in order to impose "one-world order." Glenn Beck, formerly of the *Fox News Channel*, said the U.N. pushes for "government control on a global level."[116]

Such accusations deserve examination. What is the U.N.'s contemporary agenda for the world?

Some of the U.N.'s long-term goals are outlined in a 1992 document known as Agenda 21, a non-binding action plan that takes aim at achieving global "sustainable development."

The U.S. Congress never approved America's participation in Agenda 21, but Presidents Clinton, Bush, and Obama each signed executive orders implementing it.

The U.N. bypassed national governments using Agenda 21's International Council of Local Environmental Initiatives (ICLEI) to make agreements directly with more than six hundred cities, towns, and counties in the United States.[117]

Some critics like Tom DeWeese with the American Policy Center claim Agenda 21 is a "blueprint to turn your community into a little soviet," and "it all means locking away land, resources, higher prices, sacrifice and shortages and is based on the age old socialist scheme of redistribution of wealth."[118]

DeWeese perhaps reads too much into Agenda 21. After all, both former Secretary of State Colin Powell and President George Bush endorsed the agenda. Neither man, according to one conservative writer, would ever embrace a "worldwide plot to deprive Americans of their constitutional rights to rape the land, foul the air, dirty the water, and sprawl development."[119]

Conspiracy allegations may be understandable, but they really are not fair because the ICLEI does its work in the open and with the endorsement of current and past U.S. presidents. The ICLEI convinces local U.S. governments to embrace socialist and extreme environmentalist programs with the funds and encouragement of globalists like billionaire George Soros' Open Society.[120] Where is the opposition to U.N. agents taking their extreme ideas into American communities?

The seven-hundred-page Agenda 21 is divided into four sections that illustrate the type of issues the U.N. aims to attack. Those issues include: social and economic dimensions (poverty, health, sustainable population); conservation and management of resources for development (atmospheric protection, combating deforestation, protecting fragile environments); strengthening the role of major groups (children, women, nongovernment organizations, business and industry, workers, farmers); and means of implementation (science, technology transfer, education, international institutions, and financial mechanisms).

That sounds as if there is nothing outside Agenda 21's purview. Although this is a voluntary agenda for U.N. members, virtually all nations signed up to participate.

Agenda 21 claims to promote "sustainability," a nuanced word for "environmentalism." That means the U.N. will promote vis-à-vis the ICELEI socialist goals that erode our liberties in order to push to reduce fossil fuel consumption in the name of protecting the environment. Perhaps the best way of thinking of the threat is to envision Obama's Environmental Protection Agency's agenda for America that seeks to reduce private property ownership and private car ownership and put in place population control and Obama's "social justice" reforms, which include calls for redistribution of wealth.

The ultimate threat of Agenda 21, according to a columnist in *Townhall*, is "using government to heavy-handedly accomplish vague goals of caring for the earth" which "goes contrary to our free market capitalism."[121]

International Corporate Players

Global corporations wield great influence in the nonpolar world through their economic activities like creating jobs, lobbying governments, and using mass media to reach citizens directly and bypass their governments.

Exxon Mobile Corporation is a very powerful global energy company involved in virtually every segment of the energy sector from coal mining and electricity to the production, exploration, marketing, and transportation of oil and gas. It manufactures and markets petrochemicals and conducts its operations and projects in the U.S., Asia, Africa, Canada, South America, Europe, and Australia/Oceania. It has a massive global workforce (86,300 in 2010) not counting other hundreds of thousands of contractors supporting the global energy giant.[122]

Exxon Mobile is a very wealthy company that holds 49.72 percent of its company shares valued at $161 billion.[123] That means the aggregate company stocks are worth north of $312 billion, roughly equal to the 2015 GDP for a country like South Africa.[124]

World power player Exxon Mobile uses its clout and money to influence politics in America and across the world. For example, in 2014 it made significant political contributions and ran a large ($12, 650,000) lobbying effort that focused on five primary issues on Capitol Hill in 2014: taxes; energy and nuclear power; environment and superfund; government issues; and fuel, gas, and oil.[125]

Is the Nonpolar New World Order a Sign of the End Times?

You will be hearing of wars and rumors of wars…for nation will rise against nation, and kingdom against kingdom. (Matthew 24:6–7, KJV)

Bible prophecy is very specific about end-times world politics. During the end times, Israel is reestablished (Ezekiel 37:21–22) and nested in the middle of hostile neighbors threatening the Jewish nation's annihilation (Ezekiel 35:1–36:7). This much is already fulfilled since the Israeli declaration of independence in 1948.[126]

Now consider the nonpolar new world order as part of the possible end-times prophecy. Specifically, the prophet Daniel said the Roman Empire would be revived (Daniel 2:36–41), which some argue was fulfilled with the formation of the European Union, one of the six big nonpolar world players according to the NIC. The EU does appear to fit the prophecy in spite of its ongoing problems. Then there is the reemergence of Russia, another nonpolar new world player.[127]

The Bible speaks of a great power in the land of Magog in the north, which most eschatologists associate with Russia or at least the former southern, Muslim-dominated republics of the former Soviet Union. Magog/Russia, according to the prophecy, will form a coalition with Persia, modern Iran, and other Islamic states in the region, and will ultimately invade the Jewish homeland (Ezekiel 38:1–39:16). Russia's ongoing involvement and alliance with the likes of Syria and Iran— sworn enemies of Israel—certainly supports this view.[128]

Another indicator of the end-times prophecy is unparalleled war as indicated by Matthew 24. Certainly the new nonpolar world with so many contrary "players" makes the possibility of war increase exponentially.[129] But there is more.

The war-wracked nonpolar world could become so chaotic that globalists like Soros endorse a return to unipolar rule, but this time a true one-world government. One indication of that possibility is suggested earlier in this chapter by globalists who aim to "fix the global financial architecture" and establish a world monetary authority (which sounds like the great authority mentioned in Revelation 13:2), and that a future world leader, the Antichrist, would have authority over "every tribe, people, language and nation" (Revelation 13:7, ISV). It's logical that controlling the world's wealth—all currency—through a centralized "global financial architecture" and a single monetary authority is indicative of a future one-world government.

Conclusion

Failure to grasp the nonpolar golden ring after the Soviet Union's demise has left us in a multipolar world. The twenty-first century poses a daunting future in which this new nonpolar international system likely will lead to more instability among a growing list of power players. Further, the high-tech war-making capabilities in the hands of an increasing number of actors and the uncertainty of globalized financial networks make a compelling case that the new world order is a clear sign of the end times. Major wars in the Middle East are on the immediate horizon, and the world seems ripe for the appearance of the one man who will take charge of the world systems and bring peace and stability for all. Many eschatologists believe he is alive today waiting to grab the opportunity.

Chapter 5

Terrorist Threats and Risks

TERRORISM IS ON THE RISE in part due to the instability associated with the new nonpolar world order. Terror groups will grow in number, capabilities, and influence to threaten further instability in part by radicalizing more of the world's growing unconnected population. Their weapons are sophisticated social media campaigns with messages of extremist ideas backed by actions that will lead to unremorseful killing, even with weapons of mass destruction.

This chapter defines terrorism, profiles contemporary terrorist groups, and identifies terrorist tools and their motivation. It concludes with an analysis of the future face of global terrorism and answers the question of whether terrorism is another prophetic sign of the end times.

What Is Terrorism?

What is terrorism? It can be easily defined as "a crime meant to terrorize," such as the kind that makes us afraid to send our kids to school after mass shootings in places like Columbine, Colorado, or Sandy Hook, Connecticut, or to join a footrace in a major city after the 2013 Boston

Marathon bombing, a 2015 Christmas party in San Bernardino, California, or to fly home from the Brussels, Belgium, airport in 2016.[130]

Perhaps terrorism is more than simple crime; rather, it is an infraction committed for a political reason. Was the Reverend Martin Luther King Jr. a terrorist for illegally marching in the Birmingham, Alabama, streets in 1963 to protest the rank discrimination of African-Americans solely based on their race?

Although Dr. King and other civil rights leaders denounced violence at the time, it did occur. Does the fact that violence occurred in the course of a legitimate protest make civil rights marchers "terrorists"?

The Federal Bureau of Investigation defines terrorism as "the unlawful use of force and violence against persons or property to intimidate or coerce a government, the civilian population, or any segment thereof, in furtherance of political or social objectives."[131]

King's civil disobedience was technically a "crime," according to Birmingham's law, and the march was done for political purposes. Perhaps there is something missing in the definition, such as the would-be terrorist's affiliation with a certain group that makes the "crime" terrorism. Consider the case of Eric Rudolph, also known as the Olympic Park Bomber, who was convicted for a series of anti-abortion and anti-homosexual bombings that killed two and injured 120 people. Were Rudolph's bombings acts of terrorism?

Terrorism is often referred to as a tactic used by the weaker side in a conflict to coerce the stronger adversary such as a government to abandon an activity, and at the same time the event calls public attention to the "terrorist's" political objection(s).

Unfortunately, U.S. government agencies don't share a common definition for terrorism, which might explain some of the public confusion. The Defense Department defines terrorism as "the calculated use of unlawful violence or threat of unlawful violence to inculcate fear; intended to coerce or to intimidate governments or societies in the pursuit of goals that are generally political, religious, or ideological." The key elements in this definition are violence, fear, and intimidation.[132]

The U.S. Department of State uses a different definition for terrorism: "Premeditated politically-motivated violence perpetrated against non-combatant targets by sub-national groups or clandestine agents, usually intended to influence an audience."[133]

The United Nations has its own definition for terrorism: "An anxiety-inspiring method of repeated violent action, employed by (semi-) clandestine individual, group or state actors, for idiosyncratic, criminal or political reasons, whereby—in contrast to assassination—the direct targets of violence are not the main targets."[134]

These definitions appear to agree that terrorism is a planned criminal violent act that seeks to influence a populace and/or government regarding a cause important to the act's perpetrator. This consensus definition is illustrated by the 1972 terrorist attack against Israeli athletes at the Munich Olympics. The Black September Organization killed eleven Israeli athletes at Munich's Olympic Village while drawing a global population's attention to the plight of their cause, Palestinian refugees.

Two decades later, on April 19, 1995, American Timothy McVeigh, angered by FBI actions at Ruby Ridge and Waco, Texas, earlier that decade, used a truck bomb to kill 168 people at the Alfred R. Murrah Federal Building in Oklahoma City.

Ruby Ridge was a deadly confrontation between the FBI and the U.S. Marshals Service and Randy Weaver, a northern Idaho man and his family. That incident resulted in the death of Weaver's son and his wife. Similarly, the 1993 Waco siege (a raid of the Branch Davidian compound to serve arrest and search warrants involving firearms and explosives), involved the same agencies, which fueled the widening of the militia movement.[135]

Both foreign and domestic fanaticism and violence continue today, but seemingly at a great and more deadly pace thanks to groups like ISIS and their use of car bombs, beheadings, hostage-taking, and now coordinated attacks in the West.

These terrorist events have two common factors. First, they were motivated by a specific belief or ideology to act by targeting innocent persons, and second, the authorities were surprised by each event.

Terrorists are effective because they are unpredictable. Like the former

Yankee baseball player Yogi Berra said, "It's tough to make predictions... especially about the future." That's why we must know more about the terrorist.[136]

For the purposes of this book, I will define terrorism as does the Department of Defense: "The calculated use of unlawful violence or threat of unlawful violence to inculcate fear; intended to coerce or to intimidate governments or societies in the pursuit of goals that are generally political, religious, or ideological."

Terrorist Organizations

Americans face many terrorist threats at home and abroad. The FBI identifies six domestic terror threats: eco-terrorists and animal rights extremists, the sovereign citizens' movement, anarchist extremism, militia extremism, and white supremacy extremism.[137] However, because most terrorist targets are related to global issues, the focus here will be on foreign terrorist groups that obviously have an impact on U.S. citizens and our national security.

Those entities are designated by the U.S. Department of State as Foreign Terrorist Organizations (FTO). Specifically, the State Department makes this designation using three legal criteria under Section 219 of the Immigration and Nationality Act (INA): (1) a foreign organization, (2) that engages in terrorist activities as defined in the INA law and (3) the group's actions must threaten U.S. nationals and America's national security.[138]

The 2014 State Department report, *Country Reports on Terrorism 2014*, identifies fifty-nine FTOs using the INA's three criteria. Forty-five (76 percent) of those FTOs have an Islamic affiliation and they are mostly active in the Middle East, North Africa, and Central Asia. The balance of the groups roughly fits into two categories: nationalist or communist.

These fifty-nine FTOs employ a variety of terrorist tools such as bombings to make their so-called political statements. The U.S. Department of Homeland Security operates the Global Terrorism Database, which identifies more than 140,000 discrete terrorist events that took place around the world between 1970 and 2014—of which 58,000 (41 percent) are identified as "bombings."[139] Bombs are evidently the terrorists' weapon of choice, because they tend to be inexpensive, small, and hard to detect. They also tend to draw the most blood, having killed over thousands and injured tens of thousands, and earned headlines that brought attention to their grievances.

Terrorists also use kidnappings and hostage-taking primarily in order to use the victim for bargaining or to elicit publicity for their cause. Kidnapping has an additional benefit in that it can earn ransom money for fiscally strapped terrorist groups such as ISIS, which made millions of dollars from hostage-taking.[140]

Armed attacks and assassinations constitute 11 percent of all terrorist events, according to the Global Terrorism Database.[141] Assassination is used to kill select persons either via bombs or drive-by shootings, and often the motivation is to create a psychological effect.

Arsons and fire bombings are an easy and cheap terrorist tool for groups ill-equipped for more sophisticated attacks. These devices are often used to target hotels and government buildings, with the perpetrators' intention of creating fear and a lack of confidence in that government's ability to provide security.

There are other types of terrorist events such as hijacking, skyjacking, cyberattacking, maiming, beheading, torching humans, burying humans alive, and using weapons of mass destruction. No doubt, future terrorists will find more ways of killing, instilling fear, and leveraging power players.

Terrorists' Motivation

There are two distinct terrorist motivation categories: secular and religious. The secular, non-religious groups tend to attempt "highly selective and discriminate acts of violence to achieve a specific political aim." Exercising such restraint reduces "the risk of undermining external political and economic support" because the terrorist group may need an ally to help its effort.[142]

The majority of the FTOs are religiously oriented and with few exceptions are Islamic. More often than not, these Islamic groups have an apocalyptic frame of reference that makes the loss of life irrelevant for them when associated with a terrorist act. But the selection of the terror target is key—which, for example, is why the Sunni ISIS group often selects Shia religious sites for maiming large numbers of innocents.

By contrast, secular political groups tend to attack objectives that are highly symbolic of authority, such as government buildings. While religiously focused terrorist groups like ISIS may pay attention to symbolism, they are equally seeking physical devastation. They also select objectives, as in the case of many Islamic groups, Christian individuals, their worship services, and church properties.

Anyone who observed the post 9/11 attack anniversaries will recall that those dates have special meaning to terrorist groups such as al Qaeda. Terrorists may also try to commemorate such anniversaries with new operations, such as the attack on the American consulate in Benghazi, Libya, on September 11, 2012. There also may be special days that mean something to the terrorists' enemy that warrants special attention as well.

Blame Islam for a Lot of Terrorism

Islamic terrorist groups warrant special attention because they account for the vast majority of all terrorist groups in the world, according to the U.S. State Department.

There is a mistaken view that some of the worst of these terror groups are not Islamic. President Obama said as much on September 10, 2014: "Now let's make two things clear: ISIL [ISIS] is not 'Islamic.'" "No religion condones the killing of innocents, and the vast majority of ISIL's victims have been Muslim."[143]

Mr. Obama is very wrong on this issue. Journalist Fadel Boula wrote in the Iraqi newspaper *al-Akhbar* that ISIS is motivated by an extremist Salafi ideology and claims that its atrocities represent Allah's will and directives. The Salafist movement is an ultra-conservative reform movement within Sunni Islam that embraces the doctrine known as Salafism.[144]

Mr. Boula wrote that ISIS' terror "is perpetrated by people who enlist [because they are] inspired by a religious ideology. [These people] advocate enforcing and spreading [this ideology as a set of] dogmatic principles that must be imposed by the force of the sword, and which [mandate] killing, expulsion and destruction wherever they go."[145]

> Since its inception, this movement of terror has espoused a Salafi ideology that champions religious extremism, and brainwashed people of all ages have rallied around its flag, [people who were] trained to kill themselves and kill others in order to attain martyrdom.

Mr. Boula concludes his article by stating:

> When the terrorists blew up the World Trade Center and several airplanes, killing thousands of victims, Osama bin Laden, surrounded by his people, said on television: "This is a victory from Allah." And now ISIS is bragging about killing innocent people in Paris, saying that it was "done with Allah's approval," and threatening that the next attack will be in the U.S., Allah willing. And [Sheikh Yusouf] Al-Qaradhawi [an Egyptian Islamic theologian and chairman of the International Union of Muslim Scholars] and others like him pray and hope that, in the wake of this terrorist momentum, a day will arrive when Muslims inundate Europe and subdue it to Islam. Is this not enough to convince [us] that terror [does] have a religion?

Islam is the root of many of the world's terrorist groups. Islamists use Islamic texts from the Koran and the Hadith to justify their immoral killing, raping, and capturing infidel slaves.

Further, Westerners should understand that Islam is not like Christianity in that for the Muslim, faith is the primary basis for one's identity and loyalty. Every aspect of the Muslim's life—political, economic, sexual, war-making, worship, and even dealings with non-Muslims—is dictated by Islam. No wonder Westerners have a difficult time understanding the Muslim world and especially how it views and treats non-Muslims.

I explore the Muslim faith in my book *Never Submit: Will the Extermination of Christians Get Worse Before It Gets Better?* and explain in detail Islam's view and treatment of non-Muslims such as Christians.[146] Here are a few examples of that sanctioned treatment.

Muslims can enslave the Kafir [unbeliever or infidel]: When some of the remaining Jews of Medina agreed to obey a verdict from Saed, Mohammed sent for him. He approached the Mosque riding a donkey and Mohammed said, "Stand up for your leader." Mohammed then said, "Saed, give these people your verdict." Saed replied, "Their soldiers should be beheaded and their women and children should become slaves."[147]

A Muslim may rape a Kafir: On the occasion of Khaybar, Mohammed put forth new orders about forcing sex with captive women. If the woman was pregnant she was not to be used for sex until after the birth of the child. Nor were any women to be used for sex who were unclean with regard to Muslim laws about menstruation.[148]

A Muslim may behead a Kafir: When you encounter the Kafirs on the battlefield, cut off their heads until you have thoroughly defeated them and then take the prisoners and tie them up firmly.[149]

Many of the Islamic FTOs are unquestionably orthodox and therefore strictly follow the seventh-century prophet's extremism. Understandably, these groups may attract psychopaths and invite the world's disaffected, but Islam drives these radicals to commit terrorist acts and denying the obvious is dangerous for world security.

It is also important for Westerners to understand that fundamentalist Muslims believe like ISIS and al Qaeda and are driven to terrorism by an apocalyptic theology. They want more than anything else to ultimately bring about a final military conflict so their version of the messiah (what the Shia call their Mahdi, "guided one") can come and establish worldwide sharia (Islamic) rule on behalf of Allah. In other words, they seek war and conduct terrorism to bring about worldwide chaos as the perfect conditions to bring forth their messiah.[150]

Terrorism's Future

Terrorism has a bright future and will be a major factor on the nonpolar, unstable future world stage. What then should we expect? Expect terror groups to rise up in places where there are weak or failed governments like in present-day Yemen, Syria, Libya, Nigeria, and Iraq. Expect those groups to employ aggressive tactics and seek to inflict massive damage to bring attention to their causes, including overthrowing governments.

ISIS-like groups are the way of the future. There will always be a segment of the globe's population vulnerable to radicalization and recruitment by ISIS-like Salafist groups using social media platforms (YouTube, Facebook, and Twitter) to disseminate their messages broadly and with translations in many languages.[151]

These savvy Islamist groups will demonstrate increasing abilities to adapt to counter-terrorism measures and run to exploit every political failure. They will also develop new and as yet unimagined capabilities of attack, and will improve their efficiency at delivering those deadly means.

Expect our homeland to experience more terrorist attacks orchestrated from overseas groups like ISIS. This situation will become the new "normal" for Europe, especially now in the wake of the March 2016 Brussels, Belgium, and July 2016 Dhaka, Bangladesh, attacks, which proved ISIS is capable of striking beyond its traditional arena of the Middle East. ISIS-like groups will also move aggressively toward "weaponizing" its Western recruits, leading to a major increase in the level and scale of the terrorist threat. And future terrorist operations will be far more complex, something akin to what happened in Paris and at least as deadly—especially if we can believe a leading Saudi cleric who pledged allegiance to Abu Bakr al-Baghdadi, the leader of ISIS.[152]

Saudi jihadi cleric Nasir al-Fahd said, "If Muslims cannot defeat the kafir (unbelievers) in a different way, it is permissible to use weapons of mass destruction, even if it kills all of them and wipes them and their descendants off the face of the Earth."[153]

Groups like ISIS are actively seeking weapons of mass destruction and the personnel with technical experience capable of expanding their programs. A March 2016 *New York Times* article reported an ISIS detainee and specialist in chemical weapons held by Americans in Erbil, Iraq, told interrogators that the Sunni terrorist group plans to use banned substances in Iraq and Syria. A U.S. Defense official said the ISIS detainee provided details about how ISIS had weaponized mustard gas into powdered form and loaded it into artillery shells.[154]

Chemical weapon attacks appear to be more frequent than previously reported. The Syrian American Medical Society released a report March 14, 2016, claiming chemical weapons have been used at least 161 times through the end of 2015 and caused 1,491 deaths in Syria. The report indicates attacks launched by both the Syrian government and terrorist groups like ISIS are increasing, with a high of at least sixty-nine attacks in 2015, and 14,581 people injured in all.[155]

Further, General Joseph Votel, the former commander of U.S. Special Operations Command and current commander of U.S. Central Command, warned that we must "come to grips with new types of terrorists, such as the computer-savvy individual who knows how to exploit rapid technological advances and the ubiquity of the Internet."[156]

General Votel warned that "connected youth are becoming more and more desensitized to unacceptable and violent behavior." He explained that "computerized traffic and public safety systems and electronic banking will be among the new terrorist targets."[157]

The general concluded his warning with a sobering consideration: "We should think of disparate and isolated 'lone wolves,' still independent, anonymous, and elusive, but now connected to each other in cyberspace—forming 'wolf packs.'" These packs, the general explained, can share tactics, techniques, and procedures with one another, and can instantaneously move resources across the Web anonymously—all while they collectively plan and execute their attacks."[158]

Terrorism's Prophetic Sign

Early in this chapter, I embraced the Department of Defense's definition of terrorism: "The calculated use of unlawful violence or threat of unlawful violence to inculcate fear; intended to coerce or to intimidate governments or societies in the pursuit of goals that are generally political, religious, or ideological." There will be more terrorist events as defined above in the future, and those attacks—combined with the dysfunction of the new nonpolar world order—will make it seem as if prophetic Bible verses are being fulfilled because lawlessness will reign.

Timothy describes the last days as "terrible times" in 2 Timothy 3:1–7, then he lists as evidence how lawless man will be:

> People will be lovers of themselves, lovers of money, boastful, proud, abusive, disobedient to their parents, ungrateful, unholy, without love, unforgiving, slanderous, without self-control, brutal, not lovers of the good, treacherous, rash, conceited, lovers of pleasure rather than lovers of God.

Matthew 24:12 describes the product of such lawlessness:

> Because of the increase of wickedness [lawlessness], the love of most will grow cold. (NIV)

The future lawlessness across the world in part will be due to the increasing terrorism that the nonpolar world can't contain. This is a sign of the coming end times.

Conclusion

Terrorism is here to stay and will play a significant role in the nonpolar future. These groups will grow in number, capabilities, and influence, and will take advantage of the general global instability and especially of the growing disenfranchised young that populate the megacities of tomorrow.

Chapter 6

Transhumanism and Super-Soldiers

FUTURE WAR will be like today's science fiction movies and video games. Technologically enhanced, part-automaton super-soldiers—stronger, smarter, faster, and with a dulled conscience that allows them to viciously kill on command—will fight as individuals and highly coordinated "wired-in" teams. They will fight anywhere, anytime, night or day, employing incredibly advanced, effective stand-off weapons using data transmitted directly to their brains from multiple sources: drones, satellites, and fellow super-soldiers. They will excel in urban fighting, the close-in battle necessary to defeat terrorists and other super-soldier enemies. They will be hailed as the ultimate soldier, the dream of generals and small boys. In the end, they will be nightmares.

This chapter introduces the subject of "transhumanism," which seeks to combine the "best" of man and machines and a concept embraced by the Pentagon's science arm in order to field the twenty-first century's super-soldier. However, the "creation" of "cyborgized" soldiers—beings with both organic and bio-mechatronic body parts—raises very serious ethical and moral issues as well as provides good reason to believe these super-soldiers are yet another sign of the end times.

Transhumanism: "Best" of Man and Machines

Transhumanists want to play God. They are dissatisfied with human frailties and want to use modern science to remake man into a twenty-first century, Homo sapien-like machine that attains a modern fountain of youth and superhuman physical capabilities as well as Einstein-like big brains.

The transhumanist movement is a class of philosophy that promotes what believers define as an "interdisciplinary approach to understanding and evaluating the opportunities for enhancing the human condition and the human organism opened up by the advancement of technology."[159] Still others refer to the movement as techno-utopianism, "the idea that humans can create a progressively better future through the rational mastery of nature."[160] Both transhumanist subgroups attend to technologies like genetic engineering and emerging science on molecular nanotechnology and artificial intelligence.

These ambitions raise a number of important questions. Does possessing a thread of deoxyribonucleic acid (DNA) consistent with that of Homo sapiens make one human? Can't science transfer human DNA to another species, and does that changed being then become at least part human with all the rights and privileges of a human person? What if one replaces most of a human's parts—hands, eyes, heart—with efficient and forever-lasting machines/devices; is that being still a human person? And, on the theological side, at what point does the resulting being lose or gain a soul? Is the being redeemable?

Perhaps our humanity isn't defined by our DNA but by our psychology. Are humans, even those equipped with the latest man-made replacement parts hung on their remaining DNA, defined primarily by their psychological traits such as empathy and emotion? Is a "man" any less human if he lacks common human psychological traits because he is programmed to eschew fear or behaves in a nonstandard, atypical way?

Many readers will remember Mr. Spock, played by Leonard Nimoy, on the popular 1960s television series *Star Trek*. Spock displayed no human-like emotion, and was always totally "logical" even amidst the most trying circumstances. Was Mr. Spock from the fictitious planet Vulcan really human, and is that really the personality transhumanists seek for mankind—emotionless zombies?

Transhumanists believe we can and should improve humanity by harnessing advanced technologies. For example, they want to extend our lives through genetic engineering, nanotech, and cloning. They want to boost our physical, intellectual, and psychological capabilities beyond that of mere natural humans, thus helping us to morph into allegedly superior transhumans.

Most of us would like to be smarter, no doubt. But our thinking machine (the brain), that three pounds of gray, jelly-like matter that rests inside our skulls, has a lot of shortcomings, like forgetting simple things. Who wouldn't want to overcome lapses in memory or have more brainpower? The transhumanist believes modern technology can help us overcome cognitive limitations promising to boost our brainpower with microchips that rocket our IQs to over 300, genius level, or help us upload all the data stored in our brains to a computer and then transmit all that data using electrons to anywhere in the world at the speed of light? That's certainly a scary thought (or data dump)!

Science is pursuing something worse—artificial intelligence (AI). In March 2016, software built by Google defeated the world's best player of the Asian board game "Go" in a five-game match. Go, according to Google, is the most complex game played at a professional level and could have more potential moves than there are atoms in the universe. Google's mastery of Go suggests a promising although frightening future for AI.[161]

The sky is the limit, according to David Gelernter, a professor of computer science at Yale and the author of *Tides of Mind: Uncovering the Spectrum of Consciousness.* "Once we have figured out how to build artificial minds with the average human IQ of 100, before long we will build machines with IQs of 500 and 5,000.[162]

AI's potential for good and bad consequences is staggering. "Humanity's future is at stake," Gelernter warns. Specifically, "Robots with superhuman intelligence are as dangerous, potentially, as nuclear bombs." Then he soberly explained, "Technologists are in business to build the most potent machines they can, not to worry about consequences."[163]

The same can be said of transhumanists, and they will inevitably try to harness AI to humans to make them incredibly intelligent by implanting AI-fueled chips in the brain to supercharge our IQ. Gelernter then explains that building super-intelligent computers is the "natural outcome of the human need to build and understand." Similarly, the transhumanist will inevitably harness that technology to mankind, and I believe DARPA technologists are already researching the potential.[164]

Transhumanists also seek to make us much healthier, much stronger, and able to live almost forever. Children's cartoon characters often portray a transhumanist-like lifestyle by depicting a never-ending series of phenomenal physical feats by superheroes like Superman or Spiderman, which inevitably influence impressionable young boys to act in daring and foolish ways.

The transhumanist also promises to make us forever healthy. Who wouldn't like to do away with the physical aches and pains of aging? Specifically, the transhumanist offers hope that by harnessing modern medical technology, our weak bodies will eventually be able to miraculously heal themselves and stay healthy, thus live much longer than today's normal seven or eight decades, such as the biblical character Methuselah, who lived 969 years (Genesis 5: 21–27).

Maybe the transhumanist really believes humans can live forever with the right technological fix! In fact, transhumanist advocates have already advanced the idea that our enhanced brains can be downloaded into a computer and then rebooted in a new body, thereby guaranteeing immortality.

The live-forever, superhuman-seeking transhumanists create some very serious dangers and ethical pitfalls, however. Do these transhumanists expect to stop people from reproducing if everyone in the future can live as long as Methuselah? Then if we don't die, does civilization face the possibility of overpopulation and socioeconomic disaster due to life-support limitations? Does government then mandate sterilization or even termination (abortion) to control world population? Or what if these life-extending technologies are only available to the rich? Does the availability of super-human, life-extending technology eventually lead to violent protests among the underprivileged masses demanding access to such technology as a human right?

We humans have five senses that help us interact with the world, which the transhumanist wants to dramatically improve. Who wouldn't like to sharpen those sensors—maybe to see infrared radiation, listen to radio signals without a box of electronics in hand, or control that vacuum cleaner like a robot around the house with a simple mental thought? The transhumanist believes these and other "enhancements" are possible, but are such futurist human modifications really any different than present-day "enhancements" like what Lasik surgery does for those who once wore thick glasses but now enjoy 20-20 vision, or for the deaf person who now hears clearly thanks to the use of surgery and electronic hearing aids? Aren't these improvements just the product of the evolution of medical science? The transhumanist also wants to change our psychological senses of mood, energy, and self-control. They favor altering our sense of well-being, joy, and energy—a radical move to put mental therapists out of business and shutter mental institutions. After all, altering our self-confidence might also help us lose weight or tackle a bad habit like smoking. Might such changes take place through the implantation of microchips on certain parts of our brain or through drug therapies?

These issues and questions may seem bizarre, but the transhumanist welcomes the opportunity to overcome mundane, simple humanity by harnessing our bodies to modern technology. Isn't such change rational and just keeping up with the times and the natural progression of man adapting to a better and new reality? Remember, past periods in human history evidenced significant technological change that earned the disdain of many naysayers. Our great grandparents, for example, were skeptical at first about the introduction of the radio and television, as were those who lived in the early part of the twentieth century and rode their horse-drawn carriages past the very noisy early automobiles.

Such change can be unsettling, especially if you lose your livelihood to someone who's made artificially smarter thanks to microchip implants or someone who is physically stronger—better able to perform a physically demanding job—because of drug-induced or other types of enhancements. Accepting these technologies in one's person could become socially difficult, if not impossible, especially if the government endorses enhancements like it does vaccinations.

The other part of the above is about the natural progression. Who hasn't experienced how young children quickly embrace the latest technology—video games, high-tech devices—and then end up teaching their Neanderthal parents? To children, what transhumanists promote may seem logical, the natural evolution of life, and those of us old fogies with objections are just "too human."

Christians should be troubled by the transhumanist agenda because this radical new social movement expresses great faith in the advancement of technology, believing human advancement will usher in a new era in which we transcend our bodies to conquer the universe. According to biblical prophecy, this movement evidences some of the very characteristics of human arrogance and activity that appear just prior to the Second Coming of Jesus Christ.

Before exploring the end-times signs evidenced by transhumanism, consider how the transhumanist movement has infected our societal thinking and especially how the Pentagon is harnessing technology to create the "enhanced" super-soldier and the resulting ethical and moral issues.

Untethered Military Biotechnology

The Pentagon's Defense Advanced Research Projects Agency or DARPA is a very secretive, congressionally created military science agency.

DARPA's mission is to create revolutions in military science to ensure that the U.S. military maintains technological dominance over the rest of the world.[165] DARPA is the brain behind America's new transhumanist super-soldier.

The carefully scripted public release of information about DARPA's secretive activities provides a glimpse into the agency's Orwellian research programs. Many DARPA programs are kept totally secret, while others are revealed incrementally over many years, and they demonstrate significant contributions to our modern society such as the Internet, the Global Positioning System, and stealth technology.

DARPA's scientific stable of experts work in secret on national security-related programs that often extend decades into the future to harness emerging technologies. They conduct what seems to be sci-fi, far-out research (such as brain-controlled drones) and engineer new life forms.

DARPA's newest unit is devoted to studying the interaction of biology and engineering, the Biological Technologies Office. That office investigates manufacturing biomaterials, turning living cells, proteins, and DNA into a genetic factory.

DARPA pushes the envelope on creating life at another biology office, the BioDesign program that studies how to create synthetic beings genetically engineered to be immortal and programmed with a kill switch. BioDesign's work is announced in a 2015 Pentagon budget statement.[166]

> BioDesign will employ system engineering methods in combination with biotechnology and synthetic chemical technology to create novel beneficial attributes. BioDesign mitigates the unpredictability of natural evolutionary advancement primarily by advanced genetic engineering and molecular biology technologies to produce the intended biological effect. This thrust area includes designed molecular responses that increase resistance to cellular death signals and improved computational methods for

prediction of function based solely on sequence and structure of proteins produced by synthetic biological systems.

On a related topic, a 2016 news report that doesn't finger DARPA, however, indicates that the Pentagon and U.S. intelligence agencies are looking into "synthetic biology." Evidently, the Pentagon's Office of Technical Intelligence, according to the news report, produced a study on synthetic biology, which is described as a "merging field in which scientists modify or 'engineer' DNA to improve their ability to understand, predict, design, and build biological systems." The Pentagon study states that "[synthetic biology] may be possible to enhance physical, cognitive, and socioemotional (or interpersonal) performance."[167]

DARPA has a long history of creating mechanical robots, and now, it seeks to engineer a biological robot—a humanoid. Meanwhile, until that outcome is realized, DARPA scientists, who include some of America's best minds, will continue to try to figure out how to harness emerging technology to improve on the Homo sapiens (the military service member). These efforts closely parallel the transhumanists' agenda to make our troops into super-soldiers. Never forget, DARPA exists to tap into high technology that may benefit society, but in the end it must demonstrate successful results for military use.

Let's be very clear. There is absolutely no reason to accuse either the current Pentagon leadership or the managers and scientists at DARPA of anything unethical or immoral. However, numerous historical cautions must be considered when very bright people are given great powers and operate in secret with the endorsement of their government and a fixation on scientific discovery. The one historical example of abuse that comes to most minds is what happened in Germany.

The Nazi regime of the 1930s–1940s wanted to make its military personnel stronger, healthier, and smarter. That mission was given to the Nazi Party's Schutzstaffel (SS), which ran experiments at the Auschwitz concentration camp, a network of camps in the Polish areas annexed by Nazi Germany.

At those camps, the SS conducted three categories of medical experiments intended to advance the legitimate health interests of Axis service members. First, there were experiments aimed at facilitating the survival of military personnel operating at high altitudes by using a low-pressure chamber to determine the maximum altitude from which crews of damaged aircraft could parachute to safety. Second, the Nazis experimented with prisoners to develop and test pharmaceuticals and treatment methods for injuries and illnesses such as malaria, typhus, and yellow fever. The third category of experiments was performed by the infamous Josef Mengele which involved twins to determine how different "races" withstood various contagious diseases.[168]

The German military and by association the Nazi Party's SS had legitimate interest in finding medically feasible remedies to address physical challenges facing field soldiers. The problem with the Nazi approach, of course, is the SS conducted unethical medical experiments on thousands of concentration camp prisoners as part of its "scientific" investigations. Those experiments almost always led to the death of their prisoner "guinea pigs." History has shown that not one useful discovery was made from all the experiments and thousands of lives lost.

Nazi Germany wasn't the only nation that conducted unethical medical experiments. A website called Best Psychology Degrees profiles "The 30 Most Disturbing Human Experiments in History."[169] Consider the experimental work on human "guinea pigs" of three other governments to help their militaries: the former Soviet Union, North Korea, and Japan.

Soviet Secret Service's Poison Lab

Beginning in 1921 and continuing for most of the twenty-first century, the Soviet Union employed poison laboratories known as Laboratory 1, Laboratory 12, and Kamera as covert research facilities of the secret police agencies. Prisoners from the Gulags were exposed to a number of deadly poisons, the purpose of which was to find a tasteless, odorless chemical that could not be detected post mortem. Tested poisons included mustard gas, ricin, digitoxin, and curare, among others. Men and women of varying ages and physical conditions were brought to the laboratories and given the poisons as "medication," or part of a meal or drink.[170]

North Korea's Horror Chambers

Several North Korean defectors have described witnessing disturbing cases of human experimentation. In one alleged experiment, fifty healthy women prisoners were given poisoned cabbage leaves—all fifty women were dead within twenty minutes. Other described experiments include the practice of surgery on prisoners without anesthesia, purposeful starvation, beating prisoners over the head before using the zombie-like victims for target practice, and chambers in which whole families are murdered with suffocation gas. It is said that each month, a black van known as "the crow" collects forty to fifty people from a camp and takes them to an unknown location for experiments.[171]

Japanese Army's Human Experiment Lab

From 1937 to 1945, the imperial Japanese Army developed a covert biological and chemical warfare research experiment called Unit 731. Based in the large city of Harbin, Unit 731 was responsible for some of the most atrocious war crimes in history. Chinese and Russian subjects—men, women, children, infants, the elderly, and pregnant women—were subjected to experiments that included the removal of organs from a live body, amputation for the study of blood loss, germ warfare attacks, and weapons testing. Some prisoners even had their stomachs surgically removed and their esophagus reattached to the intestines. Many of the scientists involved in Unit 731 rose to prominent careers in politics, academia, business, and medicine.[172]

Let me disabuse the reader of the notion that such horrible experiments as those conducted by the Nazis, the Russians, North Koreans and Japanese in the name of national security are just the domain of foreign militaries. Some pretty horrendous military-related experiments have also been sponsored by the U.S. government in the name of national security. This should be a warning that when our government acts in secret—like DARPA—with unsuspecting and vulnerable people, volunteers or not, terrible things can and do happen.

Our Central Intelligence Agency (CIA) leads the U.S. government with a long history of some of the most unethical human experiments in the name of national security.

CIA's Mind-Control Experiments

In 1954, the CIA developed an experiment called Project QKHILL-TOP to study Chinese brainwashing techniques, which they then used to develop new methods of interrogation. Leading the research was Dr. Harold Wolff of Cornell University Medical School. After requesting that the CIA provide him with information on imprisonment, deprivation, humiliation, torture, brainwashing, hypnoses, and more, Wolff's research team began to formulate a plan through which they would develop secret drugs and various brain damaging procedures. According to a letter he wrote, in order to fully test the effects of the harmful research, Wolff expected the CIA to "make available suitable subjects."[173]

Project Artichoke on Interrogation Methods

In the 1950s, the CIA's Office of Scientific Intelligence ran a series of mind-control projects in an attempt to answer the question, "Can we get control of an individual to the point where he will do our bidding against his will and even against fundamental laws of nature?" One of these programs, Project Artichoke, studied hypnosis, forced morphine addiction, drug withdrawal, and the use of chemicals to incite amnesia in unwitting human subjects. Though the project was eventually shut down in the mid 1960s, the project opened the door to extensive research on the use of mind control in field operations.[174]

Operation Midnight Climax—Mind-Control Research

Initially established in the 1950s as a sub-project of a CIA-sponsored, mind-control research program, Operation Midnight Climax sought to study the effects of LSD on individuals. In San Francisco and New York, unconsenting subjects were lured to safe houses by prostitutes on the CIA payroll, unknowingly given LSD and other mind-altering substances, and monitored from behind one-way glass. Though the safe houses were shut down in 1965, when it was discovered that the CIA was administering LSD to human subjects, Operation Midnight Climax was a theater for extensive research on sexual blackmail, surveillance technology, and the use of mind-altering drugs on field operations.[175]

Project MKUltra—Mind Control Using Drug

Project MKUltra is the code name of a CIA-sponsored research operation that experimented in human behavioral engineering. From 1953 to 1973, the program employed various methodologies to manipulate the mental states of American and Canadian citizens. These unwitting human test subjects were plied with LSD and other mind-altering drugs, hypnosis, sensory deprivation, isolation, verbal and sexual abuse, and various forms of torture. Research occurred at universities, hospitals, prisons, and pharmaceutical companies. Though the project sought to develop "chemical…materials capable of employment in clandestine operations," Project MKUltra was ended by a Congress-commissioned investigation into CIA activities within the U.S.[176]

The U.S. Atomic Energy Commission warrants special mention for its radiation experiments on humans.

Atomic Energy Commission's Castle Bravo "Radioactive Fallout" Study

The 1954 "Study of Response of Human Beings exposed to Significant Beta and Gamma Radiation due to Fall-out from High-Yield Weapons," known better as Project 4.1, was a medical study conducted by the

U.S. of residents of the Marshall Islands. When the Castle Bravo nuclear test resulted in a yield larger than originally expected, the government instituted a top-secret study to "evaluate the severity of radiation injury" to those accidentally exposed. Though most sources agree the exposure was unintentional, many Marshallese believed Project 4.1 was planned before the Castle Bravo test. In all, 239 Marshallese were exposed to significant levels of radiation.[177]

Atomic Energy Commission's "Great Balls of Fire" Experiment

Between 1963 and 1973, dozens of Washington and Oregon prison inmates were used as test subjects in an experiment designed to test the effects of radiation on testicles. Bribed with cash and the suggestion of parole, 130 inmates willingly agreed to participate in the experiments conducted by the University of Washington on behalf of the U.S. government. In most cases, subjects were zapped with over 400 rads of radiation (the equivalent of 2,400 chest x-rays) in ten-minute intervals. However, it was much later that the inmates learned the experiments were far more dangerous than they had been told. In 2000, the former participants settled a $2.4 million class-action settlement from the university.[178]

The U.S. military services hosted unethical medical experiments to help prepare for the possible effects of weapons of mass destruction.

U.S. Army's Biological Warfare on American Cities

In 1956 and 1957, the United States Army conducted a number of biological warfare experiments on the cities of Savannah, Georgia, and Avon Park, Florida. In one such experiment, millions of infected mosquitos were released into the two cities, in order to see if the insects could spread yellow fever and dengue fever. Not surprisingly, hundreds of [subjects] contracted illnesses that included fevers, respiratory problems, stillbirths, encephalitis, and typhoid. In order to photograph the results of their experiments, Army researchers pretended to be public health workers. Several people died as a result of the research.[179]

U.S. Navy Tests Mustard Gas on Sailors

In 1943, the U.S. Navy exposed its own sailors to mustard gas. Officially, the Navy was testing the effectiveness of new clothing and gas masks against the deadly gas that had proven so terrifying in the First World War. The worst of the experiments occurred at the Naval Research Laboratory in Washington. Seventeen and eighteen-year old boys were approached after eight weeks of boot camp and asked if they wanted to participate in an experiment that would help shorten the war. Only when the boys reached the research laboratory were they told the experiment involved mustard gas. The participants, almost all of whom suffered severe external and internal burns, were ignored by the Navy and, in some cases, threatened with the Espionage Act. In 1991, the reports were finally declassified and taken before Congress.[180]

Experiments on humans to either confirm the effectiveness of vaccines or other medical treatments, to test deadly toxins, or even to find out whether certain drugs help soldiers cope with the stresses of combat are well documented.

Years ago I sat through meetings in which the PhD speaker (a man I know but haven't asked permission to use his name) alleged that American prisoners of war from the Korean and Vietnam wars were used as guinea pigs for medical experiments in both North Korea and Russia. He shared extensive primary and secondary evidence to prove the validity of his research, and even wrote a book on the topic that was never published. I have no reason to question either the briefer or the reliability of his sources.

Then in 1987, I read a book by Dr. Richard Gabriel, *No More Heroes: Madness and Psychiatry in War* (Hill and Wang, 1987), which documents, based on his interviews of Soviet psychiatrists and other medical personnel who dealt with combat stress victims. As a result of his investigation, Gabriel recommends in his book that the U.S. military use battlefield prophylactics—chemicals that can be used to alter behavior—to enable the soldier to cope with extremely stressful conditions.

Evidently, Gabriel's "chemical" soldier would then be given to extreme risk-taking and would function only on the cognitive plane. Ethics, for the chemical warrior, would no longer be appropriate.

At that time, in a review of Gabriel's book published in *Infantry* (May–June 1987), I cautioned: "The military professional should become familiar with this book because, unfortunately, it may well capture the imagination of many militarily naïve civilian readers. Those people may accept his rationale and revelations as being technically and doctrinally accurate. The military professional must be prepared to discount such a blatantly misleading set of recommendations and conclusions."[181]

Those conclusions still reflect my views. However, until researching for this volume, I had not reconsidered the issue. Now, I'm concerned that the likes of DARPA scientists and its civilian Pentagon bosses whom I labeled in 1987 as "militarily naïve civilians" are perhaps investigating a future chemical warrior as outlined above.

Then to my surprise, on February 23, 2016, I sat through an unclassified briefing given at the Pentagon about the Ukraine-Russia war. It became clear at that point that Gabriel's chemical soldier proposal had attracted attention in Russia. The briefer, a civilian consultant, said the Russians used drug-induced "soldiers" taken from prisons in Moscow's ongoing campaign in Eastern Ukraine, and has done so "dozens of times." The briefer, who I won't identify here, said the Russians gave these "soldiers" vodka to help swallow a pill that he called "speed," and which he claimed included Demerol and lithium to reduce shock. Subsequently, those "soldiers," according to the briefer who has visited Ukrainian battlefields dozens of times, said thousands of "speed"-treated "soldiers" walked like "zombies" into Ukrainian machine gunfire similar to American Civil War soldiers marching upright into volleys of fire at the Battle of Gettysburg.

Mankind can be incredibly cruel. These examples of unethical medical experiments or mind-altered combatants illustrate the worst of man in the name of national security. Unfortunately, the very same type of thinking that is evidenced by the above examples could exist today behind the secret doors of the Pentagon's own labs—all in the name of national security.

Pentagon's Super-Soldier Crusade

DARPA's work to improve the soldier's fighting capabilities fits the transhumanist agenda and warrants special attention given the aforementioned abuses as well as the contemporary legal and ethical implications of using so-called soldier enhancements to prepare them for the twenty-first century battlefield.

DARPA's public work on soldier "enhancements" comes in two varieties. There are enhancements for the soldier's exterior and others that work inside the soldier's body; collectively, they are designed to improve performance, appearance, or capability as well as to achieve, sustain, and restore health. The internal "improvements" raise most eyebrows, but first consider some of DARPA's work on improving the soldier's capabilities via external mechanisms.

DARPA has a number of research organizations like Lockheed Martin's HULC and Raytheon's XOS developing exoskeletons to increase human strength and endurance. These "skeletons" strap to the soldier's body to give him the ability to carry payloads of two hundred pounds and run at speeds up to ten miles per hour.

The "Geckskin" is a product of DARPA's Reconfigurable Structures Program. This gear is a bio-inspired adhesive fabric that enables humans to climb walls like geckos and spiders—think Spiderman without the web-shooting finger for swinging from building to building.

DARPA's Cognitive Technology Threat Warning System is a fancy name for a fantastic pair of binoculars. Code named "Luke Skywalker," they enable the soldier to see as far as six miles using a 120-megapixel camera, a 120-degree field of view that is in tandem with an electroencephalogram cap worn by the soldier to read his mind while he scans the horizon. These binoculars radically improve the soldier's ability to detect the slightest enemy movement.[182]

These and other exterior enhancements potentially improve soldier performance, but that isn't enough for DARPA. The agency wants to give our soldiers souped-up bodies that provide characteristics advantageous on the battlefield such as being smarter, sharper, more focused, and more physically strong. Theoretically, our super-soldier will be capable of performing telepathy, running faster than an Olympic champion, re-growing limbs lost in combat, possessing a super-strong immune system, and going days without food and sleep. He will also be the perfect killer, devoid of mercy and fear.[183]

DARPA has a host of programs that use brain implants for a variety of desirable outcomes. The agency awarded grants worth $40 million to California and Pennsylvania Universities to develop memory-controlling implants.[184]

The agency is working on something like a pacemaker for the brain that monitors, analyzes, and responds to real-time information on the brain. This brain implant will help deliver more effective treatments for neurological and psychiatric disorders.[185]

DARPA'S Human Assisted Neural Devices program intends to produce drugs and treatments that can erase memories such as the events and tragedies that trigger post-traumatic stress disorder (PTSD) in soldiers. And the agency's so-called Peak Soldier Performance Program intends to boost human endurance, both physical and cognitive, and not necessarily just with stimulants.

There is also a DARPA project that implants chips into soldiers that can heal their bodies and minds. The program's goal, ElectRX, is the development of chips that are injected into the soldier with a needle. The chips then act as pacemakers for the nervous system. They stimulate nerve endings with electrical signals, thus helping relieve pain associated with chronic inflammatory diseases like rheumatoid arthritis and mental illnesses. DARPA wants to expand the use of chips to mental illnesses like PTSD.[186]

A March 2016 CNN report, "U.S. Military Spending Millions to Make Cyborgs a Reality," indicates DARPA also hopes to implant a one cubic centimeter "chip" to allow a human brain to communicate directly with computers. The goal is to "open the channel between the human brain and modern electronics," according to DARPA program manager, Phillip Alvelda. The project is funded at $62 million and part of DARPA's Neural Engineering System Design program.[187]

DARPA insists the implants are not intended for military applications. Rather, they are part of a new therapy to help people with deficits in sight and hearing by "feeding digital auditory or visual information into the brain." Of course, not everyone buys the official medical position on the implants.[188]

Conor Walsh, a Harvard biomedical engineer, speculates that DARPA's implant would "change the game," adding that "in the future, wearable robotic devices will be controlled by implants." He insists the "clips" have the potential to enhance a soldier's capabilities in combat such as allowing the combatant to operate his exoskeleton suit more effectively as well as wearable robotic devices.[189]

Let's pause here to reflect on what appears to be evidence of medical science helping veterans with brain maladies. That is likely a cover, as Harvard's Walsh suggests, for something far more sinister. Annie Jacobsen reminds us in *The Pentagon's Brain* that "DARPA is not primarily in the business of helping soldiers heal; that is the job of the U.S. Department of Veterans Affairs. DARPA's job is to 'create and prevent strategic surprise.' DARPA prepares vast weapons systems of the future. So what are the classified brain programs really for? What is the reason behind the reason?"[190]

Hold that thought for now and continue reading.

DARPA awarded $9.9 million to the Institute for Preclinical Studies at Texas A&M University to develop a means of surviving significant blood loss. This innovation would eliminate the wounded combatant's need for immediate life-saving medical care.[191]

DARPA is working on human red blood cells to produce and deliver protein antidotes and other antibody-based medicines to neutralize biological agents acquired on the battlefield.[192] The agency has also experimented with "blood pharming," which uses umbilical cords to generate universal donor blood, a vital help for war-zone trauma care.[193]

DARPA has a number of programs that deliver treatments, before and after exposure to deadly toxins, viruses, and radiation. One agency program found a treatment that protects soldiers exposed to radiation up to a day after exposure.

Millie Donlon, DARPA's program manager for this effort, said that the ability to treat radiation sickness a day after exposure is significant "because most of the existing treatments we have require they be administered within hours of exposure to potentially lethal radiation—something that might not always be possible in the confusion that would likely follow such an exposure event."[194]

DARPA wants to make humans mouse-like in their ability to resist diseases as well. Researchers found the average rat in New York City carries dozens of pathogens, yet they aren't affected. That observation led DARPA to begin investigating how to alter humans so that we can host deadly diseases without getting ill ourselves.[195]

"We should look beyond the pathogen—it's really about the host, and the host's responses to infection," Colonel Matthew Hepburn, program manager of DARPA's Biological Technologies Office, said at a recent conference. "Rats have all these things and they survive just fine—isn't that a great opportunity for us?"[196]

The agency is performing a number of genetic studies. One program seeks to genetically modify soldiers to hibernate throughout the winter. Evidently, the study of metabolic flexibility and suspended animation discovered that squirrels produce an enzyme that enables hibernation, and DARPA thinks our troops need to hibernate or at least get healthy sleep during intense combat.

DARPA is making progress overcoming cellular death as well. Evidently gene therapy with the insertion of artificial genes into an organism has proven to extend lifespan by 20 percent in rats. Of course, that work draws criticism from the 2009 Synthetic Biology Projects, which stated:[197]

> The concern that humans might be overreaching when we cre-
> ate organisms that never before existed can be a safety concern,
> but it also returns us to disagreements about what is our proper
> role in the natural world (a debate largely about non-physical
> harms or harms to well-being).

All the good DARPA is doing to improve soldier health rapidly moves into the shadows with its life-extending experiments and creating new life. Jacobsen writes in *The Pentagon's Brain* that biology is in fact driving toward immortality as suggested by DARPA's rat research. She cites an April 2014 announcement that scientists in the U.S. and Mexico have successfully grown a complex organ, a human uterus, from tissue cells in a lab. Almost simultaneously in England, scientists there claim to have grown noses, ears, blood vessels, and windpipes in a laboratory using stem cells.

"The same biotechnology will allow scientists to clone humans," Dr. David Gardiner told Ms. Jacobsen. Gardiner is a professor of developmental and cell biology at the University of California, Irvine. He maintains a laboratory where he does regenerative engineering.[198]

Jacobsen asked Gardiner, "Do you think the Defense Department will begin human cloning research?" The scientist answered, "Ultimately, it needs to be a policy decision."[199] Gardiner worked on DARPA projects but would not elaborate on his cloning comment.

DARPA might be working on a cloning program, but that would be classified. However, you can be certain that if DARPA thought the Russians or Chinese were working on human clones, the agency would push hard to do the same albeit in secret. But it gets more bizarre.

No doubt DARPA's team is excited about a new discovery that allows scientists to edit genes to determine the biological features of future humans. This breakthrough is all over the scientific literature, and there was a conference in Washington in December 2015 that exclusively focused on the new and relatively inexpensive gene-editing method named CRISPR-Cas9.[200] You can bet DARPA staffers were at the conference taking down names and plenty of notes.

CRISPR-Cas9 allegedly can permanently alter the genetic endowment of later generations to introduce new traits that can be quickly propagated through all of humanity. This discovery will excite transhumanists and will certainly attract the attention of the agency's cyborg soldier-related programs. This is the sort of breakthrough that could give DARPA scientists the tool to modify the human genome to combat genetically based disorders and, best of all for the agency, to enhance desirable traits in future super-soldiers.[201]

Soldier enhancements should be a good thing. It makes infinite sense to increase the soldier's capacity to endure pain, to heal quickly, to be sharp all the time, to run faster, and to lift heavy loads. These are all desirable outcomes and will give our troops an edge on the battlefield.

However, our future enhanced super-soldiers will create risks for their families and others who come into contact with them in America and on the battlefield. Those risks go well beyond the physical.

Issues with Cyborgized Soldiers

The transhumanist seeks to "cyborgize" humans by artificially "improving" them beyond their original (read "inferior")—biological characteristics given them at birth. The transhumanist envisions creating cyborgs that are organisms with enhanced abilities due to the integration of modern technology.

What the transhumanists have in mind is something similar to what the pop culture creates in comic books and movies. Those creations portray cyborgs as visibly mechanical like Darth Vader from *Star Wars* or they are indistinguishable from humans like *The Six Million Dollar Man*, as portrayed in the 1970s television program as a bionic man.

Obviously in the mind's eye of the pop culture, cyborgs are superior physically and mentally to humans, and the military versions inevitably have built-in weapons that are visible or they can think their adversaries into submission like *Star Wars'* Mr. Spock.[202]

Just how close is DARPA to creating a real cyborg? Closer than you may realize. DARPA's work with brain-wounded warriors extends to healthy soldiers in research called "augmented cognition," or Aug-Cog. This effort is on the frontier of human-machine interface, what DARPA calls human-robot interaction, and in fact, DARPA's scientists have created animal-machine bio hybrids that are remotely steerable. Also, through AugCog programs, "DARPA is creating human-machine biohybrids, or what we might call cyborgs."[203]

That is where we are heading with DARPA's transhumanist efforts, according to Keith Abney of California Polytechnic State University, one of three authors of a report entitled "Enhanced Warfighters: Risks, Ethics, and Policy." Abney considers the ethical and legal issues of DARPA's effort to enhance human warfighters and provides a sobering warning.[204]

"Too often, our society falls prey to a first generation problem—we wait until something terrible has happened, and then hastily draw up some ill-conceived plan to fix things after the fact, often with noxious unintended consequences," Abney explained. "As an educator, my primary role here is not to agitate for any particular political solution, but to help people think through the difficult ethical and policy issues this emerging technology will bring, preferably *before* something horrible happens."[205]

Cyborgizing soldiers "brings up a bunch of complicated issues," Abney admits.

The first issue is informed consent.

Does the Pentagon have a responsibility to earn the consent of soldiers before enhancing them for battle?

Abney explains there are at least three possible consent models to consider: medical, research, and public health. He suggests the best model for DARPA's cyborgizing experiments is the public health model, which is used in emergencies when the goal of protecting the public allegedly outweighs individual rights.[206]

The report, however, suggests a "hybrid framework" that might be necessary given the peculiar nature of the cyborg project. Specifically, the Pentagon should address a host of issues such as military purpose and necessity, the soldier's dignity, consent, and transparency.[207]

A second issue to consider when cyborgizing soldiers is whether the Geneva Convention forbids enhanced combatants. Abney admits, "There are no explicit rules against enhanced warfighters" because when the document was drafted in the late 1940s such an issue like cyborgs wasn't anticipated. However, Abney argues, Article 36 prohibits inhumane new weapons, which he understands could include cyborgized humans or super-soldiers.[208] Article 36 states:

> New Weapons. In the study, development, acquisition or adoption of a new weapon, means or method of warfare, a High Contracting Party is under an obligation to determine whether its employment would, in some or all circumstances, be prohibited by this Protocol or by any other rule of international law applicable to the High Contracting Party.[209]

At what point does a cyborgized soldier become a military killer robot prohibited under Article 36? Perhaps the red line of prohibition is crossed when the super-soldier no longer exercises free will, artificially loses fear (psycho-cognitive enhancements like mentioned above on the Russian-Ukrainian front), and has super strength that ruthlessly and indiscriminately kills.

Another issue for the ethicists is whether the cyborgized super-soldier is disqualified under the Biological and Toxin Weapons Convention (BTWC). That convention states:

> Each State party to this Convention undertakes never in any circumstances to develop, produce, stockpile or otherwise acquire

or retain: (1) microbial or other biological agents, or toxins whatever their origin or method of production, of types and in quantities that have no justification for prophylactic, protective or other peaceful purposes; (2) weapons, equipment or means of delivery designed to use such agents or toxins for hostile purposes or in armed conflict.

Abney's report suggests the cyborgized soldier could indeed be classified as a "biological agent." Obviously, the BTWC originally meant agents like the smallpox virus or anthrax as opposed to a chemical like mustard or an explosive bomb, but why not a living cyborg pumped up on a psychotropic agent? Abney argues that an enhanced warfighter free of inhibitions—fear or killing—may be a biological agent under the BTWC and just as dangerous as any toxin.

The final issue is whether cyborgized soldiers constitute a war crime. A war crime is an act that violates the law of war such as the intentional killing of civilians or prisoners, torture, wanton destructive of property, and using weapons that cause unnecessary suffering.[210]

Is a Terminator-like cyborg capable of showing the necessary mercy and judgment professional soldiers are expected to demonstrate on the battlefield? Just because the cyborgized soldier can keep killing is not justification to continue. That's a human value that must be preserved.

Abney warns that "if the physical abilities of warfighters become too great, it may take great restraint on the part of their commanders and even their peers to see them, not as a killing machine, but as a person— even if they are enhanced."[211]

Are Cyborgized Soldiers a Sign of the End Times?

Technology can be both a blessing and a curse. Luke 21:26, BSB, speaks of a future time when "men will faint from fear." Is that a future nuclear holocaust that delivers the carnage of the seal and trumpet judgments in Revelation 6 and 8 that ushers in the Antichrist? Or is it the transhumanists' dream of a super-soldier, a future agent of the Antichrist who is controlled in battle—no fear, great strength, self-healing? Might such cyborgized soldiers lack the soul to resist the orders of a commander who knows no bounds such as the Antichrist?

There is possible biblical precedence for such cyborgized-like soldiers that do the bidding of the Antichrist. The Bible is crystal clear that inhuman beings once roamed the earth before and after Noah's Flood (Genesis 6:4). The original Hebrew word for these inhumans is *Nephilim,* commonly translated as "giants" in English Bibles. The Nephilim were the children of human women and the fallen angels. They grew to enormous size; some accounts calculated them as being as tall as forty-five feet! Numerous prophecy experts have written volumes on why and how they will reappear in the end times.[212]

Jesus stated that the end times would be just like it was in the days of Noah (Luke 17:26–27). While this is commonly thought to mean that life was normal then as it would be when Christ returns, eschatologists point out that the Genesis account of the pre-Flood times speaks of widespread lawlessness and violence. But what was the source of the violence and lawlessness? Some eschatologists believe it wasn't just man's depravity but the Nephilim.

Why then was Noah and his family, but no others spared from the Flood? Genesis 6:9 states that Noah was "blameless," but the best translations use the phrase "perfect in his generations." An analysis of the Hebrew lends itself to an even more revealing interpretation: Noah and his family were perfect in their genetics. Perhaps the entire human race had been genetically corrupted by the Nephilim. Noah and his immediate family appear to have been the only humans left who could carry the genes necessary to produce the future Messiah.

Keep in mind that when the Nephilim were on the earth before and after the Flood, according to various Scriptures, they were widespread and numerous. The existence of the giants is recorded by primitive tribes around the world. There is good evidence that "giants" occupied the Ohio Valley in America and roamed the areas of South America, China, Russia—the entire planet. Native American tribes have recorded the presence of giants in America, whom they claim were here before their tribes appeared.

It may seem as if the account of the biblical Nephilim is pure science fiction; yet, recalling the efforts of DARPA in this chapter, the transhumanists and their scientific colleagues want to resurrect a Nephilim-like creature or a "cyborg" that is super human. The future creatures could be both, but one thing for sure is that something inhuman is coming this way and they will not be our pets, servants, or nannies. We could be on the verge of unleashing an evil force we cannot control.

Chapter 7

Outer Space Threats and Risks

I GREW UP smitten by the space exploration bug ever since watching my first rocket launch at Cape Canaveral, Florida. I recall as a small boy watching with amazement as those mighty rockets blasted into the Florida sky and wondering where they were going and whether someday I, too, might become an astronaut.

Today, Americans are at a crossroads regarding outer space; they weigh the threats and risks in the balance. Will we lead the world in space exploration and benefit from the science associated with such travel while accepting the risks? Or will we abandon that opportunity and pass the mantle of exploration to others that may find new technologies to improve society and create threatening situations for our future? Further, where does space adventure fit into God's plan for mankind—and is the exploration of outer space another prophetic sign of the end times?

As a teenager, I lived near Huntsville, Alabama, and my stepfather worked with computers at the Marshall Space Flight Center. My next-door neighbor worked on the Saturn V booster rocket that eventually blasted the Apollo spaceship toward the moon. Two doors down was the home of one of the scientists who came to America after World War II with the German chief rocket scientist Wernher von Braun. No wonder

I was very familiar with and excited over America's space program. It was my dream to go well beyond this earth to explore the heavens.

Years later, I was gratified when as a West Point plebe I sat in South Auditorium at the United States Military Academy watching the black-and-white, grainy, live television pictures of man's first walk on the moon (July 21, 1969). Those military officers who became astronauts inspired me and certainly made me proud to be an American.

Today, in the wake of half a century of exploring space, mostly in earth's orbit, and a few disasters like the 1986 loss of the Space Shuttle Challenger crew, I can't help but wonder like other Americans whether our space exploits should continue or rest on our past accomplishments and save our dollars. After all, based on recent national opinion surveys, there isn't a great appetite for spending more money on space adventures—e.g., only 51 percent believe increased spending on the space program is a good investment.[213] Understandably, Americans are focused on our earthly concerns—jobs, terrorism, and the latest pop culture craze—and thus they fail to grasp the true potential and importance of future space exploration.

Yes, America spent a lot of money (estimated at $170 billion in 2005 dollars) reaching the moon, but we gained tremendous insights in the process, and new scientific discoveries came from decades of investing in space programs.[214]

Some argue that our very survival may depend on future space exploration. They believe mankind is quickly using up this earth; America lacks the land to grow, and our natural resources such as water, food, and energy are dwindling.

The counter argument is that such people obviously haven't traveled to the vast empty spaces across the American West and seen the many sparsely populated regions across our expansive globe. No, to you doubters, there is plenty of open space left for more earthlings out in the American West and across this terrestrial ball. We don't have to blast to another planet to find open sky and a new home. Further, I don't accept that life on some asteroid tumbling through the solar system with one-tenth of earth's gravity, or inside some moon crater—much less on some other planet like Mars with its extreme temperatures and little oxygen—is at all preferable to living in the Utah desert or the Wyoming Mountains.

Renowned astrophysicist Stephen Hawking disagrees with my view, and he sounds alarm bells about future life on earth. Hawking favors pushing for colonization of outer space because he believes life on earth will become difficult "to avoid disaster in the next hundred years."[215] He explained:

> We are entering an increasingly dangerous period of our history. Our population and our use of finite resources of planet Earth are growing exponentially, along with our technical ability to change the environment for good or ill. But our genetic code still carries the selfish and aggressive instincts that were of survival advantage in the past. It will be difficult enough to avoid disaster in the next hundred years, let alone the next thousand or million.
>
> Our only chance of long-term survival is not to remain lurking on planet Earth, but to spread out into space.

Hawking favors human space exploration and encourages study into how to make space colonization possible. However, his statement is as much about his theology and the nature of sinful man as it is about practical science.

I'd argue, as pointed out above, we have plenty of open space here on earth, and science has already given us the ability to significantly increase food production. Earth can sustain a far larger population than the current 7.1 billion, but, as Hawking alludes, future man-made disasters—read "wars"—potentially puts living here on earth at risk. I agree on that point.

There are scenarios that make earth less habitable, such as significant nuclear events. Those could portend the prophetic end times, but for now, prior to such a scenario taking place, Hawking's argument from a resource and scientific discovery perspective appears as prudent and deserves serious consideration.

It is seldom disputed that space exploration has been especially helpful to modern man. Consider a sampling of the tangible benefits derived from past science-related space exploits.

We can thank the early Apollo moon missions for improving our healthcare. Specifically, National Aeronautics and Space Administration's (NASA) Jet Propulsion Laboratory created digital image processing that is used today in Magnetic Resonance Imaging and CT scans.[216]

NASA's work on protein crystals in space led to making albumin, a human protein that led to the development of a cancer drug combination approach and skin-care products.

Our modern life benefits tremendously from NASA's Earth Observing System Data and Information System, which required massive storage and led to the creation of a high-tech software program that can hold large amounts of information. Today, we enjoy the benefits of those scientific breakthroughs that now provide efficient information storage for hospitals, cellphone providers, the military, and businesses.

Satellites are instrumental in how we communicate and navigate. NASA and other space programs built and placed satellites into earth's orbit to give us Global Positioning Systems, worldwide television and radio programming, and access to anyone in the world with a cellphone.

Our automobile tires are more durable, thanks to the science that developed parachute shrouds for landing the Viking spacecraft on Mars. That fibrous material is now used in automobile tires such as radials that are five times more durable than steel and extends tread life ten thousand miles over a conventional rubber tire.

America's return on the space investment appears to be pretty good. One comparison found that 1.2 percent of a taxpayer's total payments have gone to science, space, and technology programs while national defense and education receives 26.3 percent and 4.8 percent of taxpayers' dollars, respectively. No one disputes the importance of our national security or education, but by comparison our space investment paid great dividends as well.

The fact is if we don't continue our efforts in space, other nations will, and they will benefit economically from those investments and possibly threaten our futures with their inevitable gains. However, as you will see in the coming pages, space is becoming a very dangerous place, and modern man's use of it may in fact be another sign of the coming end times.

Risks Related to a Future Space Program

There are plenty of risks going forward with our involvement in space exploration, such as high costs and the loss of life. But the risks of not continuing our space programs are offset by the good. The U.S. must maintain a robust effort or be left behind by the rest of the world. Specifically, space is the emerging platform for military forces, global communication, and a sphere for technological advances.

Future space work holds out promises of significant technological advances. Consider some emerging space-related technological developments, plans, and continuing endeavors.

Tough-as-steel ceramics. DARPA is doing work to make ceramic parts strong, lightweight, and able to handle heat better than many metals. Similar heat-shielding tiles were on the space shuttle, but now researchers have used a three-dimensional printer (an additive manufacturing process for making three-dimensional solid objects) to make customized ceramic parts that do not crack. This research could help a new class of ceramic-body or ceramic-engine jets, perhaps making possible hypersonic flight from New York to Tokyo in just a few hours.[217]

Mining asteroids and space service stations. A Washington state firm plans to mine asteroids in order to establish in-space gas stations. This isn't a sci-fi ambition, but has a growing technological means. The firm in question, Planetary Resources, intends to begin looking for water on asteroids, which as every science student knows is made from hydrogen and oxygen—the chief components of rocket fuel. Once mined, the asteroid would become an in-space "gas station" for spacecraft to top-off. The company also plans to mine platinum and other rare metals from space rocks.[218]

Near-earth object defense agency. The Chelyabinsk meteor explosion over Russia in 2013 prompted NASA to start up the Planetary Defense Coordination Office in Washington, D.C., to lead in efforts to find and characterize natural impact hazards such as asteroids and comets that pass near the earth's orbit. The office will also lead the coordination of interagency and intergovernmental efforts to plan response to any potential impact threats. Keep in mind there are at least 13,500 known near-earth objects of all sizes discovered to date, with 1,500 near-earth objects detected each year.[219]

Europeans plan to provide global Internet access via a network of satellites. The European satellite telecommunications industry wants to provide global Internet access using a constellation of more than seven hundred satellites in low earth orbit. Right now, much of the world lacks Internet service, and most of those who do have that service rely on high-cost terrestrial broadband. European governments are establishing new funding sources to help their satellite industry be competitive. It is noteworthy that Europe's Arianespace launch consortium relies on mainly Russian Soyuz rockets to catapult their satellites into orbit.[220] That may prove to be an Achilles heel given Europe's ongoing rumble with Russia.

No place on earth invisible. "Satellite imagery is available to basically everybody now," explained Walter Scott, founder of Digital Globe. That's one of three major changes from just a decade ago, when only a small number of organizations had access to quality on-demand imagery. "Now billions of people around the world use satellite imagery as part of their daily lives." The second major change is the degree to which "commercial satellite imagery has taken on an increasingly larger role in serving the U.S. government." The third major change is the advent of the compute cloud, which makes it possible to compute data without the end user having to own or operate a vast computer infrastructure. That phenomenal news means "there has been a major change from simply viewing a picture to actually converting that picture into useful information."[221]

NASA's *Spinoff* **leverages space for America.** NASA claims its space technologies are making life better on earth through such as providing a way to locate underground water in dry places, building more fuel-efficient airplanes, and creating shock absorbers for buildings in earth-quake-prone areas. NASA annually produces a publication, *Spinoff,* that highlights technologies like these that originated in space exploration. The 2015 *Spinoff* featured a coliform bacteria test that monitors water quality in rural communities, as well as cabin-pressure monitors that alert pilots when oxygen levels are dangerously low. NASA insists that its space exploration spinoff technologies produce commercial and societal benefits across the economy that generate billions of dollars in revenue and create thousands of jobs.[222]

The above technological developments, plans, and ongoing endeavors are just the beginning, if NASA has its way. The space agency has plenty of plans for the future.

NASA'S future plans. NASA is busy making critical advances in aerospace, technology development, and aeronautics. Consider the organization's specific plans for the long-term, subject, of course, to funding. NASA intends to send humans farther into the solar system, including to an asteroid and to Mars.[223]

NASA is developing the Orion spacecraft to carry astronauts to missions beyond the earth's moon. Some of the preliminary research for that human journey to Mars is already taking place 250 miles overhead, aboard the International Space Station. Astronauts there work in a microgravity environment that has led to breakthroughs in understanding earth, space, and the biological sciences. The station is also used to explore technologies like autonomous refueling of spacecraft and human/robotic interfaces.

NASA technology drives space exploration. The agency is developing, testing, and flying transformative, cutting-edge technologies for future exploration. This work will help humans reach an asteroid and Mars, and pays off with evolving technologies like advanced solar electric propulsion, large-scale solar sails, and composite cryogenic storage tanks for refueling depots in orbit.

NASA is working on four-propeller-driven robot explorer drones to conduct future planetary exploration. "Swamp works," a NASA facility at the Kennedy Space Center, is designing these flying probes capable of reaching hard-to-access spots.

These autonomous drones would make decisions while flying around alien environments that are "light minutes" away from mission control on earth. When operating in airless environments such as an asteroid, the drones are powered by cold-gas jets using steam or nitrogen gas, and they return to their larger base station for periodic recharging. Future missions could mine for vital resources such as water ahead of manned missions. After all, "every kilogram mined on site is a kilogram that astronauts wouldn't have to bring with them," said Mike Dupuis, co-investigator for the Extreme Access Flyer (EAF) Project.[224]

NASA continues an array of earth-specific missions to improve our lives as well. The agency studies challenges facing earth: climate change, sea-level rise, freshwater resources, and extreme weather events. NASA also studies our sun, helping us understand its variations in real-time to better appreciate its influence on space weather that impacts technology on earth, such as electromagnetic pulses.

Space Becoming a Killing Platform

"The military exploitation of space will be the defining characteristic of the twenty-first century," states STRATFOR, a Texas-based intelligence and security consulting organization.[225] We arrived at this point through a series of steps over many years, beginning with the German rocket technology in World War II. Until the early 1980s, the only weapons to enter space were intercontinental ballistic missiles (ICBM), which made an arc through the atmosphere carrying warheads aboard reentry vehicles to their targets. President Ronald Reagan's 1983 Strategic Defense Initiative (the so-called Star Wars Program) changed our view of space and simultaneously helped collapse the economically strained Soviet Union, which couldn't afford to make heavy investments in space like America was doing.

Over the subsequent years, the U.S. built a space-based war-support infrastructure. Although it didn't deploy weapon platforms in space, it did create a major network of satellites to enable its global ground, air, and sea operations. That network became critical to America's precision-guided munitions prowess and its evolving reliance on drones like those used in Afghanistan.

America's military wasn't alone in space over the past few decades, either. Many countries—especially China and Russia—built their capabilities, and in some cases are outmatching American systems.

China poses perhaps the biggest threat to America's network of military satellites. Specifically, Chinese anti-satellite tests demonstrate a growing proficiency to down satellites. Their most recent anti-satellite test took place October 30, 2015. It was a success, and this time the debris field (unlike the previous test in 2007) was negligible, which demonstrates an improved capability. That instrument of national power is growing in real terms among other countries as well, and it has understandably alarmed Washington because the Pentagon relies almost completely on its network of satellites for global operations.

Another aspect of the space race is preparation to use ballistic missile defense systems outside of earth's atmosphere. Specifically, the U.S. Ground-based Midcourse Defense System, located at Fort Greely, Alaska, and Vandenberg Air Force Base, California, is able to reach into space to attack incoming ICBMs. The "killing" is done with exoatmospheric kill vehicles, which are like a bullet that strikes the incoming missile or another "bullet."

The U.S. is scrambling to defend itself against an array of ballistic missile threats. U.S. Navy Vice Admiral James Syring, the head of the Missile Defense Agency (MDA), said in early 2016 that a new kill vehicle will fly in 2018 and be tested in 2019. Admiral Syring said his plan is to field the new kill vehicle around 2020, which he admits is "a very quick schedule."[226]

The new MDA kill vehicle came about after a string of intercept failures by the Boeing-built Ground-based Midcourse Defense System, America's primary territorial shield. That missile was designed to destroy incoming missile warheads by force of direct impact.

Syring said the new kill vehicle is more reliable, more producible, more effective, easier to maintain, and cheaper than the current vehicle. It is also a bridge to the multi-object kill vehicle, which promises to destroy multiple objects simultaneously, a capability that distinguishes between missile warheads and other objects.

Meanwhile, the MDA plans to have thirty-seven Ground-based Midcourse Defense interceptors ready to launch by the end of 2016, Syring said. Currently, the MDA has thirty deployed interceptors at the two locations and plans another fourteen at Fort Greely to counter the North Korean missile threat.

Missile defense systems are expensive, and even our likely adversaries hesitate to deploy systems of their own. For example, China has not decided whether to build a ballistic missile defense, according to a study by the Federation of American Scientists.[227] However, it is very serious about countering America's use of space for military operations. That's in part why the regime in Beijing recently created a fifth service that puts space military operations at the forefront.

On December 31, 2015, China's Central Military Commission announced the creation of a new Strategic Support Force (SSF), which will be comprised of space, cyber, and electronic warfare forces. The global times reports the SSF will be critical to China's ability to maintain information dominance in wartime and will integrate joint operations outlined in China's 2015 Defense White Paper.[228]

Military theorist Song Zhongping, a former Second Artillery Corps officer, described the SSF as a fifth service that will "focus on reconnaissance and navigation satellites." That view was echoed by Yin Zhou, a retired Chinese admiral, who said the SSF's space mission includes managing space-based navigation and reconnaissance.

Evidently, one interpretation of the reorganization in concert with the Chinese Defense White Paper is China's strategic intent to strike at its adversary's vulnerable areas of modern "high-tech" networks. That's an American vulnerability—over-reliance on its global C4ISR (command, control, communications, computers, intelligence, surveillance, and reconnaissance) infrastructure for expeditionary military deployments. Some Chinese analysts consider America's C4ISR as our Achilles heel; they term the U.S. a "no satellites, no fight" military force.[229]

All public evidence to date indicates only the former Soviet Union ever deployed a weapon to a space platform. The 1970s Soviet-era Almaz Space Station had aboard an R-23mm Kartech, a 23-millimeter cannon designed for the Tupolev Tu-22 Blinder supersonic bomber. That weapon remained in obscurity until last year.

A Russian news report indicates the Soviets were terrified that an American spacecraft would inspect Soviet military satellites, thus the space cannon. The thirty-seven-pound weapon could fire from 950 to 5,000 shots per minute and could pierce a metal gasoline canister from a mile away during its ground tests. Once in space, the cosmonauts would have to use an optical sight and turn the entire twenty-ton space station to point the cannon at the target.[230]

Unfortunately, it appears that many countries are now exploring the weaponization of space. Weapons could be based on the moon or even asteroids and used to target earth with precision munitions. Whether space-based or using space as a medium through which weapons pass to reach a target, the available science will impact the nature of future warfare and the science is here.

For years, many futurist warfare observers warned of arming space orbiters with nuclear or non-nuclear electromagnetic pulse (EMP) weapons. Detonating those weapons at a high altitude could easily obliterate an enemy's electrical grids, satellites, command and control systems, communications, computers, intelligence, surveillance, and reconnaissance operations. A large EMP weapon could even shut down an entire country, thus ending war before the first shot.[231]

A space-based EMP platform is more desirable than an earth-born system like an ICBM with an EMP warhead because the ground-launched missile is vulnerable to interception during the launch phase. Therefore, countries threatened by space-based EMP can either develop ground or air-to-space-based anti-satellite capabilities or space-based weaponized orbiters.

A real alternative to killing a missile with another is the space-based laser, which is in the works. A laser would target ballistic missiles fired during the boost phase, which is best because that's when the missile is traveling its slowest. This is an important alternative to the Standard Missile-3 launched from America's Aegis Cruiser or our ballistic missile defense systems to attack missiles launched by rogue states like North Korea and Iran, which have already demonstrated the ability to deliver warheads accurately atop ICBMs.

The U.S. Air Force has a missile-zapping Boeing 747, which proved to be too expensive—$5 billion over sixteen years—but did destroy one missile in a test. That test proved the concept, but the aircraft had to be too close to the launch site, which negated the crew's security. Today, the lasering jetliner sits in an Arizona desert, but the Pentagon now has its sights on a higher-flying drone successor. Admiral Syring says the laser drone's time may have come.[232]

"We have significantly ramped up our program in terms of investment and talking about it more of what else needs to be done to mature this capability," Syring said.

Such drones will fly at sixty-five thousand feet or higher, in all weather, and will stay aloft for perhaps weeks and engage enemy missiles as they lift off.

A drone may resolve the staying power and altitude issue, but the challenge is getting more power from lighter-weight lasers, according to Syring. "If you can balance that range, altitude, power and number of boosters you need to defeat to help augment our kinetic capability," he said, "you're thinking about the problem exactly right."

Lockheed Martin announced in March 2016 that its directed energy weapons—lasers—are ready for the battlefield. "The technologies now exist," said Paul Shattuck, the Lockheed Martin leader for Directed Energy Systems. "They can be packaged into a size, weight, power and thermal which can be fit onto relevant platforms, whether it's a ship, whether it's a ground vehicle or whether it's an airborne platform." Shattuck said their 30 KW weapon can bore a hole through a two-inch piece of steel in seconds, which is enough to disable an incoming rocket.[233]

Moscow isn't sitting on the sidelines watching as China and the U.S. fight over primacy in space, however. In 2015 Russia established its aerospace forces out of fear of being outgunned by the United States. Reportedly, the U.S. is working on a Prompt Global Strike capability, which could launch a thousand missiles toward Russia at Mach 20 arriving with pinpoint accuracy atop the Kremlin's nuclear missiles, military radars, and submarine bases. In one strike, the vast majority of Russia's ability to strike back would allegedly be destroyed.[234]

Russia's Aerospace Defense Forces are responsible for space and missile defense with a fortified operations center in Moscow described as the bridge from the reimagined *Battlestar Galactica*.[235]

The new aerospace force is pushing war preparations in space planning to knock out satellites with precision weapons. "It will be a comprehensive system, which will help detect and eliminate targets even at distant approaches," a Russian defense ministry spokesman told *Interfax* news agency. "It can be viewed as our response to the Prompt Global Strike concept being implemented by the U.S."

Russia is serious about its new space command. It already upgraded its optoelectronic satellite-monitoring station in Tajikistan and has blasted three suspected anti-satellite weapons into orbit.

"If Russia acquires the capability to destroy U.S. military satellites in low orbits, this will entail rendering the American army blind and deaf, and precision 'smart' weapons will be turned into scrap metal," said the editor of a Russian agency as reported by *Nezavisimaya Gazeta*. That view, as previously stated, is shared by the Chinese military.

In late 2015, U.S. Secretary of Defense Ash Carter expressed concern about Russian activities in space. He drew attention to Russia's "provocations," including new systems in space. He cautioned that the Russians were "flouting" the principles that underlie the "principled international order."[236]

"At sea, in the air, in space and in cyberspace, Russian actors have engaged in challenging activities," the secretary told the annual Reagan National Defense Forum. Specifically, Carter mentioned, Russia's "innovation in technologies like electromagnetic railgun, lasers, and new systems for electronic warfare, space and cyberspace, including a few surprising ones that I really can't describe here."[237]

In January 2016, Russia's defense ministry announced its intention to test launch sixteen ICBMs in 2016, including flight-test procedures for advanced weapons (read "nuclear"), and control of the technical readiness of new missile systems. Colonel-General Sergey Karkayev, commander of Russia's strategic missile troops, said fourteen of the sixteen launches are part of the development of new missiles and warheads.[238]

Further, *The Diplomat* reports:

> Russia is in the middle of modernizing its strategic and nonstrategic nuclear warheads. According to the Bulletin of the Atomic Scientists, Moscow has currently 4,500 nuclear warheads, of which roughly 1,780 strategic warheads are deployed on missiles and at bomber bases. An additional 700 strategic warheads are kept in storage along with approximately 2,000 nonstrategic warheads. "Russia deploys an estimated 311 ICBMs that can carry approximately 1,050 warheads," the Bulletin of the Atomic Scientists further notes.

There is a serious race to weaponize space and at this point there is no clear leader. America depends on space as a toll-free highway, but that is quickly changing for the worse.

Is Space Another Prophetic Sign of the End Times?

There are really two biblical issues to consider. Should man explore space, and is the exploration and development of space travel a sign of the end times?

One can argue that the Bible is on both sides of the space exploration issue. On the side against space exploration, advocates cite Psalms 115:16, arguing the heavens (space) are for the Lord and earth is for His children. They also cite the construction of the Tower of Babel, which God stopped by confusing human language because man was trying to get to "the tops reached in heaven" (Genesis 11:4).

The other side argues for space exploration because God gave his people the whole heaven (Deuteronomy 4:19). Therefore, we have the right—privilege—to explore what God created.

Harnessing technology to conquer space perhaps is a sign.

People will faint from terror, apprehensive of what is coming on the world, for the heavenly bodies will be shaken. (Luke 21:26, NIV)

Space can play host to all sorts of weapons that could lead to the people "faint from fear" due to the "the heavenly bodies will be shaken." Just imagine satellites being blown up, ballistic missiles raining down onto the earth, and EMPs exploding miles above big cities followed by a total blackout. These could easily make one say the "heavily bodies will be shaken."

Chapter 8

Cyber Threats and Risks

TOP AMERICAN OFFICIALS indicate we are losing a global war. Last fall, James Clapper, director of National Intelligence, told Congress that "cyber threats to U.S. national and economic security are increasing in frequency, scale, sophistication, and severity of impact."[239] "The demand for our cyberforces far outstrips the supply," stated National Security Agency Director Admiral Mike Rogers. Similarly, "Make no mistake, we are not winning the fight in cyberspace," Senator John McCain (R-Arizona) said in response to their testimony.[240]

The cyber battlefield is exploding with many actors, methods of attack, targeted systems, and victims. All information and communication technology (ICT) networks remain vulnerable to espionage and disruption to include our critical infrastructure such as the electric grid, water supplies, and national secrets.

Such attacks have already taken place. CIA cybersecurity analyst Tom Donahue said, "We do not know who executed these attacks or why, but all involved intrusions through the Internet." The reported attacks were against computers of foreign facilities owned by U.S. power utilities.[241]

We also know that cyberspies have penetrated the U.S. electrical grid and left behind software programs that could be used to disrupt the system, reports the *Wall Street Journal*. "The Chinese have attempted to map our infrastructure, such as the electrical grid," said a senior intelligence official, and "so have the Russians."[242]

Although Director Clapper dismisses a "cyber Armageddon" scenario that cripples America's infrastructure, he does foresee ongoing lower level cyberattacks that seriously hurt our economic competitiveness and national security.[243]

This chapter begins with a "101" introduction to cyberwarfare, identifies many of the risks, identifies the leading threats, and concludes with a prophetic view.

What Is Cyber and Cyber Warfare?

You can be excused for not understanding the world of cyberspace, much less the weaponization of that environment. There seem to be so many strange terms associated with cyberspace—bot, botnet, denial of service attack, hacker, Internet service provider, network, ping, script, worm, World Wide Web, zombie, and literally hundreds more. Perhaps before going forward, it is useful to catch up with cyber history to include a review of the thing called the Internet and then the emergence of the phenomenon known as cyberwar.

The Internet originated in the 1950s as computer networks that shared research data, as well as military radar installations sharing Cold War-era air surveillance data, on Russian Tu-95 Bear bombers (best known as the U.S. Semi-Automatic Ground Environment). Those primitive networks formed the basis for a major leap forward launched by the Pentagon's Advanced Research Projects Agency. That organization later added the word "Defense" to its title to become the now-familiar DARPA, which did the preliminary work on what became the Internet.[244]

DARPA funded the RAND (a contraction of the term *research* and *development*) Corporation to work on a distributed network architecture that would survive a possible Soviet nuclear attack. Eventually, that network matured with better programming, technology, and infrastructure, and by the 1990s, the World Wide Web architecture emerged.[245]

The network known as the Internet is an environment to which many discrete users across the globe communicate. Many modern users are linked by social networks such as Facebook, Twitter, and YouTube, others for business, and yet many others for government purposes. Collectively, each user creates a virtual extension of themselves and their purposes.

The Internet's varied connections create a structural vulnerability, however. The modern world's demand for information and communications makes the Internet more critical to everyday life, or the more correct terminology—the "cyber" world. ("Cyber" is defined as "relating to or characteristic of the culture of computers, information technology, and virtual reality."[246]) This means our cyber connections can be exploited by individuals, groups, and governments, which led to coining the term "cyberwar" or "cyberwarfare," "actions by a nation-state (or others) to penetrate another nation's computers or networks for the purposes of causing damage or disruption." Other definitions include non-state actors such as terrorist groups, corporations, political and ideological groups as well.[247]

STRATFOR defines "cyberwar" as a term that encompasses "significant geopolitical conflict in cyberspace usually involving at least one nation-state or its critical infrastructure." Cyberwarfare can be an independent line of attack or can be used in a supporting role, but in either case, it evidences five characteristics: (1) it features an extremely dynamic new battlespace, (2) distance or range is not a factor, (3) operations tend to be decentralized and anonymous, (4) it is primarily an offensive weapon, and (5) it requires low-cost investment with potentially great pay-off.[248]

Even though the term "cyber" suggests the non-physical, clearly cyberwarfare involves real people and equipment. Fighting a cyberwar requires computers, servers, fiber-optic cables, and the World Wide Web.

The cyberwarrior musters resources—a global network often created by infecting and hijacking other computers—and in an instant strikes the target without warning.

Cyberwarfare, like traditional warfare, depends on experience, skill, and the weapons brought to the battlefield. For the cyberwarrior, the growing Internet with broadband connections and ever more powerful computers makes this new battlefield incredibly complex and potentially rewarding for the victor.

What Are the Cyber Risks?

The risks and recent costs associated with cyberattacks are high. Director Clapper indicates that we are experiencing "an increase in the scale and scope of reporting on malicious cyber activity that can be measured by the amount of corporate data stolen or deleted, personally identifiable information compromised, or remediation costs incurred by U.S. victims." Some of the most noteworthy recent costs include:

- In 2015, the Office of Personnel Management (OPM) discovered their computer systems that contained the background investigations of current, former, and prospective federal government employees were compromised—hacked. OPM said that 21.5 million personal records were stolen. That means all key personal data such as Social Security numbers, family associations, financial information, and much more—a criminal's lottery for exploitation.[249]
- American commercial entities suffered serious losses as well. JP Morgan Chase (JPMorgan) suffered distributed denial of service attacks that virtually shut down the firm, and now it is pouring at least $250 million into computer security to protect itself.[250]
- Virtually the same happened to Home Depot in 2014, exposing the records of fifty-six million credit/debit card customers. That breach cost the box store operator $62 million.[251]

- A $101 million cyber heist left bank officials pointing fingers of blame at one another for the most audacious bank raid in history. Evidently, in early March 2016, hackers breached the Bangladesh central bank's security system and then pretended to officially request a large tranche of cash from the New York Federal Reserve. In fact, $101 million was transmitted before raising suspicions. Bangladesh banking officials said the cyber-criminals were ultimately stopped by a spelling mistake in its transfer instructions.[252]

The NSA chief expects such attacks to continue, and they will become more sophisticated. The OPM hack, for example, demon-strated that "data is increasingly a commodity of value all on its own," Rogers said. He continued, "What you saw at OPM…you're going to see a whole lot more" of in the future.[253]

The issue here is integrity of information, according to Director Clapper. Up to now, cyberattacks focused on the "confidentiality and availability of information; cyber espionage undermines confidentiality, whereas denial-of-service operations [like at JPMorgan] and data dele-tion attacks undermine availability." But the nature of the risks is chang-ing, Clapper explained. We anticipate seeing more "cyber operations that will change or manipulate electronic information in order to com-promise its integrity (i.e., accuracy and reliability) instead of deleting it or disrupting access to it." That type of attack affects decision-making for governments, corporations, investors, and others, because they won't trust the information needed for making decisions.[254]

Consider the cyber risks to our maritime operations. U.S. Coast Guard Assistant Commandant Rear Admiral Paul Thomas testified before a House Committee on Homeland Security. Admiral Thomas said the Maritime Transportation System (MTS) relies on information technology systems to manage the global supply chain are at risk to cyberattacks. When the MTS is tampered with, the results can include injury or death, harm to the marine environment, or a disruption of vital trade activities. Keep in mind that the MTS, Thomas explained, directly influences operations ranging from container terminal operations ashore to offshore platform stability. Today, virtually every aspect of maritime operations is tethered to cyberspace.[255]

The maritime risks associated with cyberattacks to the MTS will continue to increase as shipping firms incorporate more information technology systems into their operations. We must recognize the potential consequences, said Thomas, not only for shipping firms but for the 360 sea and river ports that handle more than $1.3 trillion in annual cargo, which our nation relies upon for much of its freight transportation.

Our commercial sector faces some very serious emerging trends because most companies are connected to the Internet and therefore vulnerable to cyberattacks. Consider five significant cybersecurity risks.[256]

First, many consumers like the flexibility of mobile devices, which makes a significant part of the connected population vulnerable to hacking. Candid Wüest, threat researcher at Symantec Security Response, said at a cybersecurity conference sponsored by Georgia Tech in 2015: "Although this would require cyber criminals to target individual cards and wouldn't result in large scale breaches or theft like we have seen in the U.S., the payment technology used won't protect retailers who aren't storing payment card data securely, and they will still need to be vigilant in protecting stored data."[257]

Second, there is an explosion in the Internet of Things (IoT), which creates a very large cyberattack target. The IoT refers to "networks of objects that communicate with other objects and with computers through the Internet." Those "things" may include remote communication, data collection, or control devices such as vehicles, appliances, electric grids, and manufacturing equipment. IoT devices tend to lack safeguards because the designers were never incentivized to think about security.[258]

IoT is morphing into the "Internet of vulnerabilities," according to some cyber experts. One expert, Jamison Nesbitt, founder of Cyber Senate, said that the IoT is "the main cybersecurity risk [and] IoT presents unique security challenges in terms of the number of connected devices present." For example, "The IoT will be integrated into every market you can think of—from healthcare to the energy industry and transport network—but it hasn't been designed with security in mind. There are millions of hackers out there that could compromise these interconnected systems. We have sacrificed security for efficiency."[259]

Third, our rapidly growing, digitized economy lacks security savvy workers. This presents a major security training requirement for the corporate world where success using the Internet is a stark choice of sink or swim.

Capgemini Consulting, a global strategy and transformation consulting organization, states that the "shortage of digital skills in the current marketplace is unprecedented. "It is estimated that over 4.4 million IT jobs will be created around Big Data by 2015; however, only a third of these new jobs will be filled."[260]

Fourth, the cyber espionage risk is not abating. Over recent years, the corporate world saw the effects of the cyberattacks on OPM, Target (2014),[261] and Sony (2015),[262] which explains the rush to bolster corporate cyber defenses.

Jamison Nesbitt, founder of Cyber Senate, said, "The next world war will be fought on a keyboard," and we should "expect cyber espionage attacks to increase." He predicts that more small nation-states and terror groups will use cyber warfare.[263]

Fifth, there is a growing use of "ransomware," a type of malware that restricts access to the computer system that it infects. "We predict ransomware variants that manage to evade security software installed on a system will specifically target endpoints that subscribe to cloud-based storage solutions such as Dropbox, Google Drive, and OneDrive. Once the endpoint has been infected, the ransomware will attempt to exploit the logged-on user's stored credentials to also infect backed-up cloud storage data," McAfee's report on 2015 cyber risks noted.[264]

"Ransomware victims will be in for a rude shock when they attempt to access their cloud storage to restore data—only to find their backups have also been encrypted by the ransomware," the experts said. Further, ransomware may be used by hackers to extract ransom payments from victims to release their encrypted data.[265]

Sixth, easy-to-crack passwords continue to be a risk. There are weaknesses with passwords, a problem the industry is hoping to phase out and replace with a number of different multi-factor options.

The whole password issue is flawed, especially the password recovery process—and who hasn't at one time or the other forgotten one of the dozens of their passwords? "The traditional method of password recovery is asking questions that only you, the real owner, should know. Unfortunately, answers to these questions often can be deduced based on information that can easily be found online—especially given people's proclivity for 'over-sharing' on social media sites."[266]

Cyber Threats

The DNI provides a mixed message regarding threats. "Although cyber operators can infiltrate or disrupt targeted ICT networks, most can no longer assume that their activities will remain undetected indefinitely," Clapper said. That's of little consequence if the cyberattack is part of a larger operation, and secrecy after the fact isn't important.[267]

Clapper testified to illustrate his point that in May 2014, the U.S. Department of Justice indicted five officers from China's Peoples' Liberation Army (PLA) on charges of hacking U.S. companies. Does anyone expect Beijing to extradite those officers to the U.S.? Of course not, and neither will any other nation. The fact is the damage is done, and whether we know the perpetrator only matters if we take action to stop them in the future. At this point that's doubtful.

No matter the course of the cyberattack, once the damage is done, it's virtually impossible to put the genie back in the bottle. Then Chairman of the Joint Chiefs of Staff General Martin Dempsey put cyberwarfare in perspective, stating that cyberattacks that target civilian infrastructure and businesses "present a significant vulnerability to our nation."

Dempsey explained, we are vulnerable because the U.S. faces a "level playing field" against our cyber enemies.[268]

Dempsey went on to say, "I'd like the enemy to play uphill and us to play downhill." He ranked deterring cyber threats among his highest priorities.[269]

The general said he was very concerned about the weak link, the commercial networks. "We have authorities and capabilities that allow us to do a pretty good job of defending ourselves," Dempsey added. "But the vulnerability of the rest of America is a vulnerability of ours, and that's what we have to reconcile."[270]

At least twenty national militaries have cyberwarfare units, according to the former chairman. He worries that our adversaries will exploit our weak link infrastructure rather than the tougher military.[271]

Director of National Intelligence Clapper identified an array of attackers. "Politically motivated cyber-attacks are now a growing reality, and foreign actors are reconnoitering and developing access to U.S. critical infrastructure systems, which might be quickly exploited for disruption if an adversary's intent became hostile." Those attackers target U.S. government, military, and commercial networks on a daily basis.[272]

The range of actors come in four groups, according to Clapper: (1) nation-states with highly sophisticated cyber programs (such as Russia or China), (2) nations with lesser technical capabilities but possibly more disruptive intent (such as Iran or North Korea), (3) profit-motivated criminals, and (4) ideologically motivated hackers or extremists like ISIS.

It is often very difficult to distinguish the cyber effects delivered between state and non-state cyber actors, however. That's because in some instances, these actors collaborate or employ similar cyber tools.

Consider what we know about some of the major cyber actors.

Russia: Russia's cyber prowess became globally evident when it attacked the Republic of Georgia in August 2008. Russia's offensive against Georgia began not with tanks but in cyberspace. The day prior to Russian tanks rushing through the Roki tunnel in South Ossetia,

Moscow launched a cyberspace attack against Tbilisi, Georgia's capitol.

The first evidence of the attack was a "permanent DDOSs [distributed denial of service] attack" against Georgia's news web site civil.ge. That was the first of many cyber-attacks attributed to a Russian hacker network.

Russia's Ministry of Defense is establishing its own cyber command, which, according to senior Russian military officials, will be responsible for conducting offensive cyber activities, including for propaganda operations, and inserting malware into enemy command and control systems. Russia's armed forces are also establishing a specialized branch for computer network operations. [273]

The DNI indicates computer security studies assert that Russian cyber actors are developing means to remotely access industrial control systems (ICS) used to manage critical infrastructures like electric grids and commercial enterprises. For example, unknown Russian actors successfully compromised the product supply chains of at least three ICS vendors so that customers downloaded malicious software ("malware") designed to facilitate exploitation directly from the vendors' websites along with legitimate software updates, according to private-sector cybersecurity experts. [274]

China: Chinese cyber espionage continues to target a broad spectrum of U.S. interests, ranging from national security information to sensitive economic data and U.S. intellectual property. Although China is an advanced cyber actor in terms of capabilities, Chinese hackers are often able to gain access to their targets without having to resort to using advanced capabilities. Improved U.S. cybersecurity would complicate Chinese cyber espionage activities by addressing the less sophisticated threats, and raising the cost and risk if China persists. [275]

Admiral Michael Rogers, commander of U.S. Cyber Command and director of the National Security Agency, told a Senate committee he believes China to be the biggest perpetrator of the volumes of attacks barraging U.S. networks and the military is struggling to keep up, even as the services train up their forces to fight in cyberspace. [276]

The pessimistic views shared particularly by the government officials came just days after the U.S. and China announced a "common understanding" not to hack each other's economic interests.[277]

"I think we will have to watch what their behavior is, and it'll be incumbent on the intelligence community, I think, to depict [and] portray to policymakers what behavioral changes—if any—result from this agreement," Director of National Intelligence James Clapper told the Senate Armed Services Committee.[278]

When asked by committee chair Senator John McCain (R-AZ) if he was optimistic about a cyber deal with China, Clapper gave an emphatic "no."[279]

On March 16, 2016, Admiral Rogers echoed Mr. Clapper's view when he confirmed that China isn't abiding by the agreement during House Armed Services subcommittee testimony. Rogers testified, "cyber operations from China are still targeting and exploiting U.S. government, defense industry, academic, and private computer networks." The admiral explained that nations with advanced cyber capabilities like China are now taking steps to mask their cyberattacks by cooperating with non-government hackers. Using surrogate hackers makes it difficult for the U.S. to confront foreign states about cyberattacks. "And they say, 'It's not us. It's some criminal group; we don't control all that,'" Rogers said.[280]

Mr. Clapper and Admiral Rogers know that China won't abide by the cyber deal and President Obama is unlikely to take any action. In fact, China's cyber threat is likely to get much worse.

China's leader, Xi Jinping, on December 31, 2015, announced the overhaul of the organizational structure of the People's Liberation Army (PLA), establishing three new organizations: the Army Leading Organ, the Rocket Force, and the Strategic Support Force (SSF). The third department is the new home of China's cyber operations and commonly known as 3PLA.[281]

In his speech that included the restructuring of the PLA, Xi said that the SSF "is a new-type combat force to maintain national security and an important growth point of the PLA's combat capabilities." A number of Chinese press reports indicate the SSF is focused on cyber operations.[282]

An official Chinese outlet compared the SSF to the armed forces of the U.S., Russia, and other countries, indicating that it is more advanced because it involves operations that are not typical for a military unit. The Chinese report said SSF "can use its own power to damage the enemy."[283]

However, there is some speculation about the more concrete details of the SSF. The *Global Times* quotes retired 2nd Artillery Corps Officer Song Zhongping as saying that the SSF consists of three independent branches. The first is the "cyber force," manned by "hacker troops" responsible for cyber offense and defense. "The second is the 'space force,' responsible for surveillance and satellites. The final is the 'electronic force,' responsible for interfering with and misleading enemy radar and communications."[284]

Iran: "Iranian actors have been implicated in the 2012–13 DDOS attacks against U.S. financial institutions and in the February 2014 cyber-attack on the Las Vegas Sands Casino Company," according to DNI Clapper. More seriously are the alleged Iranian cyberattacks on several power utilities, reported by Democratic lawmakers in May 2013.[285]

Iran likely views its cyber program as one of many tools for carrying out asymmetric warfare as well as a sophisticated means of collecting intelligence.[286] It's not surprising, therefore, other nations respond to Iran's asymmetric ways by taking out her cyber leaders.

In October 2013 the head of Iran's cyberwar headquarters, Mojtaba Ahmadi, was found dead, shot in the heart at close range. Although Iran's Revolutionary Guards denied it was an assassination, it cannot be ruled out given Iran's accusations that Israel was behind the assassination of nuclear scientists and the head of that country's ballistic missiles program.[287]

North Korea: North Korea is another state actor that uses its cyber capabilities for political objectives. In early 2013, South Korean officials reported North Korea simultaneously hacked South Korean banks and top television broadcasters.[288] According to the FBI, the North Korean government was responsible for the November 2014 cyber-attack on Sony Pictures Entertainment (SPE), resulting in the theft of corporate information and the introduction of hard drive erasing malware into the company's network infrastructure. The attack coincided with the planned release of a SPE feature film, *The Interview*, a satire that depicted the fictional assassination of the North Korean president.[289]

Cybercriminals: Profit motivated cybercriminals rely on loosely networked online marketplaces, often referred to as the cyber underground, that provide a forum for the merchandising of illicit tools, services, infrastructure, stolen personal identifying information, and financial data. As media reports have documented, cybercriminals continue to successfully compromise the networks of retail businesses and financial institutions in order to collect financial information, biographical data, home addresses, email addresses, and medical records that serve as the building blocks to criminal operations that facilitate identity theft and healthcare fraud. The most significant financial cybercriminal threats to U.S. entities and our international partners can be attributed to a relatively small subset of actors, facilitators, infrastructure, and criminal forums.

"However," explained Clapper, "our federal law enforcement colleagues continue to have successes capturing key cyber criminals by cooperating with international partners. For example, in late June 2015, the Department of Justice and the United States Secret Service worked with their German counterparts to extradite Ercan Findikoglu, a Turkish national, responsible for multiple cyber-crime campaigns that targeted the U.S. financial sector stealing $55 million dollars between 2011 and 2013. Findikoglu was apprehended by the German Federal Police after U.S. Secret Service agents confirmed he was traveling through Germany in December 2013."[290]

Clapper indicated the FBI led an international coalition of twenty countries that dismantled an online criminal cyber forum known as Darkode. The Department of Justice, according to Clapper, indicated that Darkode "represented one of the gravest threats to the integrity of data stored on computers in the United States and elsewhere."[291]

Terrorists: Terrorist groups will continue to experiment with hacking, which could serve as the foundation for developing more advanced capabilities. Terrorist sympathizers will probably conduct low level cyberattacks on behalf of terrorist groups and attract the attention of the media, which might exaggerate the capabilities and threat posed by these actors.

With respect to ISIS, since June 2014, that group began executing a highly strategic social media campaign using a diverse array of platforms and thousands of online supporters around the globe. The group quickly builds expertise in the platforms it uses and often leverages multiple tools within each platform. ISIS and its adherents' adept use of social media allows the group to maximize the spread of its propaganda and reach out to potential recruits.

U.S. Response to Cyber Threat

The U.S. is finally on the cusp of being prepared to fight the cyberwar. Indicators of that readiness are the creation of a cyber command (at Fort Meade, Maryland), putting in place an arsenal of cyberweapons, and last but not least the start-up of a cyberwarfare school to train cyberwarriors.

The *Wall Street Journal* reported the U.S. military cyberwarfare training programs began in 2012 to prepare a cadre for conflict on an emerging battlefield.[292]

The U.S. Air Force's elite weapons school at Nellis Air Force Base in Nevada, trains airmen working at computer terminals how to hunt down electronic intruders, defend networks, and launch cyberattacks.

"While cyber may not look or smell exactly like a fighter aircraft or a bomber aircraft, the relevancy in any potential conflict in 2012 is the same," said Air Force Colonel Robert Garland, commandant of the Weapons School. "We have to be able to succeed against an enemy that wants to attack us in any way."[293]

According to the *New York Times*, cyberwarfare techniques were likely first used effectively in a joint operation with Israel to undermine Iran's nuclear program by launching the Stuxnet malware worm. Reportedly the U.S. also contemplated using cyberweapons to sideline Libyan air defenses in 2011, before the start of the U.S.-led operation. Evidently Israel used cyber techniques to hide its jet fighters in 2007 as it attacked the Syrian nuclear facility in the eastern desert.[294]

The Pentagon formally created the U.S. Cyber Command in 2008 and located it with the National Security Agency (NSA), sharing the same commander. The scope of past U.S. cyberspace operations was in part revealed by Edward Snowden's leaks and, as indicated above by the operation that allegedly sabotaged Iran's nuclear program. Arguably, given the damage to America's cyber infrastructure, the Cyber Command has generally been ineffective.

The command's general ineffectiveness is demonstrated by the U.S. government's reliance on the use of threats of economic sanctions and criminal prosecution of foreign officials believed to be behind cyberattacks on U.S. interests. Those breaches are expensive to both the U.S. government and the private sector, which is a frequent target as well.

Hopefully, Cyber Command's perceived ineffectiveness will be in the rearview mirror as it quickly matures. Understand we had a late start standing up the command and now it is going through growing pains like any major organization such as budget challenges.

Fortunately, Cyber Command's budget is annually growing, but throwing a lot of money too quickly at a new and complex set of problems doesn't necessarily produce the best outcomes. Further, in terms of spending money in the federal government, there are bureaucratic contractual procedures to follow that can take months or years.

Manpower is a problem for Cyber Command. Where does the command hire cyberwarriors with security clearances? There is a dearth of high-quality information technology (IT) experts, and many of them are drawn to high-paying jobs in Silicon Valley, not Fort Meade, Maryland. Therefore, because we can't hire cyberwarriors off the street, we have to hire potential cyberwarriors, vet them for trustworthiness (think of Snowden), and then train them, which can take years. Meanwhile, the nascent command suffers from many of the same personnel challenges in a downsizing military, as does the rest of the DoD. Today, the command has 123 cyber mission teams staffed by 4,990 personnel, but sixty-eight of those teams are only "in the early stages of development."[295]

Part of the command's perceived ineffectiveness is also due to a lack of the public's understanding that Cyber Command's primary mission is to protect the Pentagon's computer networks, not private sector victims like Sony and Home Depot or even the U.S. government's Office of Personnel Management. However, because the Defense Department's cyber networks are tethered to the 90 percent vulnerable private-sector cyber infrastructure, the department suffers from the consequences of hacker attacks as well. That explains why Eric Rosenback, the secretary of defense's principal cyber adviser, said, "The Department of Defense is not here to defend against all cyber-attacks, only the top 2 percent, the most serious."[296]

Cyber commander Admiral Rogers acknowledges all these problems as well as the fact that his command must move beyond its "reactive strategy," which for a new command is understandable. Standing-up Cyber Command is like building an airplane in midflight, and mistakes have and will be made.[297]

There is no secret that Cyber Command faces an uphill battlefield with plenty of threats and rapidly evolving challenges. Those threats are familiar: nation states, transnational terrorist groups like ISIS, criminal syndicates, and insiders like Snowden. The challenges are perhaps more daunting, however.

Cyber Command must be incredibly innovative, because once it employs a tactic against a hacker enemy, that procedure is likely compromised for the future. A new tactic must then be developed, or at least the previously employed procedure must be nuanced enough to fool the same attacker at the next engagement.

There is also the matter of limited jurisdiction as mentioned above.

The command can't dictate to its public partners under current law—and besides, it has very limited control over others with whom it must work: the interagency, interorganizational groups and our multi-national partners (allies like the British). Each of these "partners" has its own protocols, policies, legal authorities, decision-making processes, and technologies, and in terms of international partners, there are treaties, different languages, statutes, and government-approved memoranda of agreement and understanding.

Is our Cyber Command ineffective? Yes, but it is getting better and fast.

Cyber Command is acquiring more capability as it overcomes these challenges. Today it has a full range of weapons, but they are a closely guarded secret, and likely this new challenging domain keeps DARPA and some high-tech contract firms very busy. We are aware that DARPA has a cyber program codenamed "Plan X." Reportedly, "Plan X" aims to "map out a digital battlefield that would keep track of potential threats in cyberspace and also help commanders carry out attacks." Then acting DARPA director Kaigham Gabriel made it clear to *The Washington Post* that "Plan X" was intended to help Cyber Command mount offensive attacks.[298]

Is Cyberwarfare a Prophetic Sign?

When you hear of wars and rumors of wars, do not be alarmed. Such things must happen, but the end is still to come. (Mark 13:7, NIV)

Cyberwarfare is one of the newest forms of war that fulfills Jesus' prediction. Although cyberwarfare is not like conventional operations, it can be at least as destructive by stealing sensitive information, shutting down utilities, and influencing the outcome of major military operations.

Conclusion

The cyber battlefield is perhaps the most serious threat to America today, according to DNI Clapper. It has the potential to shut down our infrastructure, steel our money and sensitive personal information, damage our economic competitiveness, and rob our government's secrets. We aren't totally defenseless, but fighting this emerging war will be an uphill battle.

Chapter 9

Weapons of Mass Destruction Threats and Risks

THE MODERN WORLD has so many dangerous ways to kill, and one of the worst seems to be the pervasive availability of nasty technologies, vis-à-vis weapons of mass destruction (WMD): radiological, nuclear, chemical, and biological.

This chapter examines the WMD technologies available to state and non-state actors and how these terrible means of mass killing are rapidly proliferating across the globe. Then we will consider whether the use of WMD is in fact another biblical sign of the coming end times.

WMD Technological Threats and Advances

The use of WMD is growing exponentially with the spread of technology and the evident willingness by states and non-state actors to employ these sinister killers. In early 2016, North Korea tested an alleged "hydrogen bomb"; in August 2015, ISIS used a mustard agent against Iraqi Kurds; and in late summer 2015, the Syrian regime used chlorine-filled bombs against some of its citizens.

WMD includes chemical, biological, radiological, and nuclear weapons. Until recently, the potential use of WMD was limited to a few nation-states and allegedly to be used only as a last resort. Now, WMD threats include a wide variety of users, targets, and motivations. Some powers no longer even refuse to promise not to use them preemptively.

Nuclear Threat: Nuclear technologies are perhaps the best understood WMD because they have been used in the past and can be incredibly destructive. The U.S. used two small atomic weapons against Japanese cities (Hiroshima and Nagasaki) that quickly persuaded that "empire" to end the war in the Pacific. Then during the Cold War, nuclear weapons were central to America's strategy to defend Europe from the numerically far superior Soviet military.

Since the end of the Cold War, nuclear weapons have taken on two new dimensions. First, there is a widespread concern that these weapons might fall into the hands of "rogue" regimes like Iran and terrorist groups like al Qaeda. Second, there is a push within the arms-control community that includes President Obama to abolish nuclear weapons rather than reduce their danger.

President Obama spoke in Prague, the Czech Republic, in 2009, stating that "the risk of a nuclear attack has gone up." He went on:

> More nations have acquired these weapons. Testing has continued. Black market trade in nuclear secrets and nuclear materials abound. The technology to build a bomb has spread. Terrorists are determined to buy, build or steal one. Our efforts to contain these dangers center on a global non-proliferation regime, but as more people and nations break the rules we could reach the point where the center cannot hold.[299]

President Obama concluded, "I state clearly and with conviction America's commitment to seek the peace and security of a world without nuclear weapons." He admitted, "This goal will not be reached quickly—perhaps not in my lifetime."[300] Frankly, Obama's ambition is an admirable fantasy because the risk that some country or group will use a nuclear device is in fact growing in part due to the link between the proliferation of nuclear energy and nuclear weapons.

The process used to manufacture low-enriched uranium for nuclear fuel meant for electrical generation purposes is also used to produce highly enriched uranium for nuclear weapons. Therefore, the logic follows that as we see a proliferation in nuclear energy reactors across the world, we are likely to experience an unknown number of additional albeit secret nuclear weapons programs. Further, the risk that commercial nuclear technology will be used to construct weapons facilities, as was done by Pakistan in secret, is a widespread and legitimate fear—especially among Iran-watchers like Israel and the United States.

There is also the risk that plutonium can be used to make a nuclear bomb. Plutonium is produced in nuclear reactors and found in the rector's spent fuel rods, which if diverted or stolen could be used as a radiological "dirty" bomb (a conventional explosive used to disperse radiological material). We know that North Korea's first nuclear device used plutonium extracted from the spent fuel rods taken from its Russian-built Yongbyon reactor, located about fifty-six miles from the capitol Pyongyang. The Yongbyon Nuclear Scientific Research Center, which is North Korea's major nuclear facility, also produces enriched uranium material used for North Korea's other nuclear weapons tests.

Nuclear reactors are also themselves potential targets that, if attacked, could have the same effect as a WMD. The risk of such an event is growing as the world's nuclear generating capacity is projected to increase as much as 94 percent by 2030.[301] Remember the nuclear accident at Chernobyl, Ukraine, on April 26, 1986? An explosion inside that plant released large quantities of radioactive particles across Russia and Europe. The contamination and efforts to avert a greater catastrophe involved over five hundred thousand workers, many of whom suffered radiation deaths, and cost many billions of dollars.

Nuclear reactors are not designed to withstand attacks by aircraft like those that struck the World Trade Center on 9/11, which means a similar air or ground attack on an active reactor could have severe consequences like we saw at Chernobyl.

A study by the Union of Concerned Scientists found that an attack on the Indian Point Reactor in New York could result in forty-four thousand near-term deaths from acute radiation sickness and more than five hundred thousand long-term deaths from cancer among people within fifty miles of the reactor.[302]

Chemical threat: Another type of WMD weapon is made from toxic chemicals. Those are defined as "weapons using the toxic properties of chemical substances rather than their explosive properties to produce physical or physiological effects on an enemy." Although such weapons can be traced to antiquity, their use is most often associated with World War I, during which toxic aerosol weapons were frequently used by both sides resulting in an estimated 1.3 million casualties, including ninety thousand deaths.[303] Subsequently, chemical agents were used by Italy in Ethiopia and Japan in China. During World War II, both sides, Allies and Axis, possessed but never deliberately employed chemical weapons. The most notable use of chemical agents in modern times was the Iran-Iraq conflict of 1982–87, during which Iran suffered more than fifty thousand casualties from Iraq's repeated use of nerve agents and other toxic agents.[304]

The Geneva Protocol, which prohibits use of chemical weapons in warfare, was signed in 1925 and included the U.S. but wasn't ratified by America until 1975. Although at the time the U.S. foreswore first use, it reserved the right to retaliate in kind if chemical weapons were ever used against America.

Chemical munition technology started to proliferate in the 1950s. Soon there were sub-munitions for better agent dispersal and spray tanks for uniform dissemination. By the mid-1960s, there was a thermal dissemination and aerodynamic break-up process available for chemical weapon employment.

Many of the technologies required to produce and employ chemical weapons are available today on the Internet, but effective employment is not necessarily a simple process. Clearly, nations with a sophisticated chemical industry have the potential to weaponize chemicals, but some substances like nerve agents require considerable expertise. Non-state actors can certainly produce toxic chemicals if they have access to chemical industrial equipment. With plenty of money, they can purchase already assembled chemical warfare agents as well.

Biological threat: The threat of biological weapons is perhaps the least understood of the traditional three types of WMD. Although all three are likely to find their way into the hands of state and non-state actors in the future, biological weapons could realize the most proliferation.

Forty years ago, the world community welcomed the emergence of the Biological and Toxin Weapons Convention (BWC). However, the BWC is no guarantee these weapons won't be developed and/or used especially by non-state actors that are outside the convention.

Biotechnology is an important modern field of research and development that "harnesses cellular and biomolecular processes to develop technologies and products." It employs these processes to develop technologies and products that help improve our lives. It is responsible for the creation of bread and is used for genetic engineering, such as selectively breeding plants and animals to render them with more desirable qualities. It also helped develop genetically modified organisms, such as pesticide-resistant vegetables, biofuels, and plastics, isoprene used in car tires, and more than 250 health care products and vaccines.

There is military interest in biotechnology, especially using deoxyribonucleic acid (DNA), which contributes to the concerns of proliferation given the absence of global and transparent compliance verification systems. Specifically, the global community has seen military research into biologically synthesized explosives, so-called "cyborg insects" and neuroscience. DNA is the molecule that "carries most of the genetic instructions used in the development, functioning and reproduction of all known living organisms and many viruses."[305]

The U.S. government is among the few nations conducting such military-related biotechnology research. The U.S. and the former Soviet Union performed extensive research on biological weapons programs during the Cold War. However, to the best of my knowledge, those weapons were never deployed and were allegedly destroyed to comply with the BWC in 1972. However, as we saw in an earlier chapter, DARPA continues to use biotechnology to deliver new outcomes in fields like health and agriculture, but those efforts are vulnerable to abuse or the secret production of deliberately harmful applications.

DARPA, according to one report, is investigating the possibilities of synthetic biology for warfare at its Biological Technologies Office. Two alleged projects at that office include immortal synthetic beings with "off" switches and on-demand pharmaceuticals synthesized in the field.[306]

Likely DARPA is also using synthetic biology to reprogram genes that could perform specific war-related functions such as WMD deployment. Michael Crow, president of Arizona State University, said, "The biologically-based conflicts of the future would be wild by comparison [with nuclear warfare]: I'll wipe out your food supply, I'll wipe out your water, I'll wipe out your ability to reproduce, [sic] I'll wipe out your ability for your gene line to advance."[307]

Apparently, in October 2014, the U.S. government stopped research on vials of smallpox and the deadly avian flu. The halt came in time before research to repurpose deadly biological agents went forward. Meanwhile, China continues synthetic biology research, which creates a significant challenge for the Pentagon.[308]

"In the military context, I think it's a huge dilemma for, say, our military, because we know there's [sic] people out there who won't follow the same self-imposed restrictions," said Gary Marchant with Arizona State. "To understand what those threats are and to be able to counter them, do we need to create the monster ourselves?"[309]

There is a blurry line between legitimate research and bioweapons research. The same processes and material used for medical research also applies to bioweapons development, however. Further, monitoring dual use research is a serious problem with no means of true verification.

Compounding our concern about biotechnology research is the fact that America is "dangerously vulnerable" to biological attack, according to former U.S. Senator Joe Lieberman, a co-chair of the bipartisan Blue Ribbon Study Panel on Biodefense, which was set up in 2014 and reported its findings in October 2015. The panel's task was to assess the government's actions on epidemics and bioterrorism since the attacks on September 11, 2001.

"[As] I look back, it surprised me that we haven't, thank God, experienced a bioterrorist attack in this country of any significance since the outbreak of the war against Islamic extremism," said Lieberman.[310]

A weaponized version of a killer virus is "relatively easier to put together" than other weapons of mass destruction, Lieberman said. "This is not a threat that we're creating. This is real," he warned. "We better get ahead of it before it gets ahead of us and we're running to catch up."[311]

Nations with WMD Weapons

The rise in nation-states with considerable WMD arsenals has been made easier because the end of the bipolar Cold War facilitated the proliferation of WMD technology across the Internet and mostly former Soviet Union scientists who sell their expertise to the highest bidder.

WMD have a strong allure for many states, and especially when they are paired with long-range ballistic missiles. Such combinations—WMD and missiles—act as strong deterrents to other states.

In 2015, James Clapper, the director of National Intelligence, testified before the Senate Armed Services Committee that WMD are "a major threat to the security of the United States, its deployed troops and allies."[312] He explained the threat is very different today than the past because globalized economies and the free flow of people make WMD more available to states like Iran, North Korea, Pakistan, and Syria.

Why do states like Iran and North Korea insist on pursuing WMD? The answer is simple: security. Iran maintains and improves its ballistic missile programs and likely a secret nuclear weapons program as the country's core means of deterrence. That government perceives the need to hedge against rival states including Saudi Arabia, Israel, and Turkey. Tehran's conventional forces can't match its collective adversaries, so it relies on asymmetric capabilities—ballistic missiles with WMD warheads—to deter Riyadh and others.

North Korea pursues nuclear weapons as a security blanket as well. Its leaders believe a U.S.-led alliance is determined to remove the communist regime from power. That may appear to the West as irrational, but it isn't to those in Pyongyang. They watched as the U.S. brought down governments in Afghanistan, Iraq, and Libya in the past decade, and now they ask themselves whether North Korea is next. So, logically, North Korea equates the possession of a viable WMD arsenal to the survival of the regime.

The analytical problem with rogue states like Iran and North Korea is determining their true intentions with their WMD arsenals. Iran's aggressive rhetoric, especially against Israel as discussed elsewhere, is especially troubling, as there are many indicators that the ayatollahs fully intend to attack Israel and wipe it off the map. Prophetically, we know that Iran (Persia) will in fact be part of a coalition that will attack Israel in the end times.

North Korea's aggressive rhetoric has been around since the nation was formed and has become the norm and largely downplayed with some exceptions. However, given the unbalanced state of their leadership, their new WMD technological advances and the increasing rhetoric as of this writing, North Korea bears increased attention.

The State-Based Nuclear Threat Is Very Serious

The world community long ago put in place a nuclear proliferation treaty with provisions to identify such programs and to sanction violators. Yet the nuclear club whose membership includes several rogue regimes continues to grow.

At the height of the Cold War, there were an estimated sixty-three thousand nuclear warheads, and today those numbers, after decades of demilitarization efforts, are near ten thousand worldwide. Eight nations have successfully detonated nuclear weapons: U.S., Russia (the former Soviet Union), the United Kingdom, France, China, India, Pakistan, and North Korea. The Stockholm International Peace Research Institute's *SIPRI Yearbook of 2014* also states Israel is estimated to have eighty nuclear warheads.[313]

Concerns about the proliferation of nuclear weapons dates back decades, and many world leaders including U.S. presidents have done their best to curb the growth of these systems. Most recently, in 2009, President Obama announced his vision of a world free of nuclear weapons (as pointed out at the beginning of this chapter), but, in spite of that admirable, nuke-free vision, he pragmatically endorsed a $350 billion nuclear modernization program for the American arsenal. Meanwhile, Russia, China, Pakistan, and India each continues to spend heavily on its nuclear arsenals and others, especially some emerging nuclear states, have done likewise.

In the wake of Obama's 2009 anti-nuke vision, the president failed to follow-through to combat worrisome nuclear proliferation, according to Senate Foreign Relations Committee Chairman Bob Corker (R-TN), who addressed the issue at a hearing on March 17, 2016. "I don't think there's any doubt that today there's more potential for nuclear conflict than there was in 2009," Corker said. "The potential for a military miscalculation with regard to nuclear proliferation is higher by far—by far—by orders of magnitude than it was in 2009."[314]

Corker said the U.S.' efforts to combat nuclear proliferation is poor because its partners no long respect treaties, and the senator admits he's worried about the state of American national security. Specifically, the senator and his colleagues are concerned about the nuclear behaviors of Russia, which has violated the Intermediate-range Nuclear Forces Treaty, and Iran and North Korea, which have tested ballistic missiles.[315]

Consider the nuclear arsenals now controlled by sovereign nations.

Russia: Russia controls the world's largest nuclear arsenal. That arsenal has been reduced thanks to a series of treaties with the U.S. The latest reduction treaty, New Strategic Arms Reduction Treaty (START), came into force in 2011 and limits both Russia and the U.S. to 1,550 strategic warheads by 2018. At this writing Russia has 1,643 warheads on 528 deployed vehicles—ICBMs, submarine-launched ballistic missiles, and warheads for heavy bombers. Additionally, Moscow has an estimated eight thousand warheads allegedly awaiting dismantlement.[316]

Russia also possesses a significant stockpile of weapons-grade fissile material. One estimate indicates it has 695 +/- 120 metric tons of highly enriched uranium (HEU) and 128 +/- tons of military-use plutonium (PU).[317]

It is noteworthy that the U.S. and Russia partnered via the Cooperative Threat Reduction (Nunn-Lugar) Program to secure Russian stockpiles of HEU and plutonium and to dismantle Soviet-era delivery systems and chemical weapons. However, that agreement expired in June 2013, and following Russia's incursion into Crimea, Washington suspended those projects related to nuclear energy. Shortly thereafter, Moscow announced its intention not to accept further U.S. assistance to secure stockpiles of nuclear material on Russian territory.[318]

China: China has a survivable ballistic missile fleet topped with modern nuclear weapons. Beijing launched its nuclear weapons program in 1955 with a successful test in 1964. To date, China conducted 45 nuclear tests to include both thermonuclear and neutron bombs. It has an estimated 260 nuclear warheads, approximately 60 ICBMs, and will soon have five nuclear-powered ballistic missile submarines.[319]

China also has a growing inventory of fissile material that includes 20 +/- metric tons of HEU and 2 +/- metric tons of plutonium. It is noteworthy and widely reported China supplied both warhead design information and fissile material (HEU and/or PU) to both Pakistan and Libya. Libya was eventually coerced by the international community to abandon its nuclear ambitions.

China's proliferation activities reportedly ceased once it joined the Treaty on the Non-Proliferation of Nuclear Weapons (NPT) in 1992.

However, it is troubling that China's 2013 Defense White Paper did not use the phrase "no first-use" of atomic weapons. However, Beijing did reaffirm its alleged "no first-use" commitment in a more recent publication.[320]

Undeniably, two rogue regimes (Iran and North Korea) and one so-called partner country (Pakistan) pose nuclear proliferation challenges as well.

Iran: Iran has a long history of violating its obligations under the NPT and managed to amass materials for the likely conversion into nuclear weapons. That regime will likely continue its nuclear weapons program albeit in secret in the wake of the 2015 Joint Comprehensive Plan of Action agreement, which leaves the regime with much of its non-weapon nuclear program in-tact, sanctions relief of up to $150 billion in cash, and the assurance that only the United Nations will validate the regime's compliance with the arrangement.

There is no reason to anticipate the regime will abandon its nuclear ambitions and if it sees fit it could proliferate that technology with its allies and non-state actor proxies. Meanwhile, Iran's nuclear victory over the West guarantees other nations in the region will take steps to acquire nuclear weapons, especially Saudi Arabia, which already has a robust ballistic missile capability (thanks to China) and a rumored secret arrangement with Pakistan to provide nuclear warheads on demand.[321]

North Korea: North Korea dropped out of the NPT in 2003 and continued to expand its nuclear and ballistic missile programs.[322] The most recent developments are especially disconcerting. In early 2016, the Hermit State claims it tested a "hydrogen" bomb (its fourth atomic test since 2006), although that announcement met with some skepticism in the West. A month earlier, the regime successfully tested a ballistic missile launch from a submerged submarine. That's a very significant development, and once Pyongyang mounts that missile with a nuclear warhead, the rogue regime could then credibly threaten virtually any American city or interest.

Pakistan: Pakistan is rapidly expanding its nuclear arsenal as well. Pakistani Prime Minister Nawaz Sharif visited Washington in late 2015, at which time he refused to distance himself from talk that his country is developing tactical atomic weapons for a possible war with neighbor India.[323]

Does anyone believe small Pakistani nuclear weapons won't find their way into terrorist hands? After all, Dr. A. Q. Khan, the father of the Pakistani nuclear weapon, sold nuclear weapons plans and other nuclear technologies to Iran and likely to North Korea as well.[324]

The Nation-State Chemical Threat Continues

The Arms Control Association reports that eight nations may still have chemical weapons. Both the U.S. and Russia insist they will in time destroy their remaining legacy arsenals. But there is some question about whether Iran, Israel, and Taiwan possess chemical weapons and the association has no question that Albania, North Korea, Sudan, and Syria have chemical weapons.

Chemical weapons come in various forms and delivery systems. Those used in World War I were chemically filled standard munitions. Basically, opened canisters were filled with the active agents like chlorine, phosgene, or mustard ("classic" chemical weapons), and then fired at the enemy and the prevailing winds disseminated the agent.

Between the world wars, nation-state chemical weapons industries expanded both the available chemical agents and their delivery systems. Aircraft bombs were developed to deliver lewisite, a more potent mustard agent than the one used in the First World War. Probably the most significant development in the between-war era was the development of the German nerve gas or anticholinesterase agents. Specifically, the German chemist Gerhard Schrader discovered what he called "tabun" or GA, and later the more toxic "sarin" or GB.[325] These agents are significantly more toxic than the compounds used in World War I, but were never used even though the Nazis did produce a large number of GA-filled munitions.

Nerve agents like GA and GB are liquids, not gases like those used in

World War I, which block an enzyme necessary for our nervous system and act on humans like pesticides.

After World War II, the Allies pursued chemical-weapons programs. The Russians took an entire German GB plant back to the Soviet Union, and the U.S. designed a cluster bomb for the agent GB. The French, British, and even the Canadians built facilities to produce GB. Russia also produced soman (GD), and the U.S. worked with the British to produce a form of V-agent called VX. Further, in the 1960s, non-lethal agents were developed such as the riot control agent CS, code for a solid powder known as a riot-control agent (o-chlorobenzylmalonitrile), which temporarily incapacitates the opponent.

Soon, nations acquired binary chemical weapons that use toxic chemicals produced by mixing two innocuous compounds immediately before or during use. For example, the U.S. produced a GB (sarin) binary nerve agent weapon, the M687 projectile (a 155-mm artillery shell). The Russians allegedly developed a similar binary weapon, as did the Iraqis for both bombs and missile warheads.

This history is necessary to understand, because it is evidence that the world has considerable experience developing deadly chemical weapons. That means the technology is readily available as are the precursor materials.

Future state-on-state conflicts could involve the use of chemical munitions, as we have seen in Syria. However, the more likely risk is that one of the states above, especially rogue regimes like North Korea and Iran, might provide the technology and/or actual weapons to non-state actors.

The Nation-State Biological Threat Is Rising

Nation-states have the means to employ biotechnologies for other than the uses outlined in this chapter's introduction. Specifically, the Pentagons' DARPA reportedly is looking to rewrite the laws of evolution to the military's advantage by creating "synthetic organisms" that can live forever. But remember, DARPA's mission is to improve America's national security, not prolong our lives. Inevitably, DARPA will use any biotechnology breakthroughs to improve America's killing instruments.

Possibly one such DARPA project that could improve America's killing capability is called Biodesign, which seeks to eliminate "the randomness of natural evolutionary advancement." It intends to use biotechnology to come up with living, breathing creatures that are genetically engineered to "produce the intended biological effect." Does that mean DARPA is on a path to produce transhumanist cyborgs or animal killing machines?[326]

Of course, the biotechnology threat of super killers also includes bioweapons such as the simple anthrax incidents the U.S. suffered in the wake of the 9/11 attacks. Deadly anthrax was mailed to various Washington offices as a terrorist attack.

Our rivals are working on bioweapons programs as well. Former Department of Homeland Security Secretary Tom Ridge, the co-chair of a Blue Ribbon Study Panel on Biodefense, said, "We...know that states like Russia and North Korea either have or have had bioweapons programs. So I'd say that terrorism is a serious possibility. Emerging and re-emerging diseases are actually a certainty."[327]

Ridge believes the biological threat is "right up there with the cyber threat. Both of these threats carry with them the potential for enormous damage to the global economy and commerce."[328]

The Arms Control Association identifies eight countries with possible biological weapons programs, which includes the United States. Reportedly, the U.S. gave up its biological weapons program in 1969 and then destroyed all offensive agents by 1973. However, according to a 2010 Russian government report, the U.S. is undertaking research on smallpox and is accused of undertaking biological warfare research to improve its defenses, which is "especially questionable from the standpoint of Article 1 of the BTWC [Biological and Toxin Weapons Convention]."[329]

The other nations on the association's list include Cuba, Egypt, Iran, Israel, North Korea, Syria, and Russia.

The former Soviet Union ratified the BTWC in 1975, but it evidently violated the terms of that convention by secretly operating a large offensive BW program until the USSR dissolved in 1991. That program weaponized agents like anthrax, glanders, Marburg fever, plague, Q-fever, smallpox, tularemia, and Venezuelan equine encephalitis. These agents could be reconstituted and loaded into delivery systems in wartime.[330]

Russia may still secretly operate those exotic pathogen programs. We know former Soviet biological weapons facilities remain and Moscow continues to engage in dual-use biological research activities. Further, whether the Russian government continues to conduct secret bio-warfare programs, what's not in dispute is that scientists formerly associated with the Soviet biological warfare programs are likely helping other countries or non-state actors with clandestine efforts.

Other active bio-warfare programs may include North Korea and Iran. A 2009 Defense Intelligence Agency report indicates that Iran's bio-warfare efforts "may have evolved beyond agent R&D, and we believe Iran likely has the capability to produce small quantities of BW [bio-warfare] agents but may only have a limited ability to weaponize them." Further, the 2010 report indicates that Iran continues dual-use activities and the regime has a host of potential BW delivery systems to include short-range cruise missiles, tactical missiles, fighter aircraft, artillery shells, and rockets.[331]

The association indicates that a 2010 U.S. government report states North Korea may "still consider the use of biological weapons as a military option." What is clear is North Korea has yet to declare any of its biological research and development activities, a situation given its nuclear history should surprise no one.[332]

WMD Threats Posed by Non-state Actors

Security professionals are hard-pressed to anticipate unthinkable threats like the Russian charter airplane felled by a terrorist bomb just moments after take-off from Sharm el-Sheikh, Egypt. However, given the nonpolar new world instability, we must think about the unthinkable—black swans—the use of WMD by non-state actors, terrorist groups.

The WMD terrorist threat is growing due to an expanding network of actors with mostly Islamic motivations and the availability of technology necessary to develop and then deploy WMD, ranging from radiological dispersal devices and toxic chemicals to bio-toxins.

Until the last couple of decades, the production and employment of WMD was mostly the exclusive domain of nation-states, but that has changed. Today, terror groups have unprecedented access to WMD technologies, the material resources, and the ability to operate clandestinely thanks to a growing list of failed states and ungoverned regions.

The realist will understand the risks are high because scientific expertise and dual-use technology are increasingly available for WMD fabrication. Consider the threat posed by ISIS.

ISIS controls territory that includes access to radiological material, toxic industrial chemicals and laboratories capable of significant bio-tech work. Further, ISIS recruits across the globe those who share its radical views and are willing to use their expertise to support the group's causes.

It is highly possible some of those recruits are in the West and possess WMD-related expertise. ISIS could easily operationalize them with targeting information and the means to attack industrial plants that would disperse toxic chemicals into surrounding communities. Further, radiological facilities such as those at hospitals could be targeted with conventional explosives to create a radiological dispersal device, or dirty bomb. Either type of attack would create significant panic and environmental damage—a victory for the headline-seeking ISIS.

It's also quite possible that the world's richest terrorist group, ISIS, might succeed in buying a nuclear weapon from a country like Pakistan, something ISIS boasted about in 2015. The plausibility of such an acquisition isn't that surprising, given our long-held view that Pakistan seems to take insufficient precautions with its nukes. Specifically, it is reported that Pakistan moves its nuclear weapons in panel vans on surface roads that are crawling with resource-seeking jihadists.

Possibly more troubling is the May 2015 edition of ISIS' glossy magazine, *Dabiq*. That periodical includes an article that speaks of a "hypothetical operation" regarding ISIS' proposed acquisition of a nuclear device that is smuggled through Africa, Central America, and Mexico into the United States. Is that possible? Unfortunately, yes, and that article is a warning as to the group's ultimate intentions.[333]

The nuclear device article isn't at all far-fetched. Remember, radiological material smuggling is an ongoing proposition in Eastern Europe and especially in Moldova, a small country tucked between Ukraine and Romania with an outlet to the Black Sea.

House Homeland Security Committee Chairman Michael McCaul (R-TX) validated that possibility with an account of an actual sting operation. He said the FBI interrupted arms traffickers looking to sell WMD to Islamic extremists in the Mideast. He characterized the operation—revealed in an Associated Press report in late 2015—as proof that groups such as ISIS are eyeing weapons that could be used against Americans.[334]

It is a sobering reality that groups like ISIS will eventually "get lucky" and trigger a WMD on Tel Aviv, Ankara, London, or New York City. That view is shared by officials with deep insights into such threats.

John Cohen, former acting undersecretary for intelligence and the counterterrorism coordinator at the U.S. Department of Homeland Security, said "Today, the United States faces a terrorist threat that differs greatly from the one we faced immediately after the 9/11 attacks."[335]

That threat, according to Cohen, comes from those here at home "inspired by the ideology of terrorist groups such as al Qaeda and ISIS; willing to conduct acts of violence in furtherance of this ideology; but never directly communicate with members of the terrorist group and operate independent of its command and control infrastructure." Fortunately, to date, the homegrown attacks have been against "soft targets" using simple weapons—knives, firearms, and rudimentary explosive devices.

Cohen warns:

If we have learned anything since the attacks of 9/11, it is that the threat facing the nation will continue to evolve. ISIS has actively sought to lure individuals with technical knowledge to its cause. There are concerns that as ISIS acquires more technical knowledge regarding the construct of WMD, they could seek to share information more broadly with those inspired by its ideology.

The expanded use of the Internet, encrypted communications capabilities, and social media by groups like al Qaeda in the Arabian Peninsula (AQAP) and the Islamic State has been a game changer as it relates to their communications and recruitment efforts. Once committed to a terrorist ideology, self-radicalized individuals can acquire pre-operational intelligence in advance of an attack. Also, information that can inform the design and construction of explosive or other sophisticated weapons can be accessed via the Internet. Dangerous chemicals and other materials that potentially could be used to construct a device are obtainable via the Internet as well.

Cohen acknowledges young Westerners are increasingly becoming drawn to extremist causes. He attributes this trend to a number of factors. But first, Cohen said, homegrown terrorists are not life-long Islamists, nor do they have a common economic, ethnic, or cultural background. What they do share is evidence of the dysfunctions outlined in chapter 2, trends about family backgrounds. Specifically:

They are overwhelmingly from dysfunctional families and disconnected from others within their communities. They have underlying mental health issues and may have a history of criminal behavior. They have experienced a pattern of life failures and exhibit excessive Internet and social media usage. In short, these are disaffected young people searching for something that provides meaning to their lives—and they are finding deep and meaningful connection with the cause of groups like ISIS.

Cohen's warning is echoed by other well-informed people like former Senator Joe Lieberman, who was the co-chair with Secretary Ridge on the panel that considered the biological threat to America. The former senator said he is shocked that a jihadist group hasn't pulled off a biological attack. ISIS is a good candidate for launching such an attack given its anti-West drive to demonstrate viciousness, such as its live burials, burning captives alive, and beheadings. It seems that using toxins and deadly germs or a radiological dispersal device against a soft target would resonate with the malevolent group—and imagine the shocking pictures the terror group would use for future recruiting.

Al Qaeda would like to do the same. Keep in mind AQAP's chief bomb-maker, Ibrahim Hassan al-Asiri, is very busy and innovative. In 2009, Asiri put an explosive device inside his brother, who then blew himself up in front of Saudi Arabia's Prince Mohammad bin Nayef. Asiri is also the designer of the famed underwear bomber device and the 2010 plastic explosives stashed in printer cartridges on Western-bound cargo planes. Next time, the tenacious and innovative Asiri may become more successful.[336]

Further, AQAP and Asiri enjoy safe haven in the terrorist hosting country of Yemen. It isn't hard to imagine in that war-torn country WMD materials are easily acquired, stored, and then converted and staged for use by the likes of Asiri.

On January 4, 2016, from his hideout in Yemen, Asiri, released a chilling audio threat in response to Saudi Arabia's decision to execute a Sunni cleric and forty-six others. The terrorist bomb-maker said the forty-seven had fought the "Crusaders" (read "Western Christians") occupying the Arabian Peninsula, and he promised to deal with the Saudis in a "different way." New York Congressman Peter King labeled Asiri "extremely lethal," and noted the terrorist has always intended to launch "an attack on the United States."[337]

Western officials must recognize that we have a history of underestimating the capabilities of violent extremists like Asiri, therefore threats like the use of WMD by West-hating groups must be viewed as legitimate concern.

Further, what's rather disconcerting is the fact that our WMD detection and defense capabilities aren't up to the threat. Even though we have great counter-terror capabilities, our WMD detection community has slipped in their skills. The fact is those skills have been outmaneuvered by the pace of WMD technological advancements, which leaves the U.S. trailing emerging threats.

One approach to reverse the threat is to reduce the ungoverned spaces abroad that provide non-state actors like ISIS the freedom to develop their WMD capabilities. The other piece is for Americans to report suspicious actions such as those of the San Bernardino couple who murdered fourteen innocent people in early December 2015. Their former neighbors saw suspicious activities but failed to report their observations.

One of those neighbors, Enrique Marquez, was arrested on charges that he plotted with San Bernardino shooter Syed Rizwan Farook to conduct terror attacks in late 2011 or 2012, purchasing two assault rifles for Farook, and entering a "sham" marriage with a member of Farook's family.[338]

The nexus of ideology and increasing WMD technology represents a serious threat. This demands an aggressive effort to develop and sustain WMD defense capabilities as well as response units.

Is the Use of WMD a Prophetic Sign?

Matthew 24:7 KJV states that in the last days "there shall be famines, and pestilences." Perhaps it means infectious diseases or the plague. After all Amos 4:10 uses the term to compare with the plagues of Egypt.[339]

Another interpretation comes from the name of the first horseman, Pestilence, in Revelation 6.

When the Lamb opened the fourth seal, I heard the voice of the fourth living creature say, "Come!" I looked, and there before me was a pale horse! Its rider was named Death, and Hades was following close behind him. They were given power over a fourth of the earth to kill by sword, famine and plague, and by the wild beasts of the earth. (Revelation 6:7–8, NIV)

Perhaps we are being too literal by looking for a Scripture indicating a term revealing that WMD use is in fact an end-times sign. Rather, it could be more general. For example, it's possible the sign is the growing likelihood that Israel is targeted by WMD that is related to the coming of the Lord. After all, Iran has made no secret of its desire to "wipe Israel off the face of the map," and that regime already has all the means to make a nuclear weapon and to deploy it against Israel.

There are two other Scriptures that could be interpreted to mean the use of WMD weapons. The first is in Isaiah 17, which describes the future destruction of Damascus, the capital of Syria, and a city consisting of millions of residents today. Isaiah states Damascus experiences some sort of trouble in the evening (v. 14), and before the morning comes the entire city is completely destroyed. Except for some supernatural event, only nuclear weapons could have that effect.

The second possibility in Scripture speaks of a WMD-like attack in Ezekiel. There will be a future war involving Gog and Magog in terms that are easily interpreted as descriptions of WMD employment (Ezekiel 38–39). Israel is spared, but the invading forces attack each other leaving no survivors. The warfare description includes events resembling nuclear detonations. Then it takes months to locate and bury the dead by professionals, a process resembling the care taken for biologically contaminated bodies.

Conclusion

WMD are a growing and serious threat to human kind. Nation states and non-state actors are acquiring the trilogy—nuclear/radiological, chemical and biological—of deadly mass killing weapons and the likelihood is growing they will use them in our increasingly unstable nonpolar world.

All-America's Twenty-first Century Strategy

America's future is uncertain given the trends outlined in section I and the emerging threats and risks examined in section II. Given these sobering facts, it is likely the world faces the imminent biblical end times, especially in light of the flood of prophetic signs evident today. However, should the Lord tarry, Christians must continue to witness for the Lord, and that includes participating in their government (Romans 13:1–7).

Christian participation in government only has a meaningful effect if the biblical worldview is fully engaged. We demonstrate a biblical worldview when we believe the Bible is entirely true and allow it to be the foundation of everything we say and do. Behaviorally, that means we demonstrate biblical values and principles.

Specifically, our biblical worldview should influence our vision for the future American government, which, given the opportunity through like-minded elected officials, is then translated into government policies. The policies guide the establishment of effective and accountable strategies composed of programs that are then responsibly executed. This is the least we should do and the best hope this side of heaven we have to see God's best plans for civil government.

Is this realistic? I don't know whether the Lord will permit a spiritual revival that translates into a future American government framed by Christians living out a biblical worldview. Frankly, I'm skeptical of such an outcome, given the moral condition of America today. But our God is in the business of miracles, and even though the world seems to be racing toward the gates of hell, there still may be hope.

As discouraged as we may be about America's moral condition with the frightening trends and threats outlined in this volume, we should still exercise the salt and light of Christ (Matthew 5:13–16) in the public square by embracing a Christian vision for America and strategies necessary to address the frightening challenges ahead.

This section has six chapters that provide a Christian vision for America, followed by a series of strategies addressing our national security, national intelligence, foreign affairs, economy, and culture. Each of these strategies addresses the relevant trends examined in section I and the threats and risks developed in section II.

Christians need this twenty-first century, all-American strategy as a testimony to their obedience to remain engaged until the Lord's return. It is also a reminder that although we are in the world, as Scripture reminds us, we are not to "be conformed to this world, but be transformed by the renewing of your mind, so that you may prove what the will of God is, that which is good and acceptable and perfect" (Romans 12:2, NASV).

Chapter 10

A Biblical Vision for America's Troubled Future

IN THIS CHAPTER, I will propose a future America that embraces the founders' principles and reflects a biblical worldview for America's internal and external governance and international relationships. I am NOT advocating a theocracy, but a future based on biblical principles that will incur God's blessing. Critics will call such a vision unacceptable, naïve, biased, and even anti-American. My response to such criticism is found in the book of Joshua:

> But if serving the LORD seems undesirable to you, then choose for yourselves this day whom you will serve, whether the gods your ancestors served beyond the Euphrates, or the gods of the Amorites, in whose land you are living. But as for me and my household, we will serve the LORD. (Joshua 24:15, NIV)

The vast majority of Americans self-identify as Christian. If we begin with this as given, then we can equip this majority with a clearly stated Christian vision that offers them hope and direction on how, through their participation in the American governmental process, we can right the ship of state that most Americans admit is spinning out of control.

Others with radically different governing philosophies will espouse their vision, and those alternative views are considered by our republican system of government. Certainly, every past president outlined his vision for America, and in some cases, those past chief executives reflected Christian principles and values.

This chapter begins by defining a biblical, Christian worldview (those terms are used interchangeably here) and then identifies the values and principles that necessarily flow from that view. The best chance America has to survive and flourish is for Christian leaders at all levels to incorporate these values and principles into their own lives, work, and political activity.

First, I will discuss the concept of "vision" by examining how our founders and some of our recent presidents reflected their worldviews in their vision that became the basis for their governing policies and actions. Not surprisingly, some of our past leaders have indeed embraced a Christian worldview at least in part, while others were far more secular.

Second, this chapter will sketch a fresh vision for America based on the Christian worldview. That vision will form the foundation upon which the subsequent five chapters will build strategies that tackle the tough issues America faces amidst the future trends and challenges outlined earlier in this volume.

Finally, even if this vision for America becomes a reality and we miraculously elect Christian leaders who govern using Christian principles and values, this outcome will in no way delay or accelerate the Lord's end-times plan. We don't know that plan other than what's revealed in the Scriptures, nor do we know His timing, even though, as indicated earlier, there are plenty of signs suggesting that time appears to be on our immediate horizon. However, we do understand His mandate for us to be about His work until Christ's return.

Christian, Biblical Worldview Leads to Vision for America

The vast majority (78 percent) of Americans self-identify as "Christian," but that figure is misleading. The Barna Group, a Christian pollster, indicates the faith's influence is slowly waning.[340] In fact, according to Barna, very few American Christians take their faith seriously as evidenced by the decline in those who embrace a biblical worldview.

Barna found that only 4 percent of Americans have a "biblical" worldview, a reflection of their commitment to their faith. Barna defined a biblical worldview as:

> Believing that absolute moral truth exists; the Bible is totally accurate in all of the principles it teaches; Satan is considered to be a real being or force, not merely symbolic; a person cannot earn their way into Heaven by trying to be good or do good works; Jesus Christ lived a sinless life on earth; and God is the all-knowing, all-powerful creator of the world who still rules the universe today. In the research, anyone who held all of those beliefs was said to have a Biblical worldview.[341]

Barna indicates a biblical worldview as just defined guides the believers' moral beliefs and actions. Little wonder that in spite of a heavily American Christian population, we see so little Bible-centric behavior in our society today.

One's worldview explains their reason for being. For the Christian with a biblical worldview, he/she believes the reason for existence is to love and serve God, while the humanist sees nothing beyond the material world. Therefore, one's worldview affects every aspect of his or her life—to include politics.

Del Tackett with the Focus on the Family ministry explained that so few Christians have a biblical worldview as defined by Barna because their worldview is diluted by our non-Christian culture: television, music, Internet, and academia. The culture's pull appeals to our sinful desires, which influence our worldview.

Tackett goes on to explain: "If we don't really believe the truth of God and live it, then our witness will be confusing and misleading." For example, most Christians agree with the Scripture about sexual immorality, but too often, Bible-believing Christians give into sexual sin. Our so-called Christian worldview is therefore weak, and as Colossians 2:8 states, we are taken "captive through hollow and deceptive philosophy, which depends on human tradition and the basic principles of this world rather than on Christ."[342]

A biblical worldview is peculiar because it is based on God's view of life and the appropriate response to the issues in life. Further, possessing a biblical worldview is only helpful if it is lived out in our interactions with the world through Christian values and principles.

Core Biblical Values

Biblical values are evidenced in the work of Jesus Christ. There are many Christian/biblical values, but arguably the core values are putting God first in our lives, hope, righteousness, love, respect all people, generosity, moral life, and forgiveness. Consider these values and how they may be lived out.

God is first: The Christian's first value is putting God first in his life. Matthew 6:33 (NIV) tells us to "seek first his kingdom and his righteousness, and all these things will be given to you as well." As spiritual beings, putting God first in our lives means we seek God's counsel in every decision and trust in his provision.

Hope: A Christian's hope is in Christ's redemption and our promised eternal life (John 3:16). This hope gives us perspective in the face of turmoil, especially when we don't understand why something bad happened to us. God's silence when bad things happen is no excuse to lose hope (Romans 8:28).

Righteousness: The Christian seeks to do God's will by doing what is right in God's eyes. That doesn't mean he is sinless; rather, the Christian is a forgiven sinner. Ephesians 4:24 (NIV) states: "and to put on the new self, created to be like God in true righteousness and holiness." Doing what is right is reflected in the Christian's honesty and integrity—his God-given "righteousness."

Love: Arguably the greatest of the Christian values is love (1 Corinthians 13), or in the Greek, *agape*, God-like love. This is God's love for us that while we were yet sinners Christ died for us (Romans 5:8). The Christian should desire to provide this godly love of unmerited favor to others as evidenced in Jude 21 (NIV): "Keep yourselves in God's love as you wait for the mercy of our Lord Jesus Christ to bring you to eternal life."

Respect all people: The apostle Mark writes that the second most important commandment is to "love your neighbor as yourself (Mark 12:31, NIV)." Who can argue with Jesus' golden rule, "Do to others as you would have them do to you" (Luke 6:31, NIV)? We are to care for our enemies: "If your enemy is hungry, give him food to eat; if he is thirsty, give him water to drink" (Proverbs 25:21, NIV).

Generosity: The world needs people who share generously with others in need. Christian ministries across the world reach out a helping hand to fellow humans no matter their race, ethnicity, or faith. There are numerous Bible references encouraging the Christian to generosity and to giving (Proverbs 14:21, Matthew 5:42, Luke 6:8, Romans 12:6–8). After all, the Christian recognizes that everything we have ultimately belongs to God (Romans 11:36).

Moral life: A biblical worldview demands that the Christian live a moral life. The Bible identifies values that are immoral: evil thoughts, murder, adultery, sexual immorality, theft, false testimony, and more (Exodus 20).

Forgiveness: Matthew 6:14–15 (NLT) states: "If you forgive those who sin against you, your heavenly father will forgive you. But if you refuse to forgive others, your father will not forgive your sins." This doesn't mean the Christian is to be a doormat. Rather, he is to practice forgiveness because God first demonstrated mercy and forgiveness to us.

The above list of values provides a good guideline for living the biblical life. Scripture also provides principles—accepted or professed rules of action or conduct—that help direct our understanding of and the proper role of government. In fact, some of our founders recognized the importance of applying biblical principles to our nation's foundations.

Founding father Noah Webster said:

> The moral principles and precepts contained in the scriptures ought to form the basis of all our civil constitutions and laws. All the miseries and evils which men suffer from vice, crime, ambition, injustice, oppression, slavery, and war, proceed from their despising or neglecting the precepts contained in the Bible.[343]

When biblical principles help form the legal basis for a government, as Webster said, the "miseries and evils" of man are contained. Those principles include liberty and justice for all, the rule of law, protecting the people from attacks, enabling the free flow of exchange, not imposing confiscatory taxes, and not wasting the people's tax money.

Below are biblical principles and how they apply to civic life.

Human life: Psalm 139:13–16 addresses the importance of the sanctity of human life. God cares about life; it begins at conception and arguably applies to areas of bioethics (transhumanism) which some scientists are now violating thanks to government's sponsorship. It obviously applies to the issue of abortion.

Equality: God calls for equality of human beings. Christians should not feel superior to others and are told to eschew class distinctions among people. The apostle Paul teaches spiritual equality of all people (Galatians 3:28), which applies to race, ethnicity, and religious background as well as the sexes.

Marriage: God is the author of marriage (Genesis 2:18), which He provides for the procreation and nurture of children (Ephesians 6:1–2).

It is also a God-approved outlet for sexual desire. This principle applies to distortions such as artificial reproduction, cohabitation, and homosexual relationships.

Good citizenship: God calls for man to obey the civil authorities (Romans 13:1–7). We are to submit (1 Peter 2:13–17) and render service, and this principle applies to issues such as war, law enforcement, paying taxes, and serving on a jury.

Protect citizens: Government's primary role is to protect citizens from the bad behavior of others. Thus government is to help preserve order—the people's ability to live "peaceful and quiet lives." Therefore, the state is God's agent to contain harm that would occur in the absence of public constraints on evil behavior. Romans 13:4 (NIV) states: "For the one in authority is God's servant for your good. But if you do wrong, be afraid, for rulers do not bear the sword for no reason. They are God's servants, agents of wrath to bring punishment on the wrongdoer."

Punish evil: The state is to punish evil men. The apostle Paul wrote in Romans 3:5 that the fear of punishment is one reason men comply with the authorities. By extension, this principle applies like the one above to the security of the nation from foreign threats. America builds a capable armed forces to deter evil men.

Just and righteous: Government is to be just and righteous. In Psalm 11:7 (NIV), King David states: "For the Lord is righteous, he loves justice." God expected the kings of Judah to administer justice to protect the Israelites (Jeremiah 12), but the state must also be a neutral arbiter showing no favor for one party over another.

Restrict government: Scripture puts restrictions on government as well. The most important restriction is the first commandment given to Moses: "You shall have no other gods before me" (Exodus 20:3, NIV). The nation of Israel often turned her back on God by worshipping idols, which led to God's wrath. Even in Jesus' time, the Roman emperors demanded that the people worship them. Things aren't that much different today when the concept is put into context. Many people worship the things/idols of the world rather than the Creator.

Don't replace the church: Government should not replace the church in its obligations to care for people. However, modern American government has become a cradle-to-grave nanny state for a large segment of the population through a vast array of social services such as welfare and Medicaid. That relationship, can be argued, is one of master and slave; government is the master, and the recipient of the government-provided services is the slave.

Modern government has stepped into roles which arguably it doesn't belong, such as God's concern for the poor, the vulnerable, and weak. Taking care to protect these people, especially the widows and orphans, is the believer's duty (James 1:27).

A biblical worldview is worthless unless it compels the believer to act guided by Christian values and principles. Below is evidence of these values and principles applied by our founders and former presidents.

Visions of Our Founders and Recent U.S. Presidents

Before presenting a vision for America based on a biblical worldview, it is helpful to examine our founders' vision for America and that of some of our recent presidents. It's obvious that these men—especially the founders—were influenced by biblical principles and values, but few, if any, evidence a true biblical worldview.

Some of our founders were genuine Christian men, but there is an ongoing dispute about whether America had a Christian founding. That issue is addressed in considerable detail in a later chapter. However, for our purposes here, the evidence indicates that Christian ideas were one of the important intellectual influences on America's founders with respect to religious liberty and church-state relations. The founders did not try to create an American theocracy. Rather, they wrote our framework instruments like the Constitution to include significant Christian moral truths.

Clearly our founders identified themselves as Christians. In fact, virtually all the colonists were Protestants (98 percent), with the balance being Roman Catholics. But like today's Christians, as indicated above in Barna's contemporary surveys among self-identified Christians, there is little historical evidence that our eighteenth-century founders were all "sincere" or biblical-worldview Christians.

Dr. Mark Hall is the Herbert Hoover Distinguished Professor of Politics at George Fox University and the author of an essay regarding the faith of America's founders, "Did America Have a Christian Founding?" Hall states:

> In most cases, the historical record gives us little with which to work. And even if we can determine, say, that a particular Founder was a member, regular attendee, and even officer in a church, it does not necessarily mean he was a *sincere* Christian. Perhaps he did these things simply because society expected it of him.[344]

Dr. Robert Kraynak is a professor of political science and director of the Center for Freedom and Western Civilization at Colgate University. He appears to agree with Dr. Hall. Kraynak wrote an article in *Modern Age*, "The American Founders and Their Relevance Today," which examines the founders' moral vision. He points out that our Declaration of Independence "asserts that our liberty and rights come from 'the laws of nature and nature's God.'" Further, the natural-law doctrine of the Declaration provides the underpinning of the Constitution and the republican form of government. "It states that 'all men are created equal' in the sense of possessing certain inalienable rights that come from the Creator who put them in human nature and created a natural universe with a rational moral order," Kraynak wrote.[345]

What's not in dispute, and germane to this volume, is that an excellent case can be made that the Bible and Christianity had a profound influence on the founders. Case in point, consider what founder and primary author of our Constitution, Thomas Jefferson, said about the founding of America. Jefferson's vision for America was to address fundamental questions of government and society by establishing a nation that "shall have an organized control over the actions of its government, and its citizens a regular protection against its oppressions."[346] That's a key biblical principle.

Jefferson also believed such a government was founded on two other Bible-based principles: inherent and inalienable rights and government by the people. He wrote, "The equal rights of man, and the happiness of every individual, are now acknowledged to be the only legitimate objects of government."

Many articles and books were written about our founders and a few about their faith. It is clear in many of these writings that our founders were influenced by biblical principles. However, the issue of whether America was in fact once a Christian nation is examined in more detail in the culture chapter.

For now, it is sufficient to say that biblical values and principles played a significant role in the founding of America. The same can be said about the biblical influence on many of our presidents as evidenced in their inaugural addresses. Consider some of our most recent presidents.

President Ronald Reagan acknowledged God in his first inaugural address: "We are a nation under God, and I believe God intended for us to be free," he said.[347]

Reagan said he wanted "a healthy, vigorous, growing economy that provides equal opportunities for all Americans, with no barriers born of bigotry or discrimination." The principle of equality is evident here.

Reagan was reluctant to thrust the nation into war, but he warned adversaries that "our reluctance for conflict should not be misjudged as a failure of will. When action is required to preserve our national security, we will act." This is evidence of the principle to protect the people.

President John Kennedy addressed a number of significant biblical principles in his inaugural address. He called for helping the downtrodden, a clear biblical principle. "To those peoples in the huts and villages of half the globe struggling to break the bonds of mass misery, we pledge our best efforts to help them help themselves."[348]

Kennedy addressed the threats the nation faced and promised "for only when our arms are sufficient beyond doubt can we be certain beyond doubt that they will never be employed." The security principle is evident here, and it wasn't long into his administration that he faced the Bay of Pigs invasion—and later, the Cuban missile crisis with the Russians.

President Kennedy quoted from Isaiah 58:6 (KJV) to rally support for the oppressed—to "undo the heavy burdens...[and] let the oppressed go free." Then he called for justice and caring for the weak.

The most famous line from Kennedy's inaugural address is: "And so, my fellow Americans: ask not what your country can do for you— ask what you can do for your country." This is the application of the principle of limited government—he advocated people doing more for themselves and their country.

President Barack Obama made references to the Scripture and invoked biblical precepts in his first inaugural address as well. Arguably, Obama was being cynical when he said:

> But in the words of Scripture, the time has come to set aside childish things. The time has come to reaffirm our enduring spirit; to choose our better history; to carry forward that precious gift, that noble idea passed on from generation to generation: the God-given promise that all are equal, all are free, and all deserve a chance to pursue their full measure of happiness.[349]

Mr. Obama used references to "the justness of our cause, the force of our example, the tempering qualities of humility and restraint" to explain how America defends its citizens and will do so in the future. He also promised to help "poor nations" and "to nourish starved bodies and feed hungry minds," giving them hope. Each of these are biblical teachings embedded in Obama's vision for America.

Unfortunately, Mr. Obama cherry-picked biblical principles to influence his vision for America. Some notable biblical fundamentals like the sanctity of life was ignored as applied to unborn Americans, biblical marriage and caring for the poor.

Our founders and past presidents were influenced by biblical principles and values when formulating their visions for America. That's a positive aspect of our history. However, as the old proverb states, "The proof of the pudding is in the eating." The founders' grand vision lived up to the test while those spoken by recent presidents had mixed outcomes.

A Fresh, Biblical Vision for America

The following is a vision for America from a biblical worldview perspective as seen through five lenses with a special focus on the emerging issues identified in sections I and II of this volume. The overarching questions for the reader and by extension future American leaders are twofold: First, does the biblical worldview influence government policies, strategies, and actions? Second, how should our knowledge of the end times impact our worldview, and what difference does it make in our actions?

Read this vision as if it is being delivered as part of a new president's inaugural address. Then we will in subsequent chapters put flesh on the bones of this vision.

Security Lens

America will never ask permission to protect herself or our way of life. That means in the nonpolar world where conflict is more likely, the U.S. government must field an armed force capable of both unilateral and coalition operations to protect American interests against terrorism, wars of aggression, attacks from space and cyberspace, and the proliferation of WMD.

We must also be prepared to use our might in less traditional military arenas. There will be times our armed forces must react to save lives such when black swan events raise their ugly heads: pandemics, solar storms disrupting our national infrastructure, and natural disasters like tsunamis and hurricanes ravaging population centers.

Our national security establishment will always apply the principle of the sanctity of life. Research to improve human performance must be consistent with this moral precept, as must military medicine's treatment of the unborn in hospitals.

The Pentagon must be an equal opportunity employer while recognizing that the demands of the profession of arms will always put the readiness of the force above individual preferences. It will also protect religious freedom and never distort the God-ordained marriage relationship within the ranks.

The fight against Islamic terrorists will continue for many years, and until that fight ends, terrorists captured on the battlefield will remain behind bars at our facility in Guantanamo Bay, Cuba. We must never release terrorist killers who might return to the fight, and we must never negotiate with those killers or their sponsoring states.

America will work with like-minded nations to employ joint action and military preparedness to build trust among friends and send a clear deterrent message to our adversaries. But force will never be used unless necessary, and only the force necessary should be used according to Just War guidelines.

Our homeland defense efforts will abide by the same values and precepts. We will protect civil liberties while securing our nation from foreign and domestic powers that seek to rob our people of their fruits and threaten their safety.

Intelligence Lens

The seventeen-member U.S. intelligence community is the nation's eyes and ears to protect us from enemies overseas and at home. At least two biblical principles must apply to these agencies: protect the citizenry and limit government's role.

The intelligence community must have the means to identify threats and risks to America, whether they are homegrown terrorists, killer satellites in outer space, nuclear-tipped missiles in the hands of rogue states, or cyber-attacks from foreign hackers.

These legitimate intelligence tasks must not become an excuse for an expansion of government's powers to intrude into the lives of law-abiding citizens. They must never spy on our citizens, target them when they travel abroad, or target our allied leaders.

We must put in place the most reliable of all intelligence sources, human intelligence. Further, we must work closely with ally intelligence services and cooperate with homeland and local law enforcement as appropriate.

Foreign-Policy Lens

America's dealings with other nations will reflect our Christian values and principles. Our conduct of foreign affairs is not about popularity, but about protecting the security of our citizens at home and abroad, and defending our vital interests.

A key American value is religious freedom. We will not give our support or friendship to any country that discriminates against Christians and Jews solely based on their faith. We expect tangible changes from those who currently discriminate based on religion before any favorable action.

America is a nation of immigrants, but newcomers must abide by our laws to immigrate, and they must assimilate by becoming citizens who embrace our language and culture.

We will be a member of intergovernmental organizations that share our values for life, human rights, the family, religious freedom, and our sovereignty. We will form coalitions and sign treaties that reflect those same values and conform to our Christian principles.

America will always be a generous nation and will aid the downtrodden. We do this by meeting the physical and security needs of innocent victims, and we work to end wars that threaten the innocent.

America will work with other like-minded nations to reduce the spread of HIV and AIDS through abstinence promotion, the trafficking of humans, and the scourge of drug addiction.

Economy Lens

America supports free enterprise and open markets that lead to opportunity for all. We encourage families to prosper and an economy that provides good-paying jobs, the product of an open market untethered by artificial constraints imposed by government interference.

Americans must have sufficient wages to cover the cost of life essentials: food, clothing, shelter, and medical care. It is not government's role to provide wages to cover those expenses except in dire situations and only temporarily. Otherwise, government must stay out of the way to allow the capitalistic enterprise to fuel the economy that produces enough well-paying jobs.

Our tax structure must encourage individual and corporate investment, and America must remain competitive in the globalized marketplace. Bring home those jobs sent overseas because of high taxes.

We will do more than just bring home jobs. We will take the government off the backs of the American entrepreneur who creates jobs when markets are free. America will once again become the land of opportunity, the entrepreneurship capital of the world.

America will become energy independent. For decades, we have depended on the flow of energy from some of the most unstable regions, and as a result, crises have whipsawed our economy. This will stop, and America will do what is necessary to exploit domestic energy resources to untether our economy from foreign manipulation.

America believes in fair competition. Countries that want to trade with the United States will provide open, tax-free markets to American goods.

Today, America is living beyond its means as evidenced by an over-whelming $19+ trillion debt. We have mortgaged our grandchildren's future, and this will stop. We will tighten our belts and give our economy permission to expand. Righting our trade balance and living within our means is our number-one priority. We will balance our budget while creating a fair tax system.

Culture Lens

America's culture is morally corrupt and dysfunctional. Too many of us don't know right from wrong. We've become a nation of egotists having lost the ethic that made this nation great.

Putting America back on a morally correct course begins with the nation's top leader, the president. Our president, and by association his administration, will govern with biblical values and principles, always being above reproach. Those who disappoint us will be removed and replaced by people with unquestioned scruples.

Our children deserve the opportunity to learn right from wrong and to grow in stable homes.

Children belong first to their parents and families, not to the local school—much less to government. Parents have the primary responsibility and authority over their child's education and life decisions.

At the same time as we are seeking family stability, we must provide effective law enforcement for our communities. Criminals must be deterred, and those addicted to drugs and alcohol must be given hope. Crime is a problem, and too often the perpetrators come from broken homes—then these troubled people fill our jails to capacity. The evidence shows that kids who grow up in intact homes prosper; that is our goal for all American children. We must scrub everything government does to create environments that protect our children and give them the opportunity to succeed.

Our culture is significantly impacted by various media. Although we respect our free-speech liberty, some outlets abuse the privilege and must be restrained. This may be a fine line, but we will not tolerate the coarsening of our next generation.

Conclusion

The vision the president brings to the White House is translated into policies and strategies, and is then executed by his or her administration. That's why the foundation of the president's vision—his/her worldview—really matters.

The following five chapters develop the above vision by recommending policies and strategies. As stated at the beginning of this chapter, I don't know whether successfully turning America to Christian values, principles, and leadership will in any way impact God's plan to delay or accelerate the end times. But there is biblical precedent for nations to receive new blessings if they repent and turn toward God. Ultimately, however, the Christian's duty is to obediently fulfill the Lord's mandate to be about His work until Christ's return. The following chapters outline a prescription for the future America.

Chapter 11

National Security Strategy for the Twenty-first Century

WE SEEK A national security strategy that accounts for a biblical worldview given the future world marked by the trends, shifts, and black swans outlined in section I. That complex and frightening new world must wrestle with very significant threats and risks outlined in section II of this volume: a nonpolar unstable field of players; increased global terrorism; a lust for transhumanist super soldiers and mind-blowing killing technologies; use of outer space for exploration and war making; cyber warfare that threatens our modern way of life; and the proliferation of WMD in the hands of individuals, groups, and states with the willingness to use them.

This chapter profiles the nation's security strategies as well as how things are really done at the Pentagon. The chapter ends by proposing a national defense strategy based on the vision outlined in chapter 10 and from a biblical worldview perspective, with emphasis on the six critical threats and risks addressed in section II.

America's Four Security Strategies

Before proposing a biblical worldview security strategy for the United States, we must understand the current strategic framework and processes used to secure our country.

Four documents encompass our nation's security strategy. The National Security Strategy (NSS) is the president's security vision, which is often partially introduced in his inaugural and/or annual State of the Union addresses and was later published as a glossy booklet. Second, the National Defense Strategy (NDS) is the secretary of defense's interpretation of the president's NSS; the NDS identifies defense department missions. Third, the National Military Strategy (NMS) is drafted by the chairman, joint chiefs of staff, who drills down on the NDS with an eye on assigning supporting missions and priorities taken from the NDS to the four military services. Finally, the newest strategy is written by the secretary for the Department of Homeland Security (DHS), who interprets the president's NSS to identify key missions to secure the homeland.

Consider details regarding each of these documents.

National Security Strategy: Much of the NSS is unchanged from one administration to the next. After all, every president promises to safeguard our national interests through strong and sustainable leadership. Presidents invariably outline principles and priorities in their NSS that are supposed to advance American leadership, and they always promise to work with our allies and partners to support our mutual interests—at the same time warning our adversaries against mischief. Finally, the NSS signals to Congress how the president intends to defend the country against the shifting security landscape.

Presidents also use the NSS to promise more than they can reasonably deliver by presenting a hopeful view of our future security. For example, President Obama's 2015 NSS promises greater peace and a new prosperity, and then he outlines his guiding principles: lead with purpose, lead with strength, lead by example and with capable partners, lead with all instruments of U.S. power, and lead with a long-term perspective. Few national security experts will disagree with these enduring principles.

Most presidents then promise to advance the security of the U.S., its citizens, allies, and partners. The commander-in-chief invariably says good things about our troops, such as promising to maintain the best trained, most well equipped, and best led force in the world— and of course, presidents always commit to honoring our veterans, especially those wounded in battle and the memory of those killed. Then the president promises to work with Congress to get the best armed forces for the dollar while remaining fiscally responsible and keeping Americans safe at home from threats like terrorism and nuclear-armed rogue regimes like Iran and North Korea.

The NSS provides a unique forum for the president to weave into his security plan special issues that may seem out of place. Mr. Obama's liberal worldview is evidenced in the 2015 NSS with his emphasis on climate change, an issue he considers part of the national security landscape, and he also commits America to support intergovernmental organizations like the United Nations. Further, Obama's liberal social views creep into the NSS in terms of his call for preventing and responding to human-rights abuses such as "gender-based violence and discrimination against LGBT [lesbians, gays, bisexual and transgender] persons."[350] However, Obama fails to mention the serious persecution of Christians across the globe and the ongoing genocide in the Middle East.

National Defense Strategy: The NDS is the Department of Defense's capstone document that flows from the NSS. It provides a framework for defense planning, such as the geographical combatant commands' campaign and contingency plans, force development, global defense posture, and intelligence. It reflects the president's priorities and principles, as well as does the congressionally mandated Quadrennial (every four years) Defense Review, which delineates strategic guidance for DoD consistent with the NSS that defines force structure, modernization plans, and a budget plan to execute the full range of missions.

The NDS drives the Pentagon's budgeting process known as planning, programming, budgeting, and execution (PPBE) to allocate resources that dictate how much military force is required for the anticipated missions, our weapons requirements, and business practices. The NDS also addresses how the armed forces will fight and win America's future wars and how it will work with foreign partners to shape the security environment.

Obama's 2012 (the latest) NDS identifies ten key objectives based on the secretary's interpretation of the NSS. Those objectives are:

1. Counter terrorism and irregular warfare.
2. Deter and defeat aggression.
3. Project power despite anti-access/area denial challenges.
4. Counter weapons of mass destruction.
5. Operate effectively in cyberspace and space.
6. Maintain a safe, secure, and effective nuclear deterrent.
7. Defend the homeland and provide support to civil authorities.
8. Provide a stabilizing presence.
9. Conduct stability and counterinsurgency operations.
10. Conduct humanitarian, disaster relief, and other operations.

It is likely any team of national security experts would come up with these mission sets but perhaps in a different order.

National Military Strategy: The chairman, joint chiefs of staff's (JCS) NMS supports the NSS and translates the NDS guidance into combatant command and military service missions and reinforces the need for our military to remain globally engaged by posturing the right mix of forces across the world to shape the security environment in concert with our international allies and foreign partners. It also describes how the Joint Force, the combination of service components from each of the military services that are allocated to the geographical combatant commands, align the military ends, ways, means, and risks.

The chairman of the JCS recognizes that our Joint Force must apply the military instrument of power in a complex and rapidly changing environment marked by globalization, diffusion of technology, and demographic shifts. Given this environment, he links his strategy to enduring national security interests to suggest national military objectives. The chairman elaborates on three national military objectives in the NMS: deter, deny, and defeat state adversaries; disrupt, degrade, and defeat violent extremist organizations; and strengthen our global network of allies and partners. These objectives provide guidance and context for war planners.

There are other strategic guidance documents that provide detailed instructions to the geographical combatant commands, the military services, and defense agencies. The details of those documents are classified. However, in general terms, three of the most important classified documents are outlined below.

The Guidance for Employment of the Force (GEF) consolidates and integrates DOD planning guidance and priorities, and identifies planning assumptions related to operations and other military activities.

The Joint Strategic Capabilities Plan is a companion document to the GEF that provides guidance to accomplish tasks and missions based on military capabilities. It also addresses steady-state activities like security cooperation with foreign partners and global posture as well as contingency planning requirements.

Finally, the Global Force Management Implementation Guidance integrates complementary unit assignment, apportionment, and allocation information into a single global force management document.

Homeland Security Strategic Plan: The secretary of the DHS prepares a strategic plan based on the department's Quadrennial Homeland Security Review, which establishes a unified, strategic framework for homeland security missions and goals. The DHS has five key missions: secure our country from terrorist threats, enhance security, secure our borders, enforce our nation's immigration laws, secure cyberspace, and build resilience to disasters.

The DHS works very closely with the DoD's sixth geographical combatant command, the U.S. Northern Command. On occasion, Northern Command provides military forces to assist DHS in various domestic security operations such as disaster relief, border control, some drug interdiction, cyberspace, and antiterrorist operations that have a chemical, biological, radiological, and nuclear defense (CBRN) component.

How Things Are Really Done at the Pentagon

Now consider a bit of reality about how things really work or don't work at the Pentagon—a "Pentagon 101" for the novice. My recommendation of a biblical worldview security strategy follows this brief section.

I have seventeen years in the belly of the beast—inside the Pentagon. In the early 1990s, I spent three years as an active-duty Army officer working as an inspector general investigating general officers accused of misconduct. I retired from the Army in 1993 and joined the Family Research Council, where I rose to become the vice president for policy, all the while staying close to the defense establishment in the Pentagon and on Capitol Hill. Then in 2002, I returned to the Pentagon. For the past fourteen years, I've been a contract civilian working with the Department of the Army staff and part of that time I was also a media analyst with access to the secretary of defense and his staff.

The national defense enterprise is giant, complex, and incredibly bureaucratic. Just because a document like the NDS states we do certain things and in a certain way—that isn't necessarily true, and often it is far from reality.

The shelf life of many strategic documents like the NDS seldom outlasts the last leader's departure. That is because the Pentagon is a revolving door of uniformed officers and political appointees who are supposed to run the place. Of course, these "leaders" are at the mercy of the bureaucracy that is a permanent fixture in the five-sided limestone building on the Potomac.

Sometimes leaders get things done through force of personality—usually through intimidation—but most of the time the redundant layers of bureaucracy defeat the best of intentions. After all, redundant staffs and literally hundreds of ever-changing DoD instructions, directives, and special memoranda are complimented by like numbers of military service and defense agency regulations and other guidance documents that trickle down through the military hierarchy. We practically drown in paperwork or the electronic versions of those documents at the Pentagon.

No one should doubt the power of the little lady bureaucrat in "tennis shoes" who can stop any general or secretary in his or her tracks. Literally, many government people wear running shoes to the Pentagon and then change once at their cubicle or keep them on to comfortably navigate the 17.5 miles of hallways.

The Pentagon has redundant and competing staffs such as the Army. The chief of staff of the Army is a four-star general with a staff counting in the thousands helping him resource the nearly one million soldier force (active and reserve component). There is also the political side of the Army in the form of the secretary of the Army (an appointee) and his staff of politicos and bureaucrats who more often than not duplicate (some say check on) the uniform service's efforts.

The military and political service staffs for the Army, Navy, Marines, and Air Force take direction from the chairman, JCS, and the secretary of defense. The 1986 Goldwater-Nichols Department of Defense Reorganization Act streamlined the military chain of command, which now runs from the president through the secretary of defense to the combatant commanders, thus bypassing the chairman, JCS, and the service chiefs. The chairman, JCS, is the principal military adviser to the president, the National Security Council (NSC), and the secretary of defense. He does not command the nine unified military commands or the services.

Goldwater-Nichols didn't fix all the Pentagon's problems, especially the lack of unity of command. Even though we combined the military services under a single secretary of defense in 1947, vicious rivalries continue. Air Force General Curtis LeMay explained the problem well: "The Soviets are our adversary. Our enemy is the Navy." However, it did improve operational effectiveness at the point of the spear.[351]

Army General Norman Schwarzkopf, who commanded American forces in the first Gulf War, commented that in the shadow of Goldwater-Nichols, the effectiveness of our forces improved because of clear lines of authority in the war theater. Further, that impact became even more evident during our operations in Afghanistan and Iraq.

In 2015, Senator John McCain launched a review of the Goldwater-Nichols legislation to identify and fix the unintended consequences—a lack of unity of command in the Pentagon—that resulted. Specifically, that review attempts to answer the following questions.[352]

- Are the roles and missions of the Joint Staff, Combatant Commands, Joint Task Forces, and other headquarters elements properly aligned to conduct strategic planning, equip our warfighters, and maximize combat power?
- Does the vast enterprise that has become the Office of the Secretary of Defense further our ability to meet present and future military challenges?
- Does the constant churn of uniformed officers through joint assignments make them more effective military leaders, or has this exercise become more of a self-justification for a large officer corps?

Time will tell whether Senator McCain and others can steer the Pentagon into more efficient waters, but for now the five-sided building continues to muddle along.

The Office of the Secretary of Defense has a giant staff that produces guidance such as the NDS, and the chairman's sizable staff basically responds to the guidance of the secretary and NSC by passing directives to the unified military commands that oversee worldwide joint operations.

The DoD also has twenty-three defense agencies and field activities with many thousands of full-time employees and hundreds of part-time advisory committees and boards. The Pentagon is also crawling with contract help who are embedded in the various staffs or visiting their clients from offices nearby in Crystal City or Pentagon City.

There is plenty of fat to be cut at the Pentagon, but that's a topic for another book. What's important to understand here is that the Pentagon is incredibly complex with many overlapping staffs, many levels of written guidance, and frequent turnovers of key leaders, which explains why the twenty-four thousandish-person Pentagon workforce has the reputation for inefficiency, and as the Government Accountability Office (GAO) said in a 2013 report, it is impossible for investigators to audit the Pentagon.[353]

There is little hope the Pentagon will ever pass a simple audit. A March 2016 report by Public Integrity, a private nonprofit government watchdog organization, reported on a revealing Defense Department Freedom of Information Act request regarding how many of a particular device the Pentagon had in its possession.[354]

The official response said it would take the Pentagon fifteen million labor hours at a cost of $660 million to answer the question. Public Integrity said that's enough to buy "the Washington Nationals baseball team, a 600-acre island off the coast of Australia, twelve of the most expensive Ferrari racecars, or about as much as the Pentagon is currently spending to train Iraqi soldiers in combat."

The root of the Pentagon's problem is how poorly it tracks purchasing and contracting. Robert Jarrett, the Pentagon's director of operations, defense procurement, and acquisition policy, explained the Pentagon's database—which now includes an estimated 30 million contracts—cannot be comprehensively searched. Therefore, as Jarrett explained, someone would need fifteen million hours, or about 1,712 years, to read all the contracts to answer the question.

Given the Pentagon's miles of corridors and frenetic 24/7 activity, it should surprise no one that much of the real work gets done off the grid. Specifically, staffs meet in locked conference rooms in classified discussions to keep the outside world from monitoring their work. Some, but not all, of that secrecy is legitimate. Yes, there is a problem with over-classification of material across the entire government, and DoD is a big offender.

Political appointees and senior officers give orders, stand-up study groups, and draft memoranda to make certain their subordinates carry out the intent. For example, the military services, Joint Staff and Office of the Secretary of Defense (OSD) and its defense agencies have thousands of committees that meet regularly to study and monitor issues for top leadership. There are "red teams" to anticipate problems and seek alternatives that advise senior officials.

Of course, there are the inevitable service rivalries seeking "their fair share" of the defense budget. Personalities inevitably clash among the top brass and politicos over the proper direction and country's best interest. These rivalries and personal-agenda train wrecks show their ugly heads mostly behind closed doors, however.

I've seen some of those "train wrecks" and participated in numerous study groups, committees, and discussions with bureaucrats, senior officers, and appointees. It shouldn't surprise anyone that much of the ideological work done at the Pentagon is driven by the White House via the National Security Council (NSC), which works behind the scenes like a puppeteer manipulating the politicos. But that work seldom sees the light of day except when authors like Bob Woodward of Watergate fame writes a tell-all book with the help of some Pentagon "deep throat" insider.

The point here is the political appointees may draft an NDS or other official-sounding document that says one thing, but the appointees' internal directives and sycophant staffs may say or do something entirely different. Also, the funding for those programs is often drawn from accounts that appear to be shady. In fact, there are classified parts of the budget that rightly grant the Pentagon's leadership considerable discretion. This is important, but without oversight, things have and do go amiss. Also, remember, government money comes in many colors, which explains why the Pentagon has a battalion (read "hundreds") of lawyers to help the brass sort out what color money can be spent on any number of programs. It also partly explains why the GAO complains the Pentagon can't be audited.

Let me briefly explain what is meant by "many colors" of Pentagon money. Fiscal law governs the availability and use of federal funds, which is derived from the Constitution, legislative authorization and appropriation acts, judicial court rulings, and comptroller decisions regarding legislation's intent. The authority to spend public funds can only be authorized by Congress through legislation that defines the purpose, time, and amount. That's where the "color of money" expression comes into play.

The Army, for example, receives funding via fifteen different congressional appropriations. The Army cannot spend funds appropriated for wheeled vehicles on aircraft parts. However, there is a caveat in the law called the "necessary expense rule." An Army command can't just take money from one appropriation pot (wheeled vehicle money) and use it for an entirely different requirement (aircraft). However, the "necessary expense rule" gives the lawyers wiggle room. Specifically, appropriated money from one pot may be used for a seemingly unrelated expenditure if it is necessary and incident to the proper execution of the general purpose of the appropriation. Do you see why the Pentagon has so many lawyers?

Time is another factor in deciding on the "color" of money. DoD must obligate appropriations before the authority to obligate expires (31 U.S.C. X 1502(a)).[355] That's why near the end of the fiscal year (September 30) there is often a mad rush to spend so-called "one-year money." Otherwise, that money must be returned to the U.S. Treasury.

Most DoD spending authority is one-year operations and maintenance funds. However, some funds last much longer: procurement funds can last three years; research, development, test and evaluation funds can last two years; and military community funds can stretch to five years.

One seemly undiscernible part of the budget is the money poured into what's called Building Partner Capacity. The Pentagon uses a pot-pourri of at least seventy different authorities to help so-called allies and friendly foreign partners to confront myriad challenges across the globe such as fighting drug cartels and insurgents. In total, the Pentagon spent at least $122 billion on these efforts in the past fifteen years.[356]

Finally, since combat operations started in 2001 in Afghanistan, Congress gave the Pentagon special money exclusively to fight the war above its annual budget, which in 2017 will be approximately $600 billion. The Overseas Contingency Operations (OCO) fund is a separate pot of funding often referred to as the "Pentagon's slush fund," because invariably some of those funds get stashed in non-war "operation and maintenance" accounts. The fiscal year 2016 OCO account had $59 billion.[357]

The maze of "colored money" is almost as bad for efficiency as the petty Pentagon infighting. It is important to understand the high-ego, head-strong top brass and their politico counterparts often don't get along and won't admit to getting things wrong. A good example of missing the mark is the Bush administration's claim that Saddam Hussein had WMD, one of the key justifications used for invading Iraq in 2003. The NSC pushed the Pentagon and the intelligence community to embrace that view in spite of very vague or nonexistent evidence. There were voices that cautioned against such a conclusion, but the politicos drowned the dissenters out of the conversation. I know this for a fact, because I was one of those dissenting voices albeit while serving as a Fox News analyst who attended frequent DoD-hosted meetings leading up to the Iraq invasion.

I've also spent many years watching senior officials in the Pentagon observing how too many change their views about important issues, especially the higher they rise in the ranks. As young officers, some of those who rose to flag officer rank were very socially conservative—but put stars on their shoulders and they quickly lined up with the politicos, or some would say, with the prevailing political winds. Some of those same officers tell me in confidence that they had no choice on continuous issues, but what they really meant is they wouldn't dare risk their stars and big retirement salaries to step out on a limb. That's how this insider sees the dysfunctional Pentagon culture.

The political appointees are worse than the generals. They occupy plush E-ring offices—offices on the Pentagon's outside wall with windows looking out—for such a short time that many feel compelled to seize the opportunity to make a name for themselves. So, more often than not, they work at an exhausting pace that includes significant travel and long office hours, and they face tough decisions backed by never-ending briefings and piles of staff papers. Many of these hundreds of politicos are selfless servants, and for the most part, they do good work.

President Obama said early in his administration, "Elections have consequences."[358] That view was understood by Obama to mean he earned the right through the election to do what he wanted, and that idea empowered some of Obama's Pentagon appointees as well. At the Pentagon, I saw how some Obama political appointees sacrificed readiness to push for open homosexuality in the ranks, women in all combat roles, and an anti-Christian agenda. Those Obama agenda items had opponents among the uniformed brass, but their opposition was seldom expressed to the politicos face-to-face. This is why I named my 2013 book about women in combat *Deadly Consequences: How Cowards Are Pushing Women into Combat.*[359]

There is another piece to the Pentagon's intimidation puzzle: the role of the NSC. The NSC under Obama significantly expanded to over four hundred personnel—populated by mostly young sycophants hired from Obama's 2008 political campaign—and according to former secretaries of defense like Robert Gates and Chuck Hagel, the council was very directive, always second-guessing the chairman and secretary, and even ignored the chain of command to contact military commanders in the warzones, bypassing their seniors in the Pentagon.

Others piled on that assessment. Former Deputy Defense Secretary John Hamre, who is CEO for the Center for Strategic and International Studies, reviewed the NSC's role and found "excessive demands" that are "warping" the DoD. Former Air Force Chief of Staff General Norton Schwartz elaborated that the NSC "should not be activists. They should have a limited—if any role—n execution."[360]

General Schwartz explained what he meant by "should not be activists." He explained that our warriors in Afghanistan often have to spend their sleeping hours teleconferencing with the NSC. "I think there's a fundamental dysfunction there which is not healthy," Schwartz said.[361]

This insider's glimpse of the Pentagon's "clock-making factory" is important to understand, because it demonstrates why nothing gets done easily and sometimes not at all. Further, my proposed biblical worldview-based security strategy that follows has no chance of success without the cooperation of the Pentagon's powerful bureaucracy, supportive leadership, and the NSC—and, yes, a lot of time and prayer as well.

A Biblical National Security Strategy

Chapter 10 provides a biblical worldview vision for America in terms of six lenses. The first of those lenses is security. The following translates that security vision lens into an abbreviated national security strategy-like set of recommendations that address the emerging world environment and the threats and risks.

Each of the six paragraphs that made up the security lens (chapter 10) is translated into tangible objectives below.

#1: America will never ask permission to protect herself or our way of life.

That means in the nonpolar world where conflict is more likely, the U.S. government must field an armed force capable of both unilateral and coalition operations to protect American interests against terrorism, wars of aggression, attacks from space and cyberspace, and the proliferation of WMD.

Right-sized ground force: This means the U.S. must field a right-sized armed force capable of performing a host of missions either unilaterally or as a coalition partner. That guidance begs for answers to two questions: What are the missions that must be accomplished and what is the right sized force for those missions?

The U.S. needs a force capable of dealing with the highly unstable nonpolar world that is full of global players armed with sophisticated weapons. There are many complex and challenging scenarios our forces must prepare to confront over the next decade. They include Russian threats to its former satellites, a second Korean war, an Asian maritime conflict pitting China and Japan and/or small nations, an Indo-Pakistan war, Middle East nuclear war, biological pandemic, and civil wars in Africa and the Middle East.

The most important challenge for the future U.S. armed forces—Army, Marines, Navy, and Air Force—is correctly sizing our ground force, Army and Marines, which is capable of securing American interests at home and abroad. A large ground force is required because most of the future fights will employ ground forces in complex urban terrain. Unfortunately, the U.S. has a history of cutting our ground forces too much, then major challenges requiring ground forces quickly emerge, such as just prior to World War II we shrank to the seventeenth-largest army in the world and then again just prior to the Korean War. In both situations, the U.S. faced great peril.

The Obama administration wants to set up the U.S. for a repeat of the periods just prior to the Second World War in Europe and later in Korea. There are good cases made that such actions are intentional and cases made that Obama's actions are merely naive. Either way, Mr. Obama dismisses the need for a ground force to "conduct large-scale, prolonged stability operations," the rationale he uses to justify deep ground force cuts. However, most of the world's military personnel are in armies and most of the future perils we face will require significant ground forces.

"We are in an extraordinarily rare position in American history where our budgets are coming down but our missions are going up," then Secretary of the Army John McHugh said. In 2009, our Army was 553,000 strong and now it is on a gliding path to bottom out at 450,000 by October 2018.[362]

This figure (450,000) is endorsed by the congressionally mandated National Commission on the Future of the Army and the current Army chief of staff. However, their endorsement comes with caveats. Army Chief of Staff General Mark Milley reminds us, "We are an Army of three components in the Regular Army, National Guard and United States Army Reserve." The chief explains that he is comfortable with a smaller active-duty force of 450,000 as long as it is backed by a trained, equipped, and ready Guard and Reserve. That's a big "if" and will require a potentially giant bill to keep them ready, one Obama wants to avoid.

The other caveat is, according to General Milley, the world is increasing in "velocity of instability," requiring a force that is nimble and prepared for long fights in which advanced technology won't necessarily give soldiers upper hands. Milley's statement is consistent with the new nonpolar world outlined in section I of this volume.

Our other ground force, the U.S. Marine Corps, faces significant shortfalls to win in a major fight, according to General John Paxton, Marine Corps' assistant commandant, in testimony to the Senate Armed Services Committee on March 14, 2016. The general explained:

> In the event of a crisis, these degraded units could either be called upon to deploy immediately at increased risk to the force and the mission, or require additional time to prepare thus incurring increased risk to mission by surrendering the initiative to our adversaries… This does not mean we will not be able to respond to the call…. It does mean that executing our defense strategy or responding to an emergent crisis may require more time, more risk, and incur greater costs and casualties.[363]

Although we need a large ground force we also need capable sea and air services. Those capabilities must be robust to address the challenges posed by our many threats in the sea and air domains. Yet, those forces are shadows of their past selves.

America's Navy, according to the 2014 National Defense Panel, ought to have between 323 and 346 ships. However, as of September 2015, the U.S. Navy had 272 "deployable battle force ships" built around its 10 aircraft carriers (with another, the USS Ford, expected soon). Today's fleet compares with 550 during the Reagan administration and the 300s during the Clinton years.[364]

Another way of looking at our naval strength is to consider the tonnage. Our present Navy weighs a total of 5.1 million tons compared to 5.7 million in 1975, which at that time included 559 ships. The difference is in types of ships such as today's Arleigh Burke-class destroyers (DDG-51s) which weigh nine thousand tons compared to 1970-era destroyers which were half that size.[365]

Is our Navy the right size? That's the key issue and not numbers or tonnage. The Obama administration decided "not to maintain continuous carrier coverage in all three theaters but to gap the coverage particularly in the Mediterranean and Europe." This means Obama accepted more risk, which in a more dangerous world is not in America's best interest, but may be necessary if our budget can't sustain a larger fleet.

Similarly, our Air Force is at a dangerous level thanks to Obama. We now have the smallest Air Force ever. For example, during Operation Desert Storm in 1990, our Air Force had 188 fighter squadrons compared to just 54 today and soon to be 49. There were 511,000 Active Duty airmen in 1990; in mid-2016, there are 313,000 even though the current operational tempo requires significantly more personnel.[366]

Air Force Chief of Staff General Mark A. Welsh III put the condition of his service's aircraft fleet into perspective at a 2015 Senate Armed Services Committee hearing. "If WWII's venerable B-17 bomber had flown in the first Gulf War in 1991, it would have been younger than the B-52, the KC-135 and the U-2 are today. We currently have 12 fleets of aircraft that qualify for antique license plates in the State of Virginia!"[367]

The bottom line is the roles that the traditional ground forces (Army and Marines) play are now in jeopardy due to Obama's ill-begotten strategies. Our sea and air forces are almost as bad, and arguably in some situations worse. And the risk to national security is becoming unacceptable. Remember that in the instance of a fast-moving national emergency, the draft is unlikely to be our salvation. We will be forced to respond with the team we have and that team is ill-equipped.

Space: Our armed forces must prevail in space as well as on the land. As seen in an earlier, chapter space is becoming militarized and that domain plays a critical role in America's future national security.

On March 16, 2016, Air Force General John Hyten, commander of the Air Force Space Command, gave alarming testimony to Congress that the threat to U.S. space systems has reached a tipping point. "Adversaries are developing kinetic, directed-energy, and cyber tools to deny, degrade, and destroy our space capabilities," Hyten said. "They understand our reliance on space, and they understand the competitive advantage we derive from space. The need for vigilance has never been greater," the four-star general said.[368]

Lieutenant General Buck, commander of Joint Functional Component for Space, a U.S. Strategic Command unit, testified along with Hyten that China and Russia pose the most serious threats to space systems. "Simply stated, there isn't a single aspect of our space architecture, to include the ground architecture, that isn't at risk," Buck said. "Russia views U.S. dependency on space as an exploitable vulnerability and they are taking deliberate actions to strengthen their counter-space capabilities," he said.[369]

Two American military capabilities in the space domain warrant special attention: satellite protection and deterring space-based threats.

First, we need the means to protect our fleet of satellites and the capability to shoot down a threatening enemy satellite.

One way to protect our fleet of satellites is to make them smaller targets. Frank Calvelli, deputy director of the National Reconnaissance Office, the spy agency that operates intelligence and reconnaissance satellites, revealed in March 2016 congressional testimony that in October 2015 the U.S. launched a new satellite that carried thirteen smaller "CubeSats." The "CubeSats" are part of NRO's efforts to develop "better and faster" systems in space that provide better overall "survivability and resiliency" to operate in a future war.[370]

Dyke D. Weatherington, director of unmanned warfare and intelligence, surveillance, and reconnaissance at the Pentagon, also testified to Congress that "the rapid evolution and expansion of threats to our space capabilities in every orbit regime has highlighted the converse [to our current strategic advantage]: an asymmetric disadvantage due to the inherent susceptibilities and increasing vulnerabilities of these systems." While space threats are increasing, "our abilities have lagged to protect our own use of space and operate through the effects of adversary threats," Weatherington said.[371]

China has already demonstrated the capability to down a satellite. This is significant, given China's public statements regarding our dependence on a constellation of satellites to effectively conduct expeditionary military operations anywhere in the world. We should understand China's statement as a threat.

In 2008, the U.S. also downed a satellite as a proof of concept. An American National Reconnaissance Office's NROL-21 Radarsat was downed by a three-staged Standard Missile (SM-3) launched from the USS Lake Erie, an Aegis cruiser.[372]

Russia as well as U.S. allies like Japan and Israel are also investing in hit-to-kill systems that can be used for either missile defense or anti-satellite operations.[373]

Second, we need the capability to deter space-based threats such as that posed by nations like North Korea. In February 2016, North Korea launched an object into space that flew over the United States. That is one more step to the rogue regime having the capability of launching a devastating attack on America from space.

North Korea has tested nuclear devices and claims the ability to miniaturize those devices as recently as March 2016.[374] The threat to America will be well beyond crisis stage when North Korea places a nuclear device in orbit above the U.S. which could then be remotely detonated.

A single nuclear weapon exploded at high altitude could shut down vast areas across the United States. That explosion will interact with the earth's atmosphere to produce an electromagnetic pulse (EMP) that will have a cascading effect—knock out electrical power, fry circuit boards, and disrupt telecommunications. Most non-hardened military systems will become inoperable. Some scientists have likened the outcome for America to be a societal return to the 1800s.

Director of National Intelligence Clapper testified that North Korea was "committed to developing a long-range nuclear-armed missile that's capable of posing a direct threat to the United States." He also testified about his views regarding Iran's nuclear capability. So far, said Clapper, Tehran is complying with the terms of our (2015) nuclear agreement, but he said "we in the intelligence community are very much in the distrust-and-verify-mode."[375]

It is noteworthy that Iranian media previously addressed the EMP as a weapon. A 2001 article in Siyasat-e Sefa-i (*The Journal of Defense Policy*) includes EMP as a part of "terrorist information warfare" and an article published in Iran's security journal *Nashriyeh-e Siasi Nezami* in 1999 identified an EMP attack as a way to defeat the U.S. as a military power or as a state.[376]

Clearly, the U.S. needs a capable antiballistic missile system either earth or space-based to prevent the potential nuclear weapon threat posed by rogue and other nuclear-capable parties.

New hypersonic arms race: A new arms race is heating up between conventional arsenals and space-based systems called hypersonic missiles that promise to revolutionize warfighting in both conventional and nuclear settings.

Hypersonic missiles are both very accurate and extremely fast, and as a result, they will change the face of modern warfare and render current missile defense systems ineffective. Specifically, according to Stratfor, hypersonic missiles will travel at least five times the speed of sound (Mach

5 or 3,852 MPH), but first some technological barriers must be overcome such as the stress and extreme temperatures of hypersonic flight, maintaining the high speeds for extended periods, and guidance stability.[377]

The advantages of hypersonic missiles justify the investment. They are superfast, cover great distances in short time periods, highly accurate, better at penetrating enemy air defenses than conventional missiles, and likely can carry nuclear warheads. Further, hypersonic missiles are on the brink of becoming operational.

U.S. Major General Thomas Masiello said the U.S. Air Force plans to host operational prototype tests by 2020, and four experimental flights already successfully proved the concept.[378]

The Chinese and Russians are close on the U.S.' hypersonic development tail. In 2014, China conducted three tests of its DF-ZF hypersonic strike vehicle and Russia is developing its own hypersonic vehicle. Moscow successfully tested a short-range hypersonic missile, the 3M22 Zircon, in March 18, 2016.[379]

The consequences of this technology are very significant. Hypersonic systems threaten large naval warships like aircraft carriers and will speed up development of directed-energy weapons—both lasers and microwaves—as possible counters. It could also radically alter nuclear strategies.

Hypersonic missiles equipped with nuclear warheads render current antiballistic technologies harmless. That fact makes preemptive strikes more likely, which means world powers must totally rethink their deterrence and national security strategies.[380]

The U.S. must maintain the lead in hypersonic development and at the same time begin to develop defenses and adjust our strategy as necessary.

Cyber: On February 10, 2016, President Obama sought a surge in funding to counter cybersecurity threats. That request came on the heels of DNI Clapper's Senate testimony that cyber threats "could lead to widespread vulnerabilities in civilian infrastructures and U.S. government systems."[381]

At the same time, Mr. Obama wrote in the *Wall Street Journal* that cyber threats are "among the most urgent dangers to America's economic and national security."[382]

Similar warnings come from other leaders across the government. Yes, we must take immediate and decisive action to defend ourselves, but we must also equip the Pentagon's U.S. Cyber Command with the offensive means to shut down those who would attack our cyber infrastructure systems.

The following are urgent actions needed to protect our cyber infrastructure.

First, government and private industry must share threat intelligence about the state of our networks and the capabilities and intentions of our cyber adversaries. This information will help critical infrastructure companies and vulnerable industry to be better prepared when inevitably attacked and the sharing must be done on a timely, ongoing basis.

Second, government must establish mandatory cybersecurity standards for public utilities that are not overly cumbersome and costly. Americans deserve to know their public service providers reliably comply with industry-wide cybersecurity standards just like water treatment plants monitor contaminants for public safety.

Unfortunately, our public electricity services are especially vulnerable. A Government Accountability Office study states "the electricity grid's reliance on IT systems and networks exposes it to potential and known cybersecurity vulnerabilities."[383] Congress could vaccinate our utilities from outside hackers by requiring them to create cyber networks no longer tethered to the Internet.

Third, reduce cyber threats by working with foreign partners to create global cyberspace standards, strengthen law enforcement against cybercrime, and grant our armed forces the authority to deter adversaries with effective counters.

Finally, Congress must protect fundamental freedoms. The tendency will be to allow more monitoring and use of private information. Somehow we must thread the cybersecurity-privacy needle by protecting privacy while keeping the Internet open. Further, private industry must not become part of the government's "Big Brother" network violating user privacy while passing cybersecurity information to government entities.

Top leaders warn that America faces a cyber "Pearl Harbor." That is why Congress must quickly pass tough legislation that gives the country the readiness to confront potentially catastrophic cyberattacks while protecting civil liberties.

Weapons of Mass Destruction: Chapter 9 detailed the massive danger posed by WMD and especially biological weapons. What needs more development is the nuclear deterrence in a disordered world.

The world community is failing in its efforts to prevent the proliferation of nuclear arms. Specifically, the early January 2016 North Korean nuclear test was that regime's sixth in the past decade. Unfortunately, the nuclear landscape goes beyond the risks from the likes of Kim Jong-un's penchant for provocation. Iran continues to develop its nuclear capabilities as suggested in a February 2016 Stratfor report indicating Tehran continues to expand its nuclear test site at Parchin in spite of the 2015 agreement with the West.[384]

Pakistan and India remain hostile while both build up their nuclear arsenals and Pakistan acknowledges a range of "tactical" nuclear weapons as part of its "full spectrum deterrence."[385] That's a very frightening statement given their history of warfare.

China is unabashed in its hegemonic regional desire to modernize its nuclear fleet and diversify and harden its nuclear arsenal, to include putting ICBMs on trains hidden in mountains.[386]

Russia is demanding an expanded sphere of influence and rejects further arms control while rapidly modernizing its nuclear weapons and making highly provocative statements about its nuclear capabilities. Russia's new nuclear program includes nuclear-capable missiles tipped with low-yield weapons. Further, due to Russia's conventional military weakness, Russian President Vladimir Putin has dramatically lowered the threshold for when he would use nuclear weapons, allegedly hoping to alarm the West to avoid conflict and as one analyst put it, Putin has adopted a new doctrine that says nuclear war is winnable.[387]

What then should the U.S. do about the proliferation of WMD technologies, especially atomic arms and the growing willingness to use these weapons?

- We need a strategy that preserves stability and builds confidence through international agreements and verification.
- The U.S. must champion WMD nonproliferation and meanwhile, America must urgently modernize its nuclear fleet (launchers and weapons) and then posture them against those who might use such weapons with a credible deterrence.
- The U.S. needs to move full speed ahead with the deployment of a comprehensive antiballistic missile network against a wide range of threats.
- We need a more aggressive military response to emerging threats rather than our past practice of hand-wringing and trying to institute sanctions that never work. An Israeli-like attack on Syria's secret reactor in 2007 must be a viable option against emergent threats.

#2: Our national security establishment will always apply the principle of the sanctity of life.

Research to improve human performance must be consistent with this principle, as must military medicine's treatment of the unborn in hospitals.

Sanctity of Life: The most obvious application of this guidance is to protect the lives of the innocent, and that begins with the unborn. Military hospitals will not provide abortions on demand.

This principle also applies to the Pentagon's efforts to improve human performance. Those transhumanist efforts must not violate ethical bounds, and this is a clear message to DARPA, which was thoroughly examined in Chapter 6.

DARPA wants to dress our troops in skintight suits called "warrior web" to help them run faster, lift more weight, and be less at risk to injury. There is no objection to this form of enhancement and protection; it seems common sense. In fact, the societal benefit from the spinoffs of such research will be tremendous, just as the now-common, everyday use of technology developed for the space program.

However, DARPA is experimenting with the drug modafinil to keep military personnel awake for long periods of time, and it also is trying to alter the soldier's metabolism to operate for days with little or no food. It is trying to artificially sharpen the soldier's brain by ultrasound, electromagnetic fields, mild electric currents, and even drugs or implanted microchips. There are advantages to some of these developments, but these experiments and drug-related so-called enhancements are pushing the limits of ethical research using human subjects. We must exercise more oversight.

Chapter 6 also introduced a revolutionizing scientific gene editing technology known as CRISPR. It is noteworthy that DNI Clapper testified in February 2016 he considers CRISPR a threat posed by WMD proliferation. Although Clapper didn't outline any bioweapons scenarios in his testimony, there is speculation in the science literature that CRISPR could be used to make "killer mosquitoes," plagues that wipe out crops, and a virus that snips at human DNA.[388]

Further, DARPA's super soldier-related research is pushing a transhumanist agenda that could include altering the very genetic makeup of a human. The resulting inhuman being would be a clear breach of ethics and morality and goes against biblical principles. This is the most dangerous area of research and one that other nations are developing now.

Who is regulating DARPA to keep it from going too far?

#3: The Pentagon must be an equal opportunity employer while recognizing the demands of the profession of arms must always put the readiness of the force above individual preferences.

It must also protect religious freedom and never distort the principle of marriage within its ranks.

Three social issues warrant special attention: women in combat, religious freedom, and homosexuality.

Women in Combat: Mr. Obama set a deliberate course to change the very nature of the U.S. military and its relationship with women. It ordered with Congress' silence the lifting of all combat exclusions in early 2016. This decision erodes the military's warrior culture and its ability to defend America. The nation's top military leaders showed moral cowardice in the face of the enemy by failing to speak out against this ideological initiative that harms readiness and troop morale.

Congress was derelict in its duty to establish the laws and policies that govern the military. Rather, it kowtowed to the Obama administration's radical agenda and to feminists and accepted the mass media's illogical formulation of the issue as one of "equal rights."

Putting women in combat will seriously weaken our fighting force, discourage males who are already abandoning the all-volunteer force, encourage sexual improprieties that erode unit cohesion, inflict physical and psychological injury on young women fooled into serving in combat, and ensure that eighteen-year-old females will be subject to the draft just like men. America is being deceived by the highest levels of its government.

The debate about sending women into combat raises other questions, however. Both sides avoid them, but they are perhaps the most important questions. What kind of society sends its women into combat? Do we want to be that kind of society? Is sending our daughters, wives, and mothers into combat good for women, for men, or for children? To a certain kind of feminist, hardened by ideology, it's repugnant even to ask these questions. We don't ask if we should send *men* into combat, so we shouldn't ask if we should send *women*.

People who have blinded themselves to the profound and wonderful differences between the sexes are not open to a discussion about the consequences of those differences, and perhaps there is nothing more to say to them. But those people should not set the terms of the public debate nor should they drive decisions.

The American people need to stop pretending that sending women into combat involves questions no deeper than how far they can carry a seventy-pound rucksack.

Religious Freedom: Military personnel who face serious injury and death while serving our country often have been very committed to their faith in God. That devotion is a serious aspect of their lives and must be protected.

Over the past decade, many Christians like me have noticed a growing hostility to religion within the armed services. We see pressures to be more secular, and the antireligious culture has intensified especially during the Obama administration with the exception of Islam. Muslims are freer than ever to flout their faith within the military community and that's celebrated by the so-called politically correct liberals. However, Christians are targeted for being open about their faith. This has been especially true in the U.S. Air Force.

The Family Research Council chronicled a long list of egregious anti-Christian incidents of religion-based persecution within the military and many were evidently sanctioned by Pentagon authorities. For more information, see FRC's report, "A Clear and Present Danger."[389]

Anti-faith and especially anti-Christian bigotry must stop. The secularists' intimidation that now prevails much of the faithful in the ranks must be replaced with tolerance and not just for Muslims, something President Obama seems to support.

Homosexuality: Mr. Obama pushed to repeal the military's centuries-old homosexual ban and after little resistance from Congress. Then Obama made homosexual marriage entitled to benefits and the administration is pushing for the inclusion of transgendered people and even military-financed operations for sex changes.

I was a member of the 1993 Army Study Group that advised the Congress on the issues associated with open homosexuality in the military. I also advised the DOD committee and the staff director who eventually wrote the now-repealed policy known as "Don't Ask, Don't Tell" (Policy Concerning Homosexuality in the Armed Services, 10 U.S.C. § 654).[390]

Congress failed the men and women in the military by rolling over to President Obama's radical move. In spite of Pentagon reports that found the majority of service members in 140 focus group sessions were against repeal and 67 percent of combatants expressed negative views about the proposed repeal, Congress went along with Obama.

Congress is still constitutionally responsible to reconsider their failure. They need to conduct a true deep dive on the policy change to understand all the ramifications of the policy switch such as spousal benefits, marriage, reinstatement, medical issues associated with homosexuality, impact on recruitment, discipline issues, readiness considerations, homosexual assault within the ranks, and impact on those with moral objections to homosexuality.

It is paradoxical that few people give gender segregation a second thought due to privacy and modesty concerns, but those same people expect the military to force heterosexuals to share facilities with homosexuals. But call for coed sleeping and bathing facilities and the objections are loud. Who says homosexuals are any more perfect than anyone else?

#4: The fight against Islamic terrorists will continue for many years, and until that fight ends, terrorists captured on the battlefield will remain behind bars at our facility in Guantanamo Bay, Cuba.

We must never release terrorist killers who might return to the fight and we must never negotiate with those killers or their sponsoring states.

Keep terrorists locked up: Islamic terrorism will be a serious security challenge for many years. In February 2016, top U.S. intelligence officials testified ISIS will attempt attacks in Europe and the U.S. homeland in 2016 and ISIS was "taking advantage of the torrent of migrants to insert operatives into that flow."[391]

DNI Clapper estimated that violent extremists were active in about forty countries and that there were more terrorist safe havens "than at any time in history." Further, he said, more than 38,200 foreign fighters have traveled to Syria from more than one hundred countries since

2012. Meanwhile, ISIS is the "preeminent terrorist threat," however, "al Qaeda affiliates are positioned to make gains" such as the Yemen-based al Qaeda in the Arabian Peninsula and the Syria-based al Nusra Front the "most capable al Qaeda branches."[392]

We must maintain our facility at Guantanamo Bay given that the war on terrorism is expected to last many years. That facility serves two critical functions: It keeps the terrorist killers off our battlefields, and it keeps America's streets safer. Bringing terrorists to the U.S. as Obama wants and granting them the rights of citizens in hybrid courts based on civilian criminal standards with civilian due process rules is wrong-headed and very dangerous.

#5: America will work with like-minded nations to employ joint action and military preparedness to build trust among friends and send a clear deterrent message to our adversaries.

But force will never be used unless necessary, and only the force necessary should be used according to Just War guidelines.

Building Coalition Partners: The terror attacks on September 11, 2001, and subsequent global war on terror gave rise to DoD's security cooperation efforts with friends and partners under the rubric "building partner capacity," a term of art that came into use in the 2006 Quadrennial Defense Review.

The 2015 NMS appears to link building capable partners both to counterterrorism and alliance/coalition building with durable states and U.S. allies. The document states:

As we look to the future, the U.S. military and its allies and partners will continue to protect and promote shared interests. We will preserve our alliances, expand partnerships, maintain a global stabilizing presence, and conduct training, exercises, security cooperation activities, and military to military engagement. Such activities increase the capabilities and capacity of partners, thereby enhancing our collective ability to deter aggression and defeat extremists.[393]

The key here is working with like-minded nations. The U.S. will leverage its relationships and aid to encourage likeminded allies and partners. Allies like Israel have our trust, and that relationship is enduring. Other friends will continue to enjoy our support as long as the security burdens such as with the North Atlantic Treaty Organization (NATO) equitable.

#6: Our homeland defense efforts will abide by the same values and principles.

We will protect civil liberties while securing our nation from foreign and domestic powers that seek to rob our people of their fruits and threaten their safety.

The Department of Homeland Security (DHS) works closely with other federal and state agencies. Of course, DHS' primary focus is keeping the homeland safe from a host of future threats: terrorism, geopolitical risks such as the loss of control of a nuclear device, cyber-threats, border violations, and significant natural disasters such as Hurricane Katrina or the Fukishema earthquake/tsunami that created a nuclear reactor disaster in northern Japan.

Although each of these threats is potentially serious, our poorly defended border demands immediate action. Not only do many people illegally cross our border with Mexico, but the border is a key highway for illegal drugs and weapons into this country. Recently, U.S. intelligence officials have sounded alarms about the increase in threats from jihadist groups and growing evidence those groups are infiltrating the U.S. from Mexico.

A 2015 Texas Department of Public Safety (DPS) report revealed that illegal immigrants have been working with terrorist groups like East Africa's al-Shahab, an al Qaeda affiliate, to infiltrate personnel into the U.S. via the Mexican border. The DPS report cited 143 land border crossing encounters with watch-listed individuals in southwest border states between November 2013 and July 2014. Further, a leaked Customs and Border Patrol intelligence report indicates illegal aliens from more than seventy-five countries to include some in the Middle East attempted to enter the U.S. from 2010 to 2014.[394]

A Clarion Project report, *U.S. Mexican Border Porous to Jihadists*, cites a number of media reports that indicate Muslims are in Mexico to convert locals to Islam and work with drug cartels to subvert Mexico with the ultimate goal of establishing a base for sabotaging the U.S. Two of the cited media reports indicate Iranian proxy Hezbollah built a network in Tijuana, Mexico. "Over the years, Hezbollah—rich with Iranian oil money and narco-cash—has generated revenue by cozying up with Mexican cartels to smuggle drugs and people into the U.S."[395]

In September 2015 the U.S. Border Patrol nabbed two Pakistani men with terrorism ties at the U.S.-Mexico border. The men, both in their twenties, were caught south of San Diego and just over the international border from Tijuana. The report indicates both men had been previously processed by immigration officials in Panama and then used smuggling networks to sneak into the U.S.[396]

Lawmakers are worried about potential terrorists gaining entry to the U.S. through the Mexican border. "The southern land border remains vulnerable to intrusion and exists as a point of extreme vulnerability," Representative Duncan Hunter (R-CA) wrote to DHS Secretary Jeh Johnson in late December 2015. "The detention of the two Pakistani nationals underscores the fact that any serious effort to secure our homeland must include effective border security and immigration enforcement," Hunter wrote.[397]

The homeland is under assault, and our porous southern border with Mexico is a significant vulnerability that must be resolved. This is a principled security priority that could require U.S. National Guardsmen to supplement Border Patrol personnel.

Begin this effort by hosting "town-hall"-like meetings with law enforcement and Guard personnel from our border states. Ask them to present ideas and solutions, but make it clear the border will be closed to all but authorized traffic.

Finally, we need to give Mexico one chance to clean up the border area of Islamists, gun runners, and drug cartel, and leverage their cooperation with threats to end tourist trade and other economic incentives. Failure by the Mexican government to cooperate could invite our intervention into that land to capture or destroy the havens of criminals and Islamists.

Conclusion

The U.S. faces many serious security threats which demand immediate attention. The consequences of not acting quickly could be devastating given the global trend of instability and the plethora of dangerous threats and risks.

The above biblical worldview-based security strategy includes many priorities common across recent administrations with the exception of the ethical and social issues and a much stronger push for tougher action on several fronts, especially along our mostly undefended borders.

Chapter 12

National Intelligence Strategy for the Twenty-first Century

FEW AMERICANS understand their national intelligence community (IC), much less how vital it is to keeping the nation safe. If we were to ask the average American whether he or she believes that "big brother"—the government's IC—is monitoring their every move, they would likely agree and they might be correct!

National security organizations and missions described in the last chapter depend on the IC to serve as their eyes and ears in this unstable, nonpolar dangerous world. Before outlining a principled, biblically based worldview of an IC strategy, we will first take a whirlwind tour of that community.

The tour begins with a brief history of American intelligence, which is followed by a frightening view of the threats facing America today from the pen of the Director of National Intelligence (DNI). That is followed by an introduction to the IC—what it is, does and how it collects intelligence. Next we examine how intelligence-collecting can and has gone wrong, and the chapter ends with a fresh IC strategy.

History and Maturation of the American Intelligence Community

The nation's security demands an effective intelligence capability to anticipate threats and then to provide the national defense enterprise information to defeat those threats. Fortunately, throughout the history of America, the republic has enjoyed a range of improving intelligence capabilities that dates back to the Revolutionary War when patriot Paul Revere was a member of a secret surveillance committee, "The Sons of Liberty," that patrolled Boston streets looking for signs of impending British raids against Americans. Then on April 7, 1775, when British army activity suggested the possibility of troop movements, Revere alerted the colonial militia to the approach of British forces before the battles of Lexington and Concord.[398]

Colonel Elias Dayton fought alongside General George Washington during the American Revolution. Dayton wrote a letter to Washington on July 26, 1777, explaining the importance of intelligence to the war effort. Dayton wrote:

> The necessity of procuring good Intelligence is apparent & need not be further urged--all that remains for me to add, is, that you keep the whole matter as secret as possible. For upon Secrecy, Success depends in most Enterprises of the kind, and for want of it, they are generally defeated, however well planned and promising a favorable issue.[399]

The American Civil War (1861–65) had agents spying for both sides and the Union Army first used aerial intelligence when Thaddeus S. Lowe pioneered the use of hot air balloons for reconnaissance to track the movement of Confederate forces.[400] Intelligence-gathering became more sophisticated with technological advances such as code-breaking (cryptography) in World War II. The Cold War pitted the American and Soviet intelligence agencies against each other, which included the creation of the National Security Agency (NSA) to track Soviet nuclear developments.

The terrorist attacks on September 11, 2001, were a wake-up call for America's intelligence community. Those attacks led to the creation of a host of oversight committees, courts, boards, and the Department of Homeland Security. But even those changes evidently didn't go far enough, as evidenced by a raft of new and disturbing developments.

Over the last few years, we saw a tsunami of scary developments regarding our intelligence capabilities. Some of the disturbing and otherwise eye-opening IC capabilities and shortfalls include the maturation of the Central Intelligence Agency's drone-killing capability, which included the killing of an American-turned-terrorist in Yemen; the American contractor Edward Snowden's leaks about the NSA's surveillance activities at home and overseas; the controversies over CIA detention and interrogation practices like waterboarding; the missed analysis of the WMD threat posed by Iraq in 2003; the unanticipated disintegration of Syria; and the evident failure to anticipate the precipitous rise of ISIS. These developments have renewed calls for further reform to limit the IC's threat to civil liberty while equipping it with the tools to deal with a very complex, new nonpolar incredibly dangerous world.

The IC's sophisticated tools for dealing with the emerging threats complicate both our efforts to defend the nation and uphold American civil liberties. President Obama correctly outlined possible threats to civil liberties that accompany the IC's efforts to gather critical intelligence in a speech at the Department of Justice, "Security and Privacy: In Search of a Balance." Specifically, the same digital technological advances that allow U.S. intelligence agencies to find terrorists (email and cellphone calls) also scoop up similar communications by average citizens—thus the rub. Further, the explosion of digital information and powerful computers gives the IC the opportunity to sift through massive data—called data-mining—to look for communication patterns both here at home and abroad. Of course, these actions cannot function without secrecy, which is why the public became alarmed—especially in light of the mass-surveillance revelations by Mr. Snowden.[401]

Americans are rightly skeptical about government intrusion into their private communications, and the Obama administration did increase oversight and auditing of those programs. Those new rules were proposed by the administration and approved by the Foreign Intelligence Surveillance Court, which exercises oversight of the IC's data-mining efforts. However, it is appropriate to ask ourselves whether those actions are sufficient, or perhaps they go too far given the serious threats now facing America.

Director of National Intelligence Outlines the Global Threats

The IC is vital to keeping America safe in a very dangerous world. There is no evidence and only suspicion that our IC has sought to violate the law, but at the same time we must not become cavalier about the civil liberties of Americans.

Our IC does an extraordinarily difficult job, and their daily successes at thwarting dangers facing America must by necessity remain secret. The problem is our IC has to be always right to stop increasingly effective and dangerous adversaries, and too often it is by sheer chance that we piece the intelligence puzzle together in time to thwart an attack. Failure on today's incredibly dangerous global stage can lead to significant black-swan-like catastrophes.

We must appreciate that our threats are very real, and our IC serves a vital role in confronting them—preventing terrorism, cyber threats, and much more. Therefore, we must be careful not to disarm the IC and in fact they likely need more capabilities. Specifically, we need increased intelligence capability because globalization and the Internet make it easier for our enemies to plan and coordinate their activities. Sophisticated technology has erased international borders and empowered individuals as well as terrorist groups to accomplish their goals worldwide. The conundrum is we must simultaneously demand more from our IC and at the same time expect them to never violate our civil liberties.

The DNI provided the U.S. Senate a great assessment of the hydra of threats facing America today. The most serious are profiled in section II of this volume, but it is helpful to consider the DNI's perspective before considering how the DNI does his incredibly difficult job of protecting America.

DNI James Clapper testified before the Senate Select Committee on Intelligence regarding the IC's 2016 assessment of threats facing the U.S. today. That report outlines global and regional perils, but for purposes of this volume we will only review the eight global threats.[402]

#1: Information Technology

The DNI cautions that America's information technology is especially vulnerable and it's going to get worse. "The consequences of innovation and increased reliance on information technology in the next few years on both our society's way of life in general and how we in the intelligence community specifically perform our mission will probably be far greater in scope and impact than ever," Clapper cautions. He continued, "Devices, designed and fielded with minimal security requirements and testing, and an ever-increasing complexity of networks could lead to widespread vulnerabilities in civilian infrastructures and U.S. Government systems."

#2: Terrorism

Terrorism is the DNI's second global threat, and he emphasizes Sunni Islam violence: "Sunni violent extremism [think ISIS and al Qaeda] has been on an upward trajectory since the late 1970s and has more groups, members, and safe havens than at any other point in history." Then the DNI warned, "Shia violent extremists [think Hezbollah] will probably deepen sectarian tensions in response to real and perceived threats from Sunni violent extremists and to advance Iranian influence."

DNI Clapper paid special attention to ISIS, which he said has "become the preeminent terrorist threat because of its self-described caliphate in Syria and Iraq, its branches and emerging branches in other countries, and its increasing ability to direct and inspire attacks against a wide range of targets around the world."

#3: WMD

WMD and their proliferation is the third global threat in the DNI's report. He said, "nation-state efforts to develop or acquire WMD, their delivery systems, or their underlying technologies constitute a major threat to the security of the United States, its deployed troops and allies." He acknowledged the use of chemical weapons in Syria and the dual-use materials and technologies associated with biological and chemical threats move "easily in the globalized economy," as does "the scientific expertise to design and use them."

#4: Space

Space is a global security issue of considerable interest to the DNI. "Changes in the space sector will evolve more quickly in the next few years as innovation becomes more ubiquitous, driven primarily by increased availability of technology and growing private company investment," according to the DNI. He indicated the number of space actors now includes eighty countries and more are expected soon. Space as a new frontier is also making modern services and capabilities less expensive, such as big data analytics and additive manufacturing. Further, the DNI anticipates the rapid expansion of space services such as reconnaissance and communications and for military and intelligence applications.

#5: Counterintelligence

Counterintelligence is a global threat that is possibly one of the least understood roles of the DNI. "We assess that the leading state intelligence threats to U.S. interests will continue to be Russia and China, based on their capabilities, intent, and broad operational scope," Clapper said. Those are not the only intelligence threats, however. He said, "Iranian and Cuban intelligence and security services continue to view the United States as a primary threat." He indicates that foreign intelligence will focus on penetrating and influencing the U.S. national decision-making apparatus and the IC. Also, they will target national security information and proprietary information from American companies.

#6: Transnational Organized Crime

The DNI commits significant resources to tracking transnational organized crime and especially drug trafficking, which "poses a strong and in many cases growing threat to the United States at home and to U.S. security interests abroad." He warns that organized crime is growing more sophisticated, pointing out that Mexican drug traffickers are capitalizing on strong U.S. demand for heroin. The cartels have mastered production of the white heroin preferred in eastern U.S. cities and have boosted heroin's overall potency thirty to fifty times by adding fentanyl. The National Institute on Drug Abuse defines fentanyl as "a powerful synthetic opiate analgesic similar to but more potent than morphine. It is typically used to treat patients with severe pain, or to manage pain after surgery. It is also sometimes used to treat people with chronic pain who are physically tolerant to opiates. It is a schedule II prescription drug."[403]

#7: Economies and Natural Resources

Our IC monitors and provides the U.S. government with analysis regarding global economies and access to natural resources. The DNI indicates "global economic growth will probably remain subdued, in part because of the deceleration of China's economy [in 2016]." While advanced economies grew in 2015, developing economies "saw the first net capital outflows to developed countries since the late 1980s."

#8: Human Security

The DNI's eighth threat is what he labels human security, which includes environmental risks and climate change. These are important to the IC and by association our national authorities because extreme weather, climate change, and environmental degradation are related to human suffering, which is often a spark for instability.

The DNI wrestles with understanding and then forecasting how these and other threats will impact American security interests using the IC's significant capabilities.

Intelligence Community 101 — An Introduction

Now that we reviewed a history of American intelligence and the DNI's assessment of global threats, it is useful to review the IC's organization, mission, and processes.

The need for an office of the Director of National Intelligence dates back to 1955 when a blue-ribbon study commissioned by Congress recommended the Director of Central Intelligence focus on coordinating the overall nation's intelligence effort rather than just the Central Intelligence Agency (CIA). Finally, the attacks of 9/11 and the recommendations of the National Commission on Terrorist Attacks (9/11 Commission) pushed the government to create the DNI. President Bush signed four executive orders in 2004 and Congress amended the National Security Act of 1947, which created the need and structure for the DNI in February 2005.[404]

The DNI leads the IC, overseeing seventeen discrete organizations, which includes the DNI's office as the seventeenth organization. He directs the implementation of the National Intelligence Program and acts as the principal advisor to the president, the National Security Council (NSC), and the Homeland Security Council for Intelligence Matters. The DNI's mission is to integrate foreign, military, and domestic intelligence in defence of the homeland and of U.S. interests abroad. His authorities and duties are outlined in the Intelligence Reform and Terrorist Prevention Act (IRTPA) of 2004.

The IRTPA also established the National Counterterrorism Center, the National Counterproliferation Center, the National Intelligence Center, and the Joint Intelligence Community Council—to protect the American people.

The IC's seventeen federal organizations work separately and together to conduct intelligence activities. Those seventeen, which include the DNI's headquarters organization as the over-watch entity, are:

- Air Force Intelligence
- Army Intelligence
- Central Intelligence Agency
- Coast Guard Intelligence
- Department of the Treasury
- Drug Enforcement Administration
- Federal Bureau of Investigation
- Marine Corps Intelligence

- Defense Intelligence Agency
- Department of Energy
- Department of Homeland Security
- Department of State
- National Geospatial-Intelligence Agency
- National Reconnaissance Office
- National Security Agency
- Navy Intelligence

The DNI and his staff receive a lot of guidance from myriad agencies. The DNI reports to the president and works closely with the NSC and the president's national security adviser. Further, he receives guidance from the Office of Management and Budget regarding budgets in light of the president's policies and priorities. He also receives guidance from two boards: the President's Intelligence Advisory Board (regarding the quality and adequacy of intelligence collection) and the President's Intelligence Oversight Board (that oversees the DNI's compliance with the Constitution, applicable laws, executive orders and presidential directives).

The DNI answers to two congressional intelligence committees by keeping them "fully and currently" informed of the IC's activities. Besides the Senate Select Committee on Intelligence and the House Permanent Select Committee on Intelligence, the DNI also works with congressional committees where there is overlapping jurisdiction such as the Armed Services and Homeland Security Committees.

The IRTPA set in motion the reform of the U.S. IC in the wake of the 9/11 Commission's recommendations. President Bush described the act: "Under this new law, our vast intelligence enterprise will become more unified, coordinated and effective. It will enable us to better do our duty, which is to protect the American people." The Act is divided into eight titles:

- Reform of the intelligence community
- Federal Bureau of Investigation
- Security clearances
- Transportation security
- Border protection, immigration, and visa matters
- Terrorism prevention
- Implementation of 9/11 Commission recommendations
- Other matters

The previous section in this chapter outlined what the IC gathers intelligence against: terrorism, proliferation, WMD, information infrastructure attack, narcotics trafficking, and more. The IC gathers this intelligence across the entire world with an eye on America's security interests.

The IC has six basic intelligence sources, or collection disciplines:

- Signals Intelligence (SIGINT)
- Imagery Intelligence (IMINT)
- Measurement and Signature Intelligence (MASINT)
- Human-Source Intelligence (HUMINT)
- Open-Source Intelligence (OSINT)
- Geospatial Intelligence (GEOINT)

For a full definition of each of these six sources see http://www.dni.gov/index.php/about/faq?start=3.

Finally, the IC says it strives to exhibit three characteristics: integrated (team making the whole greater), agile (adaptive, diverse, continually learning enterprise) and mission-driven (embraces innovation and takes initiative). It also espouses America's values of rule of law, privacy and civil liberties, human rights—all to retain America's trust.

Colored Intelligence

The abuses of intelligence collection garner the most public attention. We saw this in the 1960s when the government spied on civil rights leaders and anti-Vietnam War protesters. The abuse of information in Watergate and the struggle against communism made people at that time rightly concerned about their liberties.

Many American suspicions about the IC were confirmed by a young computer contractor. The 2013 case of Edward Snowden, a computer professional and former CIA employee, copied classified information from the NSA and leaked it to journalists who then published the material in the *Guardian*, the *Washington Post*, *der Spiegel*, and the *New York Times*. Snowden is now at an undisclosed location in Russia seeking long-term asylum, while in the U.S. he is sought on charges of espionage and theft of government property.[405]

Snowden is characterized by people based on their perception of what he did: hero, whistleblower, or traitor. His leaks of NSA documents fueled a groundswell of debate about government's overreaching in the name of intelligence.

Issues of an overreaching government are understandably a hot topic for most Americans, according to Pew Research Center. While Americans support monitoring the communications of suspected terrorists, Americans don't see by a large majority (74 percent), the need to sacrifice civil liberties to be safe from terrorism. Rather, they oppose (54 percent) the government collecting bulk data on its citizens supposedly to help track terrorists, and two-thirds believe there aren't adequate limits on what types of data the government collects. Further, although most (93 percent) Americans want to control their personal information, only a few (9 percent) say they have a lot of control over that information.[406]

These strongly held views are at the center of the 2013 policy earthquake regarding NSA's vacuuming of all telecommunications (email and cellphone) data. Thanks to Snowden, many Americans came to believe and still believe the NSA has a secret "back door" into its vast databases to search U.S. citizens' email and phone records without a warrant. Unfortunately, that interpretation was encouraged by U.S. Senator Ron Wyden (D-OR) in an interview with the London *Guardian*. Wyden said the NSA data-mining program is a loophole allowing "warrantless searches for the phone calls or emails of law-abiding Americans."[407]

The data to which Wyden refers is gathered under Section 702 of the Foreign Intelligence Surveillance Act (FISA) Amendments Act (FAA), which gives NSA authority to target without warrants foreign targets—read terrorists—communications and communications of Americans in contact with foreign targets. The NSA acknowledges that purely domestic communications can also be inadvertently swept into its databases, known as "incidental collection."

Shortly after the Snowden story blew up in the media, Dianne Feinstein (D-CA), the ranking member on the Senate Intelligence Committee, said she believed the IC and the Justice Department were sufficiently mindful of Americans' privacy. However, she continued, "The intelligence community is strictly prohibited from using Section 702 to target a U.S. person, which must at all times be carried out pursuant to an individualized court order based upon probable cause."

Mr. Obama jumped into the data-mining debate to promise to right the wrongs exposed by Snowden. Specifically, Obama promised to reform programs and procedures to provide greater transparency to our surveillance activities. Specifically, his review resulted in the declassification of various opinions of the Foreign Intelligence Surveillance Court including the Section 702 program that targets foreign individuals overseas and the Section 215 telephone meta-data program. The president promised to personally review future court opinions that involve privacy issues.

Why should we care if government is monitoring our communications? One writer explained that digital liberty is currently defined by lawmakers. Does anyone believe lawmakers are going to really make it harder for government to monitor citizen communications? Our silence is consent and that ultimately will rob us of our privacy, something we will unlikely ever recapture.[408]

A lot can be done with access to our private communications. Data-mining—a powerful tech tool—allows the analyst to know about every online, cell, or other call moment. Already, Internet consumers understand the concept because of "cookies" and pop-ups, nuisances that track our Internet search interests and interrupt our routine online activities with unsolicited advertisements. Search for an item once, and for days you will be spammed with pop-ups offering similar items. Our habits, likes, friends, favorite locations, and every view or communication creates our profile and grants outsiders insights into our private lives.

Although Obama said "nobody is listening to your telephone calls," the fact is that programs like Prism, xKeyscore, and Boundlessinformant collect content and meta-data of our communications. Federal rules (outlined above) allow the NSA to target anyone within two or three degrees of a suspect—read "suspected terrorist." So, even if in a casual way you talk to a "suspect" you know or unknown, you are subject to being monitored as well.

One of the more embarrassing revelations from the Snowden leaks is the fact that we were spying on Ally leaders like German Chancellor Angela Merkel. That revelation prompted Germany's domestic intelligence agency to expand its counterintelligence operations to include friendly countries like the U.S. as well.

Obama didn't deny the allegation that the NSA was monitoring foreign leaders. He put a spin on his answer: "And the leaders of our close friends and allies deserve to know that if I want to learn what they think about an issue, I will pick up the phone and call them, rather than turning to surveillance." So why did he authorize spying on Merkel?

The president said his new presidential directive clearly indicates what can and can't be done relative to overseas surveillance that applies to foreign leaders. He promised, "I have made clear to the intelligence community that—unless there is a compelling national security purpose—we will not monitor the communications of heads of state and government of our close friends and allies." The president has total control over the IC, but he still hasn't explained why he directed the spying on Merkel. Let there be absolutely no doubt, spying on a foreign head of state is done at the president's direction.

While pulling the IC back from some overseas spying, Obama also promised: "Our intelligence agencies will continue to gather information about the intentions of governments—as opposed to ordinary citizens—around the world… [And] we will not apologize simply because our services may be more effective. But heads of state and government with whom we work closely, and on whose cooperation we depend, should feel confident that we are treating them as real partners. The changes I've ordered will do just that."

We must not be naïve, however. Everyone spies on everyone else. Every embassy in Washington is manned by trained spies who regularly report to their seniors at home. Every federal agency welcomes foreign guests into their facilities and in some, like the Pentagon, we even allow foreign military officers to work on our staffs where they enjoy great access to sensitive information.

Allies that feign disbelief know better. They monitor us and would do more if we didn't make it so difficult. They also eavesdrop on their own citizens as much or more than the U.S. monitors its citizens.

Our northern neighbor Canada collects personal information on its citizens via a program named Communications Security Establishment Canada, operated by NSA's counterpart in Ottawa. It also collects metadata similar to NSA's PRISM (surveillance) program.[409]

Even the world's largest democracy, India, snoops on its people. In 2008, India's parliament passed the Information Technology Act, making it a crime to publish obscenities or make it legal "to search premises without warrants and arrest individuals in violation of the act." That government monitors such behavior and follows with enforcement.

The president is personally responsible for the IC's actions, which include collecting meta-data and spying on allies. Clearly, there are times when this might be justified, and we obviously are not alone among the world's democracies in performing such collection. However, the average citizen has good reason to be suspicious that "big brother" is watching our every move.

A Fresh Strategy for the Intelligence Community

In 2014, President Obama spoke at the Justice Department, where he acknowledged the tension between secretly gathering useful intelligence and protecting civil liberties. He said:

> The task before us now is greater than simply repairing the damage done to our operations [in the wake of the Snowden leaks]; or preventing more disclosures from taking place in the future. Instead, we have to make some important decisions about how to protect ourselves and sustain our leadership in the world, while upholding the civil liberties and privacy protections that our ideals—and our Constitution—require.[410]

Mr. Obama used that forum to introduce reforms that his administration adopted administratively and passed to Congress to codify. These are important to understand before outlining a new strategy, because we need to take them into account.

First, he approved a directive on signals intelligence that strengthened executive branch oversight. He promised it will take into account our security requirements and those of our allies, trade partners, American companies, and basic liberties.

Second, he promised to reform programs and procedures to provide greater transparency to our surveillance activities. This was mentioned earlier in terms of restrictions on Section 702 (targeting foreign individuals) and Section 215 (meta-data programs). He also promised more effort to declassify court cases and to keep Congress in the loop.

Third, he promised other restrictions on Section 702 activities such as the government's ability to retain, search, and use in criminal cases, communications between Americans and foreign citizens.

Fourth, he promised to make the secrecy on the FBI's use of National Security Letters less indefinite. (A National Security Letter is a U.S. government-issued administrative subpoena intended to gather national security information.[411]) That will help the public to understand how the government uses this authority and whether it is abusing the letters that require companies to provide specific information without disclosing the order to the suspect.

Finally, Obama addressed the bulk collection of telephone records under Section 215. He claims the program does not include the content of phone calls, or the names of people making calls. Rather, it provides a record of phone numbers and times and lengths of calls—meta-data that can be used to link a telephone number to a known or suspected terrorist.

Obama explained that the meta-data program was established after the 9/11 attacks, but had it existed prior to those events, it could have been very helpful at preventing the attacks. For example, Khalid al-Mihdhar, one of the 9/11 hijackers, called from San Diego to an al Qaeda safe house in Yemen, according to Obama. Although the NSA had a record of the terrorist's call, it could not determine that the call originated in San Diego. The meta-data program under Section 215 now allows the NSA to see who a terrorist calls, which is a piece of valuable information in a crisis.

Clearly, much more needs to be done to protect our civil liberties while keeping America safe. Future administrations need to follow-through on these efforts and then get tougher when necessary.

Beyond picking up Obama's IC legacy future presidents must implement a strategy that supports the following biblical worldview vision.

Each paragraph of the three-part IC strategy is followed by specific goals.

#1: The seventeen-member U.S. intelligence community is the nation's eyes and ears to protect us from enemies overseas and at home. At least two biblical principles must apply to these agencies: protect the citizenry and limit government's role.

The IC is too big, complicated, and inefficient. It needs to be trimmed.

Earlier it was pointed out the IC is comprised of seventeen federal agencies assigned an array of missions—national defense, foreign relations, homeland security, and law enforcement. Each of these agencies sprawls out across an enormous enterprise that touches local, state, national, and international audiences, and as a result, there are significant, expensive redundancies.

The agencies have different communications platforms and portals including the National Counterterrorism Center, the Joint Counterterrorism Assessment Team, seventy-one FBI Joint Terrorism Task Forces, fifty-six Field Intelligence Groups, and seventy-eight state and local intelligence fusion centers, which can incorporate military and private sector participants.[412]

The IC has many redundant information sharing systems. The largest is the Pentagon's Secret Internet Protocol Router Network (SIPRnet), but there are many others: the U.S. Navy's Law Enforcement Information Exchange (LInX); the Department of Homeland Security Information Network (HSIN); the Director of National Intelligence's Information Sharing Environment (ISE); and the FBI's eGuardian, National Data Exchange (N-DEx); National Crime Information Center (NCIC); and Law Enforcement Online (LEO), and there are others.[413]

FBI and Department of Homeland Security officials operate several private sector intelligence sharing organizations as well, including the Domestic Security Advisory Council, InfraGard, and the National Cyber Forensics and Training Alliance.[414]

The IC also has close working relationships with international partners, including the governments of the United Kingdom, Canada, Australia, and New Zealand under the "five eyes" agreement. They share intelligence with other nations such as Israel and Saudi Arabia through memoranda of understanding, or other less formal agreements.[415]

The IC needs to be better controlled by the executive but within limits. The IC is adept at empire building and largely unconstrained by the political executive, according to Professor Samuel Rascoff, an intelligence expert writing in the *Harvard Law Review*. He argues that too much separation between the IC and policy "impedes the purpose behind intelligence—to generate useful and relevant insights for the policymakers."[416]

Paul Pillar writes in *Foreign Affairs* about the Iraq WMD fiasco, which illustrates Rascoff's point. That situation evidences how political pressure shapes—distorts—intelligence estimates.

Jennifer Sims expressed a contrary view to Pillar's perspective in *National Intelligence*. She wrote:

> U.S. intelligence officers often do not seem to believe they are working on behalf of policy makers or as part of their team. They tend to see themselves as a check on an administration's power and the repository of truth in a system riddled with biases.... Although policy makers do want intelligence to provide facts or "ground-truth," other branches of government have the job of checking the power of those in office, not intelligence.[417]

This loose-cannon perception has as a result led to significant diplomatic blowback and jeopardized the bottom lines of some American industries.[418] Rascoff suggests greater numbers of political appointments in the IC agencies, which entails the assertion of presidential control and better centralized review from the White House "to ensure that important bureaucratic decisions are made, or at least overseen and monitored, by presidential agents." Proceed with caution on this advice, however. Such an approach could create a dilemma as well, because much like explained in the previous chapter, vis-à-vis the Pentagon politicos, political appointees with limited intelligence background can run the risk of undermining truthful analysis.[419]

Further, newcomer politicos to the intelligence business and especially those who have a distorted perspective influenced by Hollywood's exaggerations can be especially dangerous for America. They seldom have the judgment to sit in a secure location and watch special operators on live feed perform masterful actions and then keep their tongue. America's national security and the very lives of our brave special operators are at risk. Therefore, future administrations who assign inexperienced people to intelligence over-watch positions must ensure those appointees at least have the maturity to listen to the experienced personnel rather than direct some insane action that is untethered from reality but fits the administration's political agenda.

In 2015 there were allegations swirling about that intelligence on America's progress against ISIS was "manipulated" to fit President Obama's policy.[420] Those allegations led to investigations that at this writing have not been made public.

#2: The intelligence community must have the means to identify threats and risks to America whether they are homegrown terrorists, killer satellites in outer space, nuclear-tipped missiles in the hands of rogue states or cyber-attacks from foreign hackers.

The IC has the budget and should have the means for all the capability it needs but, as outlined above, much is being wasted on duplications across the seventeen IC agencies. The president runs the IC and controls its budget via the OMB, which is divided into two components: National Intelligence Program (NIP), managed by the DNI and

Military Intelligence Program (MIP), which is managed by the Secretary of Defense. Not surprisingly, the intelligence budgeting processes remain opaque and involve bureaucratic sleights-of-hand like "reprogramming." It's past time for an administration to realign the IC's budget. The fiscal year 2015 NIP had $50.4 billion and the MIP had $16.6 billion.[421]

Back to the basics—IC needs better human intelligence: Technology cannot gather information like a well-trained human spy. Unfortunately, the IC seems captivated by technology to the point that we've significantly degraded the art of human spying.

Human intelligence (HUMINT) is critical if the IC is to meet the demands and expectations of the political leadership and ultimately the American public. This was evident especially in the lead-up to the 2003 Iraq invasion. The vast majority of the IC's efforts leading up to that invasion were focused on technical intelligence (e.g., SIGINT) and had the IC put in place trusted spies much of the unsubstantiated information about WMD, the reliability of the Iraqi army and Saddam Hussein's inter-circle might have saved our going to war.

No technical intelligence collection system can replace the understanding and verification of human intelligence. It takes many years to train HUMINT personnel and years to recruit officials in countries we need to penetrate. Those efforts cannot be turned on and off like a light switch in the budget world; it requires long-term thinking and investment.

#3: These legitimate intelligence tasks must not become an excuse for an expansion of the government's powers to intrude into the lives of law-abiding citizens.

They must never spy on our citizens, target them when they travel abroad, or target our allied leaders.

At the beginning of this section, we reviewed some of President Obama's changes intended to protect civil liberties without disarming the IC. Those must be vigorously monitored and reinforced to make certain spying on our citizens and allied leaders doesn't reoccur.

America depends on capitalistic competition, and the IC must not undermine business unless it is absolutely necessary. The economic impact on American firms being closely identified with the IC is estimated between $22 and $180b.[422] For example, the Snowden revelations allegedly persuaded the German government to transfer an important contract from Verizon to Deutsche Telekom.

Another significant threat is turning the globalized economy away from the globalized Internet. Google chairman Eric Schmidt said the aggressive NSA could lead foreign governments to "eventually going to say, we want our own Internet in our country because we want it to work our way, and we don't want the NSA and these other people in it."[423] Evidence of this concern is echoed by Google's chief legal officer, David Drummond, who said his company was "outraged" that the U.S. government would have intercepted data from Google's private networks.[424]

American spying has created a nearly impenetrable alternative and aligns the U.S. government with industry. FBI Director James Comey admits as much. However, he says the pendulum has now swung too far in the direction of protecting privacy. Specifically, Comey said the recent push toward encryption is very problematic and he warned that "Apple and Google have the power to upend the rule of law." The Obama administration promises not to try to legislate "back doors" to defeat encryption by the likes of Apple but that's an issue that hampers the IC from exploiting enemy networks.[425]

The "back-door" encryption issue came to the forefront in February 2016 with the FBI's request for Apple to unlock an iPhone used by one of the two attackers who killed fourteen people in San Bernardino, California, in December 2015. The Justice Department requested and gained a court order directing Apple to unlock the iPhone and provide the FBI with the requested information.

"This Apple case really goes right to the heart of the encryption issue," said Ira Rubinstein with the New York University Information Law Institute, "and in some ways, this was a fight that was inevitable."[426]

Apple argued the mechanism to satisfy the FBI's request does not exist but experts indicate the company can comply if it so chooses. However, Apple CEO Timothy Cook initially rejected the FBI's request. The back-and-forth between Apple and the government could eventually force the creation of legislation that grants government the means of compelling the surrender of terrorist-related information.

Companies other than just Apple and Google are reacting to the IC's data mining with a form of commercial "self-help" by employing default encryption on mobile devices, according to Rascoff. This is also a business issue, because if the global marketplace demands technology to defeat government surveillance, and if American firms can't or won't provide that capability, then others outside the U.S. will get the business.[427]

The IC must maintain a good working relationship with our best allies' intelligence services. This is true for the most part in spite of the huffing about Snowden in the press. In fact, Snowden's interviews demonstrate a close relationship among the allies' intelligence services.

Snowden gave an interview in which he said that Germany's intelligence services are "in bed" with the NSA, and that's true for most other Western countries.

Snowden's statements are echoed by intelligence experts and historians who agree the latest leaks reflect the complicated relationship among the allies. "The other services don't ask us where our information is from and we don't ask them," Mr. Snowden said in the German magazine *Der Spiegel.* "This way they can protect their political leaders from backlash, if it should become public how massively the private spheres of people around the globe are being violated." Not surprisingly, Mr. Snowden's statement is off the mark. The fact is the IC in its cooperation with foreign partners has ways of communicating the reliability of the raw information and ways of indirectly communicating the identity of its sources.

Britain is America's closest intelligence ally in Europe, which is implicated in Snowden's leaks especially regarding his claim the British enjoy access to the PRISM computer network, which shows American Internet company data. The *Guardian* suggests there is sort of a quid pro quo that benefits the IC by which the British surveillance center tapped fiber-optic cables carrying telephone and Internet traffic that is shared with the NSA.[428]

Conclusion

The IC is hard to understand mostly because it is so secretive out of necessity. However, that doesn't give it license to run roughshod over American civil liberties. Americans have every right to expect it to operate within the law to gather the intelligence needed to keep us safe.

This chapter is a primer for those new to understanding our IC and then it outlines a clear way ahead for those charged with the responsibility of keeping America safe in an incredibly dangerous non-polar world.

Chapter 13

Foreign Policy Strategy for the Twenty-first Century

THE AMERICAN PEOPLE tend to rebound to a foreign policy extreme after an overassertive presidency. They were exhausted by President George W. Bush's wars in Iraq and Afghanistan, so they chose the opposite—an apologetic, nonassertive, and "lead-from-behind" President Barack Obama. There was a similar pattern with the transfer of power from the 1970s President Richard Nixon/President Gerald Ford era to that of a feckless President Jimmy Carter.

President John F. Kennedy nailed the importance of keeping a consistently strong foreign policy in spite of our national apprehension about overseas assertiveness. In his inaugural address, Kennedy said, "Domestic policy can only defeat us; foreign policy can kill us." [429] The twenty-first century forecast outlined in section I of this volume demands a well-defined and strong foreign policy to protect America from the nonpolar, incredibly dangerous future.

This chapter builds on the previous strategies on national security and intelligence. It defines foreign policy and how it works, outlines policy drivers and its heart, compares recent presidential foreign policy programs and how they have become dysfunctional, and then advances a foreign policy for our future America from a biblically based worldview.

Foreign Policy and How It Is Supposed to Work

A country's foreign policy consists of strategies aligned with critical national interests intended to achieve goals within the international community. That policy is designed by the president and executed by his administration with the support of the Congress. Those goals are generally accomplished through peaceful cooperation across many disciplines, or, when under duress, they can also be accomplished through coercive means like military action.[430]

Sometimes, the U.S. exercises its foreign policy via humanitarian aid. It did this with North Korea and a number of countries in war-torn Africa hoping to spur the rise of democratic ideals. However, too often, aid alone in the face of tyranny fails and dictators like those who have ruled North Korea for sixty years maintain a tight grip on power.

The U.S. employed a foreign policy guided by the strategy of "containment" during the Cold War. That policy was intended to keep communism from spreading beyond the Soviet sphere. Our policy goals were to advance American national security and to promote world peace by working with likeminded nations. That policy eventually succeeded, thanks to President Reagan's stalwartness, with the fall of the former Soviet Union in 1991.

The U.S. president has considerable help crafting and executing his administration's foreign policy. The primary players tend to be his national security adviser, the National Security Council (NSC), the secretary of state, and more often than not, the secretary of defense. Although the main objective of foreign policy is to use diplomacy to solve international problems, the threat of other action such as using military force always looms in the background as a fallback option.

The president and his team shape foreign policy by working with

Congress, foreign partners, and outside advisers. The secretary of state is the chief coordinator of the government's actions that affect relations with other nations. According to our Constitution, the president leads those efforts to establish treaties with other nations, after which he must seek the "advice and consent" of the Senate.

The secretary of the state supervises political appointee ambassadors and the Foreign Service personnel who serve as the president's representatives in more than 294 physical embassies, consulates, and diplomatic missions across the world.[431] At these facilities, the ambassadors and Foreign Service officers represent American interests and maintain relationships with those countries.

Foreign Policy Driver

Most Americans are close to total ignorance about the world. They are ignorant. That is an unhealthy condition in a country in which foreign policy has to be endorsed by the people if it is to be pursued. And it makes it much more difficult for any president to pursue an intelligent policy that does justice to the complexity of the world.[432]

—ZBIGNIEW BRZEZIŃSKI, National Security Advisor to President Carter

Most Americans pay attention to foreign policy only when there is a major international incident like a terrorist attack. Otherwise, as Zbigniew Brzezinski said above, they remain oblivious to world events. Further, it's noteworthy, as pointed out earlier in this chapter, that American voters run hot and cold on foreign policy. They revolted against Bush's activist foreign policy to elect Obama, and even though Obama's policies earned consistently low marks among voters, he was still reelected in 2012.

The U.S. experiences such wild swings in its foreign policies because there is a tendency to root the foreign policy in the incumbent president's worldview. We explored this worldview in chapter 10 and won't elaborate here except to cite an important perspective from John Hillen, the chairman of *National Review* and a former assistant secretary of state under President George W. Bush.

Mr. Hillen wrote, "If the central organizing principle for the foreign policy of a great power is based on a political philosophy or worldview that is coherent but not widely held over the long term by a vast majority of the body politic, it will falter with time—especially if it requires sacrifices and offers only a long-term payoff."[433] That is a cautionary note to this volume, as this author outlines a national strategy based on a biblical worldview, which won't be popular among many and likely will require sacrifice.

Hillen goes on to suggest that America, a diverse representative democracy, will likely lack the staying power of "any ideological or philosophical basis" for a foreign policy. Rather, he proposes that America needs more of a "consistent, smartly managed, prudent, and unapologetic exertion of American power and leadership than any particular political philosophy or perspective on human nature [worldview]."[434]

He elaborates on the substance of this geopolitical lens. Specifically, the geopolitician must consider "not only geography and natural resources, but also demographics, education, technology, culture, leadership, economics, and other elements of the competitive landscape." It is noteworthy that Hillen's foreign policy algorithm fails to mention religion or faith, an important ingredient for a significant minority in America.[435]

A 2012 study in *Social Science Quarterly* found the "faith factor" is a powerful force driving American attitudes about foreign policy. "Specifically, seculars, mainline Protestants, and Catholics variously stand out as more moderate…when compared to evangelical Protestants." Further, "Evangelicals and Jews are ardent supporters of Israel who *sometimes* endorse hawkish foreign policy tools, up to and including the use of force, in the Greater Middle East."[436]

Juxtapose these faith-factor study results with Hillen's argument that foreign policies "born of a strong philosophical worldview" are fatally conceited. Then Hillen explains that "ideologues cannot understand why everyone is not getting with the program—their program." This is a fair critique of a Christian/biblical worldview strategy—those who don't share a Christian worldview won't support a foreign-policy strategy based on a biblical worldview very long-term, if ever. However, although Hillen's view is worth noting, it is of little consequence if in fact Christians believe the best governing strategy for America is based on a biblical worldview.

Although we can disagree with Hillen's proposition that a biblical worldview isn't an appropriate driver for America's foreign policy, we should seriously consider his exhortation to reboot our foreign policy. His four "reboot" proposals are instructive, and if applied within a biblical worldview framework, they have merit.

First, he suggests "policymakers [should]...avoid the seduction of trendy conceptions" and "fully appreciate the shifts in world geopolitics that are constantly reshaping opportunities and challenges." It is rational to consider the trends, shifts, potential game changers and future black swans like those outlined in section I of this volume when formulating America's foreign policy.

Second, Hillen suggests policymakers must not be so conceited to assume that "other parts of the world must be viewing things as we view them." This is a pragmatic reminder that much of the world does not understand much less embrace democracy or a biblical worldview. However, if our biblical values and principles are universally appropriate, they must be advanced by our policies no matter what others may think. Further, it is not necessary to announce to the world that our government's policies will be based on a biblical perspective. Policy drives us and our actions, but the world wants to see our actions first.

Third, Hillen's geopolitical, steady-state approach to foreign policy promises to keep the nation from "whipsawing back and forth between various philosophical views of how the world 'should' work." That's a prudent position if faith were irrelevant. However, faith is critical to mankind and "whipsawing," although a regrettable outcome of competing ideologies in elections is not justification for failing to advance biblically sound foreign policies.

Finally, "a sober geopolitical perspective allows for management of the world's toughest policy problems in digestible pieces." This is good advice for any policymaker. It answers the question: How does one eat an elephant? Answer: One bite at a time. The same is true for massive geopolitical problems. Our foreign policy must advance over time, winning one building block at a time.

Geopolitical secularists like Hillen will dismiss those of us who embrace and apply a biblical worldview to foreign policy. However, the secularist ignores the primary motivating factor in the lives of the majority of human kind, their faith and that's a major blind spot for many when formulating a nation's foreign policies.

The Heart of a Foreign Policy

The heart of a nation's foreign policy is its interests, which in turn guide strategy. Unfortunately, the Obama administration appears to not have well-defined interests to guide its foreign policy. Consider the mess it created in the Middle East in the wake of the Arab Spring. One misstep after another resulted in chaos in Libya, Syria, Yemen, and Obama's withdrawal of all forces from Iraq in December 2011 significantly contributed to the rise of ISIS. Why? Arguably, President Obama failed to clearly define our interests and then apply those to our foreign policy. What resulted was a series of serious foreign policy failures that are still having a negative ripple effect on our homeland security.

Noted American diplomat and political scientist George Kennan advised President Harry Truman on the importance of clearly defining America's vital interests. That effort, Kennan explained, will form the standard by which to evaluate threats. Presidents need a standard to sort out the critical from the peripheral threats, because in foreign policy, decision-makers need "a clearly delineated set of preferences when evaluating possible foreign policy decisions," which is an insight shared by two contemporary military science professors.[437]

Those professors write in *Space & Defense* that "the lack of a clearly defined set of national interests has made it difficult to develop effective foreign policy; a central question must be addressed; what are the nation's vital interests?"[438]

The nation's vital interests shift with administrations. President George Washington identified a vital interest in his farewell address to the nation on September 17, 1796. Specifically, he made interstate commerce a vital interest but warned otherwise to "avoid the entangling alliances of Europe."[439] Washington's foreign policy was clear: embrace commercial internationalism but avoid foreign military intervention.

Post-World War II presidents made the containment of the Soviet Union a vital national interest, which led to big budgets for an enormous nuclear arsenal and for the forward stationing of 535,000 troops until the end of the Cold War. Once the Iron Curtain crumbled, President Bill Clinton cashed in our Cold War victory by redefining our national interest as a combination of commercial internationalism and the spread of democracy and international institutions.

President George W. Bush took a page out of Clinton's foreign policy to spread democracy and to make the world safer, but used coercive means to attain that objective in Iraq and Afghanistan. Those efforts are failing today because Mr. Obama's follow-on war policies were wrongheaded. Specifically, he gave-up on Iraq and withdrew our forces, which unquestionably led to a rapid escalation of sectarian fighting. Further, in Afghanistan, there was no justification for surging fresh forces into that backward country because our vital interests were keeping al Qaeda from reestablishing a base of operations from which to launch further and not, as it became evident under Bush and later Obama, building a democratic ally in Central Asia.

The issue of "national interests" must be considered in light of the previous section in this chapter, foreign policy drivers. National interests must not be defined by the whims of politics and they must transcend short-term political objectives. Further, as Dennis Drew and Donald Snow explain in *Making Twenty-First-Century Strategy*, national interests come in three levels: vital, major, and peripheral.

Vital interests have two basic characteristics, according to Drew and Snow. A vital interest is any action that might result in an unacceptable outcome and that might legitimately trigger a war. Other scholars refer to this as a concern for which "blood and treasure" are worth expending.[440]

Less than vital interests are major and peripheral interests that do not require the nation to resort to blows. One definition of these concerns is they involve a situation in which "a country's political, economic, or social well-being may be adversely affected but where the use of armed force is deemed excessive to avoid adverse outcomes."[441] These interests are more often defended using diplomatic and economic tools.

This distinction is important especially when something horrible like genocide is happening elsewhere in the world and the nation's leaders must decide what to do. If the genocide doesn't trigger a vital national interest but a peripheral concern, then the decision-makers must weigh the costs and benefits of alternative actions. For example, does the Christian genocide in the Middle East at the hands of ISIS reach to the "vital" national interest level to warrant a significant military response, thus resulting in the deployment of military capabilities in order to destroy the terrorist group?

If Christian genocide in the Middle East at the hands of ISIS is not a "vital" national interest, then at the very least we should provide all the humanitarian aid possible, as well as put diplomatic pressure on our so-called regional partners to assist, arm Christians in the region to at least defend themselves, and increase the opportunity for Christians to emigrate from that region.

If the answer is "yes," stopping Christian genocide in the Middle East is a "vital" national interest, then we should deploy sufficient American forces to eliminate the threat and compel partner countries to join the fight. Then we should restore the remnants of the Christian community in that land through humanitarian aid and diplomatic pressure to ensure they are protected in the future.

Adam Lowther and Casey Lucius identify six vital American interests in their *Space & Defense* article, "Identifying America's Vital Interests." They provide a reasoned rationale why each is enduring and are identified below without explanation.

1. Trade and economic prosperity
2. Energy supply
3. Freedom of the seas
4. Space access
5. Cybersecurity
6. Homeland security

These "vital American interests" intuitively fit the criteria of worthy of "blood and treasure" to defend. However, other concerns might rise to that level as well. For example, the U.S. has a stake in sustaining an economic relationship with China but also in checking China's evident hegemonic actions across much of Asia and the Pacific Rim. Are we willing to go to blows with China should the communist giant block America's use of the South China Sea or it attacks Japan? Perhaps, and China, for a number of reasons, becomes an interest in itself because it potentially jeopardizes a number of vital national capabilities such as our cyber and space security.

Russian intimidation and bullying of its former satellites rises to some level of U.S. national interest. Russia's reemergence as a nuclear threat with its new possible first-use nuclear doctrine and its growing investment in new nuclear-capable platforms potentially threaten vital interest areas such as our homeland, trade with Eastern European partners, and of course, Russia's cyber-attacks against American infrastructures are well documented.

Finally, the most difficult potential "vital interest" to consider is the promotion of democracy. In chapter 4, Henry Kissinger is quoted as stating, "The search for world order has long been defined almost exclusively by the concepts of Western societies." Kissinger explained that, for America, world order is identified with the spread of liberty and democracy. Further and in agreement with Kissinger, Mark Lagon with the Council on Foreign Relations points out that one of the "few truly robust findings in international relations is that established democracies never go to war with one another."[442]

Lagon states that promoting democracy as a foreign policy goal increases stability, enhances economic development, grants more access to justice, and supports pluralistic outcomes most Americans support. Historically, the U.S. deepened and widened democracy in Europe and Japan after the Second World War, and similarly after the Cold War, Western-style democracy was encouraged and then blossomed in Eastern Europe.[443]

Alternatively, "Foreign policy 'realists,'" argues Lagon, "advocate working with other governments on the basis of interests, irrespective of character, and suggest that this approach best preserves stability in the world."[444]

President Roosevelt was a foreign policy "realist" when he quipped about Nicaraguan dictator Anastasia Somoza: "He may be a bastard, but he's our bastard." Similarly, President Obama showed his "realist" stripes when he stiff-armed Iran's democracy-promoting Green Movement in 2009, "presumably to keep a door open for dialogue on Iran's nuclear program." But then two years later, Obama favored democracy as a goal that brought to power the radical Muslim Brotherhood after ousting a long-time authoritarian ally, Egyptian President Hosni Mubarak.[445]

Going to war to win other peoples' right to democracy is a very tough decision, much different than going to war to protect our homeland. However, future American governments must weigh in the balance whether the potential outcomes of a new democracy in our troubled world is worth the sacrifice in blood and treasure as opposed to working with the world's "bastards" who happen to serve our purposes.

The Gatestone Institute points out that there is a serious downside to America's recent push for democracy in the Middle East. Specifically, "In every Muslim nation where the U.S. has intervened in the name of 'freedom and democracy,' Christian life has exponentially worsened." Further, "empowering forces hostile to Christians is synonymous with globally empowering forces hostile to America," according to Gatestone.[446]

Clearly, and this won't be popular, sometimes America must act on the Christian principle of justice, and that means we use our power and expend our blood and treasure to help others earn the freedoms we enjoy in America. However, we must exercise wisdom when choosing the venue to promote democracy—especially if doing so empowers the rise of likes of Islamic forces bent on destroying America.

Americans should demand that transparency be a vital part of foreign policy. That includes the future administration's definition of America's vital security interests. What interests are tomorrow's leaders willing to shed American "blood and treasure" to defend?

Foreign Policies Compared and How Policy Is Really Executed

Republican presidents from the 1950s through the early 1990s left behind favorable legacies on foreign policy, argues Daniel Drezner, a professor of international politics at Fletcher School of Law and Diplomacy, Tufts University. Eisenhower stabilized relations with the Soviets; Nixon opened relations with China and extricated us from Vietnam; Reagan called the Soviets an "evil empire" but then paved the way to the peaceful end to the Cold War; and George H. W. Bush extended America's liberal order to the world and ejected Saddam Hussein from Iraq without expanding the fight.[447]

These presidents were tough, but ably combined principled beliefs with prudence and flexibility, said Drezner—and they far outshone the Democrat Presidents Lyndon Johnson and Jimmy Carter. Why? Drezner says they relied on a string of steady-handed professionals, such as John Foster Dulles, Henry Kissinger, George Shultz, James Baker, and Brent Scowcroft.[448]

Unfortunately, America hasn't been as blessed with similar wise leaders and advisors since the 1990s. Consider the Bush and Obama administrations to illustrate why America's foreign policy seems so unpredictable and arguably unwise today. The two administrations demonstrate very different interests and therefore foreign policies.

President Bush was a take-charge leader who, in the wake of the terrorist attacks on September 11, 2001, saw his popular approval rating skyrocket to 90 percent as he quickly defined the threat and attacked the terrorist enemy. Soon he took the fight to terrorist leader Osama bin Laden in Afghanistan. But then some of his advisors argued to broaden the conflict to include Iraq, which Bush initially tabled.[449]

The 9/11 attacks inspired President Bush's foreign policy, which he summarized as: "Our war on terror begins with al Qaeda, but it does not end there. It will not end until every terrorist group of global reach has been found, stopped, and defeated." In his 2002 State of the Union Address, President Bush declared that the United States considered any nation that supported terrorist groups a hostile regime, and he called out an "Axis of Evil" consisting of North Korea, Iran, and Iraq, declaring them a threat to American security. This declaration was the genesis of what became known as the three-part Bush Doctrine: (1) preventive war to strike enemies before they attack America; (2) U.S. will act alone to defend itself; and (3) the U.S. will seek to spread democracy and freedom around the world.

The Bush administration soon applied that doctrine in a war of choice against the Iraqi dictator, Saddam Hussein. Iraq was a war of choice because no vital American interest was threatened with the possible exception that certain members of Bush's national security team argued with insufficient proof that Iraq posed a WMD threat. Also, there were some Bush advisers who wanted to develop Iraq into a democratic country friendly to the U.S. interests.

Eventually, Bush took his case for war to Congress, and that body consented by passing a resolution authorizing the president to go to war with Iraq. Then Bush convinced the United Nations on the dangers of WMDs in the hands of a murderous dictator, arguing that it was riskier for the world not to act. Tragically, America led a coalition into a war based on faulty intelligence. It was later confirmed that Iraq had actually disposed of its WMD stockpiles.

After the 2003 invasion, Bush's plans to stabilize Iraq proved to be just as bad as his WMD intelligence. Soon the administration dismantled the Iraqi army, which seeded the rise of sectarian violence between the two dominant strains of Islam in Iraq: Shia and Sunni. By mid 2006, a full-blown civil war raged across Iraq.

Once the situation in Iraq became desperate, and against popular opinion at home, President Bush authorized a surge of fresh troops under the leadership of Army General David Petraeus. The general brought to the fight an appreciation for counter-insurgency that slowly won the trust of the disenfranchised Sunni population in the Anbar Province. By the end of 2008, the surge paid off and casualties radically declined as sectarian tensions were contained and stability returned to that nation.

Bush's post-surge policy became known as "return on success," which meant as stability grew, more troops would return home. In 2008, Bush entered into a Strategic Framework Agreement that promised a comprehensive relationship with the new Iraqi government, and he signed a Status of Forces Agreement that outlined a plan to withdraw all forces by the end of 2011, providing stability continued.

On other fronts, Bush's Doctrine stirred controversy, especially with regard to his security and surveillance actions. The USA Patriot Act expanded domestic security and surveillance to disrupt terrorist funding and increase the efficiency within the American IC, a serious shortfall leading up to the 9/11 attacks. In late 2001, Bush signed a military order establishing tribunals to try non-U.S. citizens fighting for al Qaeda and to hold the accused terrorists at Guantanamo Bay in Cuba, without a right to a writ of habeas corpus. Those held at Guantanamo Bay were also classified as unlawful enemy combatants instead of prisoners of war to keep the Geneva Conventions at arm's length.

At that time, President Bush also authorized legal memos that justified enhanced interrogation for terrorists accused of fighting against the United States. Those interrogation methods, which included waterboarding (simulated drowning), were used in order to extract information—acts critics labeled as torture.

A far less controversial aspect of President Bush's foreign policy was his use of foreign aid, which he said was used "to do our part to help relieve poverty and despair." He radically increased America's aid to fight HIV/ AIDS and malaria. His 2003 initiative, the President's Emergency Plan for AIDS Relief (PEPFAR), was aggressive, well-funded, and made a significant difference by testing, counseling, and treating millions in Africa.[450]

President Obama was elected in 2008 as the un-Bush, who promised to end wars and take the country in a radically different direction. However, unlike Bush, Obama evaded hard decisions and was allergic to military force although he seemed to like to use killer drones against terrorists in places like Yemen and Pakistan. Further, although Obama on occasion sounded like Bush, his actions seldom matched his words. "This is a pattern," a retired four-star general said. "He issues stern warnings [like Syria must not use chemical weapons against its citizens], then does nothing. It damages American credibility."[451]

Obama never articulated a foreign policy doctrine or clear interests as did Bush. However, Obama promised during his 2008 election campaign to withdraw American troops as soon as possible from Iraq, which he did, and that contributed to the emergence of ISIS. Then he followed through with another campaign promise to increase U.S. operations in Afghanistan by sending an additional twenty-one thousand troops to that country, raising our military presence there to about sixty thousand. However, he failed to articulate our vital national interests with that directive.

Obama soon became convinced that our Afghanistan strategy still wasn't working. Then in December 2009, he accepted the advice of his

Afghanistan commander, General Stanley McChrystal, to send a surge of thirty-three thousand fresh troops into the fight but with the promise to withdraw them in July 2011. Since that time, Obama has withdrawn most of the troops, but he still keeps delaying a final departure as that war continues to rage with no end in sight.

On other fronts, Obama's lack of a clear strategy is evident across the Middle East as the Arab Spring created new foreign policy challenges, and Obama demonstrated no inclination to resolve them. His actions and inactions tended to create more chaos and distrust.

Obama's tepid response to the Arab Spring in Libya, characterized as "leading from behind," left that country in chaos and it now plays host to an ISIS province and serves as the staging area for Islamists flooding elsewhere into the Sahara. Similarly, in Yemen, Obama's efforts backfired as civil war raged, and our formerly good ally Egypt is in crisis and forced by Obama's poor decisions into the arms of a welcoming Russia. Then Syria deteriorated beginning in 2011 in spite of Obama's secretaries of state (Clinton and Kerry) who promised that dictator Bashir al-Assad was a "reformer"[452] and "someone we can deal with."[453]

Obama's decision to abandon Iraq in 2011 unquestionably contributed to the rise of ISIS and the spread of that cancerous terrorist movement across much of the Middle East, North Africa—thirteen countries in all—and in late 2015 to Europe via a tsunami of refugees infiltrated by ISIS extremists. However, history will demonstrate that Obama's Iran policy was the worst blunder of his "leading-from-behind" foreign policy, and as a result, gave a pass to that radical regime to continue its nuclear program and terrorist ways.

It was bad that Obama's premature withdrawal from Iraq contributed to ISIS' emergence, but arguably just as bad was Obama's very slow response to ISIS' actions. At first Obama dismissed the group by labeling it "the JV team." Then as the ISIS threat metastasized, Obama launched pinprick-like air strikes, started to help retrain Iraqi forces to retake lands lost to ISIS, and train so-called Syrian rebels who promised to fight ISIS as well.

On other international fronts, Obama's foreign policies backfired as well. His efforts to jump-start relations with Russia led to a New START nuclear arms agreement; however, eventually Russia took advantage of Obama's naiveté. Specifically, Russian President Vladimir Putin annexed Crimea without a shot and then seeded a revolution in eastern Ukraine, but that drew only more tough talk from Obama, little meaningful action. Today, I argue, Russia is in a very good geopolitical position because Obama consistently failed to push back against Putin's aggression.

The situation with China is arguably worse under Obama as well. Chinese actions against our ally Japan and its failure to restrain its puppet North Korea make Northeast Asia a powder keg. Further, China built islands hundreds of miles from its mainland in the South China Sea and then claimed that entire sea its sovereign territory—an obvious assault not that different from Russia's annexation of Crimea. What did Obama do? He threatened, but once again took no meaningful action.

The other piece of the American foreign policy equation is the process used to carry out that executive direction. President Bush relied on his war cabinet to set the course of his overseas strategies, holding most of the key decisions to a close-knit group of advisors. Obama's approach is very different and worth examining in some detail to understand why his foreign engagement is in such disarray.

Susan Rice is Obama's national security adviser, a Rhodes Scholar with a history of political and think-tank advocacy work. She micromanages foreign and national security policy and runs the largest ever NSC staff—nearly four hundred compared to just fifty under President George H. W. Bush. Her foreign policy decision-making style is characterized as "sclerotic at best, constipated at worse," said a defense official who works with the NSC.[454]

The Obama NSC operates like a free agent and accepts little help from the balance of the administration. For example, Deputy National Security Adviser Benjamin Rhodes handled secret talks that led to announcing an opening to Cuba, and even the secretary of state was unaware. Former CIA Director and Defense Secretary Leon Panetta complained that before he left the Obama administration in 2013, there was an "increasing centralization of power at the White House [NSC]" and a "penchant for control." Former Defense Secretary Robert Gates echoed that view to say that "micromanagement" by the NSC "drove me crazy." Others said the NSC undervalued their expertise and experience and subjected non-White House administration officials "to the whims of less knowledgeable NSC staffers."[455]

One senior state department official complained, "Any little twerp from the NSC can call a meeting and set the agenda." The problem with the NSC is too much process and no one will make a decision. One official said, "There's too much airing of every agency's view and recommendations, and not enough adjudicating.... Someone's got to be the decision-maker, who's just going to say, 'We're going to do this' and 'We're not going to do that.'"[456]

The dysfunctional NSC directly contributes to Obama's failed foreign policy. "It used to be that State ran foreign policy," said a former White House official to a *Washington Post* reporter. "Now, everyone's got a hand in it. Go around the table, and they've all got equities, they've all got personnel out in the field, and all that needs to be managed."[457]

This is so typical of Washington bureaucracies. I've worked in Washington with the Pentagon, State Department, White House, and many congressional staffs and offices. Washington bureaucrats are decision-adverse and much like think-tanks, of which I've been a part, they love to examine issues from every possible angle. What Washington lacks and doesn't like are people who make tough decisions and force action. That's why Bush, like him or not, bothered much of the Washington establishment and why Obama is so popular among the government's bureaucrats.

A Foreign Policy from a Biblically-based Worldview

A biblically-based worldview foreign policy is consistent with the six "vital" American interests identified by Lowther and Lucius in an earlier section: trade and economic prosperity, energy supply, freedom of the seas, space access, cyber security and homeland security. However, as you will see across the six-part foreign policy vision that follows, there are other vital and major interests that warrant immediate attention and require a biblical worldview.

#1: America's dealings with other nations will reflect our Christian values and principles.

Our conduct of foreign affairs is not about popularity, but about protecting the security of our citizens at home and abroad, and defending our vital interests.

The first part of the foreign policy vision encompasses the "vital" interests outlined above. The difference here is in implementation, which must reflect biblical principles such as punishing evil and acting justly while protecting our citizens. This approach must permeate every aspect of America's foreign policy which, for much of our history, has been the case. That's why America has a global reputation as a compassionate country—reflected in a quote on the Statue of Liberty: "Give me your tired, your poor, your huddled masses yearning to breathe free."[458] It's also why World War II British Prime Minister Winston Churchill said, "You can always count on Americans to do the right thing—after they've tried everything else."[459]

The Obama administration seems intimidated by China and Russia. These international bullies must be shown a tough-minded U.S. that's willing to stand up for its interests and, given those countries' recent behavior, a taste of a reassertive U.S.—and that can't happen too soon. For example, consider the fact that Moscow is using the Syrian civil war to reset its competition with the West. Russian President Vladimir Putin is a skilled opportunist who is exploiting the Syrian war to push a flood of Mideast refugees into Europe in order to increase Europe's pain, a useful lever for Russia to divide the European Continent, impact the continuing Russian sanctions, and respond to Europe's movement of troops to Russia's border.

This realpolitik gamesmanship will continue until the U.S. pushes back against Moscow. One way is to hurt Russia in its Muslim-majority North Caucasus by fueling a new Chechen war via secret military aid and helping radical Islamists in the Russian homeland. This is the same thing the CIA did in the early 1980s via Pakistan among the Taliban in Afghanistan. Eventually Moscow abandoned that war, and arguably this effort contributed to the downfall of the former Soviet Union.

Similarly, China's naval intimidation in the East and South China Seas must be met with strength. That nation must not be permitted to exercise sovereign rights over international seaways vital to American commerce and certainly key to our Asian allies.

The inevitable confrontation among major players—U.S., Russia, and China—in the new nonpolar world will be vicious at least until the U.S. reestablishes itself as a tough player that refuses to be pushed around.

Next, America must protect those persecuted for their faith.

#2: A key American value is religious freedom.

We will not demonstrate support or friendship to any country that discriminates against Christians and Jews. We expect tangible changes in their behavior before any favorable action.

Religious freedom should be considered at least a major or possibly a vital American interest because it is a life-and-death issue for those being persecuted and could come to our shores if not stopped elsewhere.

As a critical aspect of our stand on religious freedom, U.S. foreign policy will explicitly embrace Israel as a "best friend" and guarantee its security. Further, the U.S. will make clear in no uncertain terms that any nation that is hostile to the Jewish nation will be treated as an enemy of the United States.

The pervasive anti-Semitism evident in the Middle East vis-à-vis the treatment of Israel is joined by a region-wide anti-Christian discrimination, with the worst evidenced by genocide at the hands of ISIS.

I recommended the following four foreign policy actions in my 2015 book, *Never Submit: Will the Extermination of Christians Get Worse Before it Gets Better?*[460]

First, the State Department and by association all American embassy staffs across the world will make religious freedom part of every foreign government engagement.

Todd Nettleton with Voice of the Martyrs said, "The biggest thing [the Department of] State can do is make [religious freedom] part of the agenda all the time." It is important that the U.S. be seen talking about religious freedom so other nations understand it is our priority.

Retired U.S. Army Lieutenant General William G. "Jerry" Boykin and former Undersecretary of Defense for Intelligence under President Bush expects the State Department to do more than talk. He wants the American government to pressure Middle East governments to pass laws that protect Christians from persecution, and he would use diplomatic and economic pressure, withholding foreign military sales as well as all other forms of U.S. aid to compel those governments.[461]

Second, the U.S. is a party to the 1948 Convention on the Prevention and Punishment of the Crime of Genocide. Christians should push the U.S. government to declare the violence against Middle East Christians as genocide and do the same elsewhere where and when people are targeted solely because of their faith.

It took the U.S. forty years to finally ratify the genocide convention. The Genocide Convention Implementation Act of 1987 binds the U.S. to the provisions of the 1948 convention. Specifically, if a party to the convention (like the U.S.) determines that genocide is occurring such as what is now happening with Christians in the Middle East, then that party is obligated to undertake appropriate actions to prevent it and to punish the guilty parties. See the governing statute 22 U.S.C. § 8213 at the endnote.[462]

The United States Commission on International Religious Freedom's (USCIRF) 2015 annual report summarized the assault ISIS has launched against Christians—and other religious minorities—in Syria and Iraq. That report states, "ISIS issued an ultimatum that all Christians must convert to Islam, leave Mosul, pay a tax, or face death. The Christian community in Mosul dates back more than 1,700 years, with an estimated 30,000 living there before the ISIS offensive."[463]

On December 7, 2015, the commission released a statement: "USCIRF calls on the U.S. government to designate the Christian, Yazidi, Shi'a, Turkmen, and Shabak communities of Iraq and Syria as victims of genocide by ISIS."

Finally, on March 17, 2016, after a year and a half of terrible atrocities against Christians and other religious sects at the hands of ISIS, the Obama administration admitted the obvious. "The fact is Da'esh [ISIS] kills Christians because they are Christians, Yazidis because they are Yazidis, Shias because they are Shias," Secretary of State John Kerry admitted. "We will do all we can to see that the perpetrators are held accountable."[464]

Now that the Obama administration bowed to pressure and the overwhelming evidence of genocide, it is obligated to take action. Unfortunately, given the administration's past record, meaningful action is unlikely and the genocide will continue.

Third, in 1998, Congress passed and then President Bill Clinton signed the International Religious Freedom Act (IRFA), which strengthened U.S. religious freedom advocacy on "behalf of individuals persecuted in foreign countries on account of religion." The act established an ambassador-at-large for international religious freedom, a commission on international religious freedom, and an advisor on such matters to the national Security Council. The major flaw in the act is the lack of teeth.[465]

Title IV of the act outlines presidential actions available to promote religious freedom and if necessary to punish countries that engage in or tolerate violations. Unfortunately, taking action against violators is left to the discretion of the president, and in the case of President Obama, he seldom if ever uses the act to advance religious freedom anywhere in the world.

A future president should use the act to punish violators, and the 114th Congress prepared language to strengthen the act. Congressman Chris Smith (R-NJ) is leading the effort to amend the act to "advance religious freedom globally through enhanced diplomacy, training, counterterrorism, and foreign assistance efforts, and through stronger and more flexible political responses to religious freedom violations and violent extremism worldwide." Christians should encourage their congressional delegations to support a tougher IRFA. Once amended, the IRFA won't go far enough, according to some of the nation's leading Christian aid leaders, however.

Franklin Graham with Samaritan's Purse said we must adjust our attitude toward some of the worst violators of religious freedom like Saudi Arabia and Iran. He said, "The Saudis are supposedly our friends [but] they are friends who kill all Christians and they destroy all Christian churches."

General Boykin would pull out all stops on the Saudis because of their anti-Christian persecution: use economic penalties, trade agreements, and leverage foreign military sales.

Tom Farr, who worked for years on religious freedom inside the State Department, said the only way to get Middle Eastern governments to protect Christians is to use "an interest based approach." State must use "not just carrots and sticks," Farr explained, but demonstrate to the Middle East regimes that "to have religious freedom is to have a stable democracy, sustained economic growth and security." He believes such arguments might entice countries like Egypt to "opt for stable democracy," which begins by granting full equality to Egypt's large Coptic Christian population.

#3: America is a nation of immigrants but newcomers must abide by our laws to immigrate and they must assimilate by becoming citizens that embrace our language and culture.

The 2016 presidential campaign put America's immigration policy in front of the voter. This issue touches two vital American interests—security and trade—and therefore must be considered a vital issue for our future foreign policy and by association must also reflect our commitment to religious freedom as outlined above.

Today roughly 13 percent of the U.S. population is foreign-born; that's up from 4.7 percent since 1970. The post-1965 immigration surge contributed more than 60 million people to our now 325 million population, and the Pew Research Center anticipates that new immigrants will account for 88 percent of America's population growth over the next fifty years. That startling figure is due to the slowed birth rate among native-born Americans and because, logically, immigrants account for a much larger part of our emerging demographics. Further, non-Hispanic whites will become a minority of all Americans in the coming decades.[466]

The flow of immigrants to America is also a security risk in part because some have terrorist leanings. They must be stopped, but the Obama administration seems unwilling to plug the gaps and in fact appears to be encouraging the flow.

Then-Texas Governor Rick Perry warned Obama in a letter that our open-borders approach to immigration enforcement invited an evident unprecedented surge of illegal immigration, a view shared by U.S. District Judge Andrew Hanen, who wrote that "[the government] has simply chosen not to enforce the United States' border-security laws."[467]

The border-related security crisis was addressed in chapter 11 to include how suspected terror-related personnel were pouring into America. Evidently, our efforts over the last few decades have failed to stop the undocumented flood across our borders, which results in an illegal and poor, uneducated under class willing to perform menial jobs, but now, with a stagnant economy, robs jobs from the underemployed citizens.

Therefore, the immigration issue rises to become a vital interest that warrants immediate and comprehensive action. We must secure our border to stop illegal entry and hold accountable those who legally cross to abide by our laws. For those who are in America illegally but can demonstrate they are crime-free and contributing to America's betterment, then Congress and future administrations must chart a path to citizenship that insists they assimilate. However, there must be stipulations that failure to register will result in permanent expulsion.

The other part of immigration is the nearly one hundred thousand people the U.S. government invites into America annually through State Department programs such as those who seek exile from war-torn regions such as the Middle East. The Obama administration relies upon the United Nations to initially vet would-be Mideast immigrants living in refugee camps in countries like Turkey, Lebanon, and Jordan. Unfortunately, the many thousands of refugee Christians in that region avoid U.N.-run refugee camps because they are targeted inside those camps by Islamists who kill or kidnap them. Therefore, Christians are seldom given the opportunity to immigrate to the U.S. because, at least in the Mideast, they aren't to be found in U.N. camps.

Future administrations must abandon the Obama administration's U.N.-centric immigrant vetting approach and replace it with one that allows more Christians to immigrate. Further, we must only accept those Muslims who clearly demonstrate a willingness to assimilate and demonstrate no support for radical Islamic behaviors such as that evidenced by ISIS. However, if vetting of that population proves too difficult, then we must stop all Islamic immigration until a system is developed that's more reliable. In the meantime, we must help those would-be immigrants where they are to regain the means to a peaceful livelihood and grow democratic processes.

#4: We will be a member of intergovernmental organizations that share our values for life, human rights, the family, religious freedom, and our sovereignty.

We will form coalitions and sign treaties that reflect those same values and conform to our Christian principles.

The United States' interests are not served when it joins organizations that use its dues and good name to promote anti-Christian values. The most obvious example is the United Nations, which funds abortion on demand and homosexuality as a "human right," and routinely fails to hold accountable anti-Christian and terrorist-sponsoring states like Iran. Further, the United Nations castigates our good ally Israel for defending itself against Islamic terrorists and consistently fails to condemn Israel's enemies like Iran that call for the Jewish state's destruction.[468]

America's foreign policy must also include a top-to-bottom reexamination of all treaties and abandon those that demonstrate the benefit is no longer worth the investment. One of the best cases is the North Atlantic Treaty Organization (NATO). The U.S. carries a giant share (up to one quarter) of the alliance's costs, while very few of the twenty-seven other members provide for their own defense, as demonstrated by their failure to satisfy the 2 percent GDP-mandated defense investment.[469] Why should America pay the toll to defend Europe when the vast majority of those countries fail to do their part?

The United States is a member of seventy-nine international organizations, which includes the United Nations and NATO. Do any of these organizations share our values for life, human rights, the family, religious freedom, and our sovereignty? Are the "dues" we pay for membership a wise investment of the taxpayer's money?

The truth is that most of these international organizations don't endorse American, much less Christian, values—and in fact the U.S. more often than not becomes a punching bag for the blowhards inside these organizations who sit on fat salaries funded in part by our tax dollars. The U.S. must establish strict conditions for joining international organizations and exit those that fail to endorse our values.

#5: America will always be a generous nation and will aid the downtrodden.

We do this by meeting the physical and security needs of innocent victims, and we work to end wars that threaten the innocent.

America and Americans are very generous. The U.S. government provided $35 billion in economic aid to over 140 countries in 2014[470] while in the same year the American people gave a record $358 billion to charity. Most (72 percent) of that giving came from individuals.[471]

These government aid figures include numerous international programs. For example, our foreign assistance includes President George W. Bush's initiative on HIV/AIDS, PEPFAR ($3.1 billion in 2014) and the extensive help the U.S. Agency for International Development provides to other countries ($2.7 billion in 2014 for economic development). The Departments of State and Defense provide significant assistance to foreign security partners via programs intended to build capabilities in those security forces to defend themselves and or be prepared to fight as coalition partners with U.S. forces such as the Foreign Military Finance program, which grants Israel more than $4 billion annually.[472] Other monies such as that for Overseas Contingency Operations administered by the Pentagon amount to tens of billions of dollars annually, but doesn't show up as aid, although they help pay for development in places like Afghanistan.[473]

America's foreign aid must continue, but refocus it on those most in need. At the same time, it must be used to encourage freedom of religion and promote our message on respect for other human rights and advance our views on democratic reform.

#6: America will work with other like-minded nations to reduce the spread of HIV and AIDS through abstinence promotion, to stop the trafficking in humans, and to end the scourge of drug addiction.

This is a continuation of programs advanced by the Bush administration in Africa and Southeast Asia like the clinics we run for HIV/AIDS patients in Thailand. That policy was an aggressive helping hand to the sick and included a compassionate message that abstinence (not condom use) is the only true way to prevent the transmission of the deadly AIDS virus.

The trafficking in humans and the scourge of drug addiction that beset millions of Americans are long-term issues that warrant ongoing investment. The Defense Department works closely with the State Department to fight these serious problems, and so must our foreign partners.

Our foreign policy must provide clear incentives and disincentives that encourage all foreign partners and less friendly nations to stop human trafficking and the production and distribution of illegal, dangerous drugs.

Conclusion

The American people are rightly tired of being "whipsawed" by administrations with dysfunctional foreign policies. A biblical worldview foreign policy represents the values of most Americans and will consistently represent our interests across today's nonpolar, dangerous world.

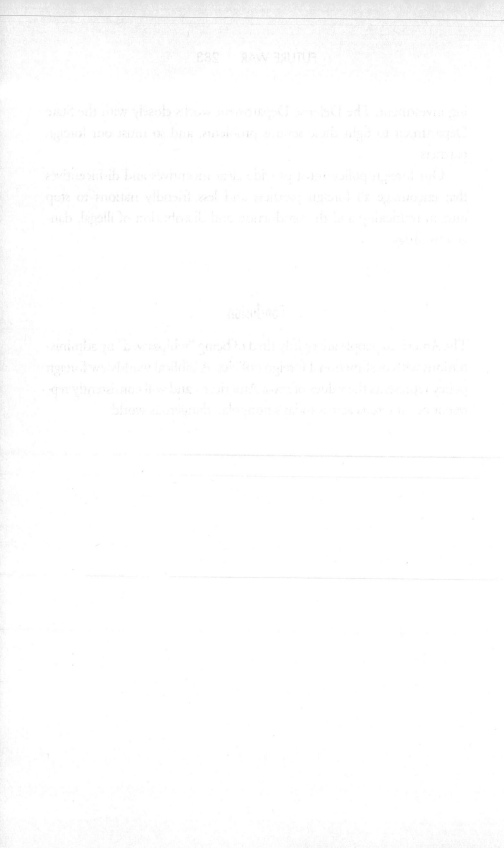

Chapter 14

National Economic Strategy for the Twenty-first Century

THE UNITED STATES is a federal republic with a free enterprise economy based on capitalism. Government and business are inextricably linked economically, which means it is in government's best interest to keep the economy healthy by encouraging a pro-business environment marked by fair trade, low interest rates, tax incentives that encourage business activity, and reasonable personal and corporate tax rates.

This chapter will examine the six-part economic vision outlined in chapter 10. It begins with an examination of a biblical view of economics.

A Biblical View of Economics

The Bible has a lot to say about economics, which is why this chapter reviews that foundation at the start.

David Noebel, the former director and founder of Summit Ministries, provides an outstanding introduction to Christian economics in his book *Understanding the Times: The Collision of Today's Competing*

Worldviews. Noebel holds to the view, as does this author, that the Bible encourages a capitalistic economic system of private property and individual responsibilities (Isaiah 65:21–22; Jeremiah 32:43–4; Ephesians 4:28).[474]

"The Bible as a whole supports an economic system that respects private property and the work ethic," Noebel wrote. He illustrates that conclusion with Proverbs 31, Isaiah 65:21–22 and Acts 5:1–4. Then he quotes author Rodney Stark, who defines capitalism as "an economic system wherein privately owned, relatively well-organized, and stable firms pursue complex commercial activities within a relatively free (unregulated) market, taking a systematic, long-term approach to investing and reinvesting wealth (directly or indirectly) in productive activities involving a hired workforce, and guided by anticipated and actual returns."[475]

Stark makes the case that capitalism traces its origin to early Christian monasteries that promoted property rights, free markets, free labor, management, and a strong work ethic. Further, the Bible teaches about private property, good stewardship, and that "our right to own property stems from our duty to work." This is backed up by Genesis 3:17–19, in which Adam and Eve were cast out of the Garden of Eden to scrape out a life from hard work.

The fruit we acquire from our labor is to be used wisely. That means that as God prospers us, we are to use that prosperity for His glory because everything we have really belongs to God (Psalm 24:1). Further, as we understand from the parable of the talents (Matthew 25), we are but stewards of God's property, and it's our responsibility to use it by serving others. In that parable, the two servants who made a profit were rewarded while a third who hid his master's money was punished.

The Bible instructs us to reward the worker but not the one who is lazy (Proverbs 10:4; Luke 10:7). This instruction applies to the capitalist system whereby competition leads to fruitfulness—work hard and you can get ahead economically.

Noebel explains, "Competition encourages cooperation in a capitalist society when we act in accordance with the principle of comparative advantage." We work hard where there is a good chance of success, and that effort benefits not just ourselves but society as a whole.

Therefore, "The Christian worldview embraces a form of democratic capitalism that allows for the peaceful and free exchange of goods and services without fraud, theft, or breach of contract as the biblical view," Noebel writes. He makes three final observations about biblical economics.

- The Bible grants us the right to private property and calls us to be good stewards of our resources.
- A free enterprise system affords the greatest opportunity to steward our resources responsibly by creating wealth and opportunity.
- Competition in a free market system works according to the principle of comparative advantage, which affirms our inherent worth as individuals.

Now let's apply this biblical worldview of economics to a seven-part future national fiscal vision with solutions.

#1: America supports free enterprise and open markets that lead to opportunity for all.

We encourage families to prosper and an economy that provides good-paying jobs, the product of an open market untethered by artificial constraints imposed by government interference.

This is a great goal, but what does it really mean?

Let's make certain we understand this part of the vision by appreciating the economic terms used here before exploring our fiscal problems and some possible solutions.

Free enterprise: The American economy is based on the principles of free enterprise, which is an economic system wherein there are few restrictions on business, and competition is based on the factors of price, quality, and service. Further, free enterprise is based primarily on capitalism, where private ownership involves few restrictions on trade and limited government intervention. The true free-enterprise economic system rewards hard-working people who demonstrate initiative, and it provides a higher standard of living for all while narrowing the gap between the common people and the wealthy.[476]

Our free enterprise system depends on open markets, which means the same rules and conditions are applied to all participants and in which the competitive action of some does not harm the ability of others to compete. Unfair competition means that the gains of some participants are conditional on the losses of others when the gains are made in ways that are illegitimate or unjust.[477]

Open markets: Some U.S. manufacturers say it isn't fair competition when labor is so inexpensive in a trading partner country, which results in a similar yet cheaper competitive product. Too often, that disparity earns calls for tariffs (taxes imposed by the receiving partner's product(s), but such an outcome not only is unfair to the producing partner but also to millions of U.S. consumers who must pay a higher price. This type of "fairness" is dubious. The uncooperative American company ought to either adopt new technologies to cut production costs to become competitive with the foreign partner country or shift to a different product. Neither outcome harms the American consumer, because he/she continues to have access to the least expensive and best product.

Fair competition also fosters innovation. Remember the personal computers in the 1980s? They were very expensive and slow. Today, because of competition, the personal computer price is less expensive by far, and the consumer gets a far superior product in terms of reliability, computing power, and capability. The same is true for cell phones, iPhones (really mini-computers), and even air travel. The options today are marvelous, and the prices are much better all because of fair competition.

A "living wage": We will explore this term in detail in this chapter. However, for now, we translate a "good-paying job" as a "living wage," which is one "a full-time worker would need to earn to support a family above the federal poverty line, ranging from 100 percent to 130 percent of the poverty measurement." Communities vary in terms of a mandated "living wage" from $6.25 in Milwaukee, Wisconsin, to $12 in Santa Cruz, California, and beginning in 2016 Seattle, Washington, and San Francisco, California, bumped up their wage to no less than $13 per hour.[478] Calculations of a "living wage" are imperfect, but typically include: food, health care, clothing, housing, utilities, transportation, education, child care, taxes, telecommunications, and the unforeseen such as lost job and car accident.

Government policies: Government impacts our free enterprise system in at least four ways.[479]

First, the Federal Reserve System (the U.S.'s central banking system) directly influences the prime interest rate (the lowest rate of interest at which money may be borrowed commercially), which, if it drops, can stimulate lending to businesses and consumers. This results in business expansion and consumers increase their discretionary income.

Second, government lowers corporate income taxes to encourage business activity and can also stimulate specific industries by providing incentives to attract consumers to patronize those products. President Obama poured billions of dollars into green energy programs to stimulate that industry. However, many rightly argue that was a waste of the taxpayers' money because green energy isn't ready to compete with fossil fuels.

Third, government changes foreign trade policies, such as tariffs and import quotas, to encourage, reduce or eliminate foreign trade. It can also permit manufacturers to lower production costs by sourcing cheaper materials or outsource labor abroad.

Finally, government uses grant and loan programs to encourage entrepreneurs to use that tax revenue to stimulate business activity, which then creates more jobs. Similarly, the Small Business Administration helps start-up entrepreneurs in a variety of ways such as loans and mentor programs. For example, according to the Government Accountability Office, the Obama administration approved nearly $30 billion for green technology programs. Some of those efforts like the infamous Solyndra stimulus project defaulted.[480]

Now that we understand some of the economic terms, consider some specific economic challenges and possible solutions for future governments.

#2: Americans must have sufficient wages to cover the cost of life essentials: food, clothing, shelter, and medical care.

It is not government's role to provide wages to cover those expenses except in dire situations and only temporarily. Otherwise, government must stay out of the way to allow the capitalistic enterprise to fuel the economy that produces enough well-paying jobs.

Scripture teaches that the worker deserves his wages (Luke 10:7). However, does the worker deserve to be paid a living wage? That's a question for government policymakers and begs the question whether the federal minimum wage should be scaled to deliver a "living wage."

Before answering that pregnant question, we ought to put into perspective the federal minimum wage and further examine the "living wage" as well as explore how much public assistance is already given to the working poor who rely on the federal minimum wage.

The American minimum wage earner is rare in the United States. The U.S. Bureau of Labor Statistics reported in 2014 that 77.2 million workers age sixteen and older were paid at hourly rates, or 58.7 percent of all wage and salary workers. Among those paid by the hour, 1.3 million (1.6 percent) earned the prevailing federal minimum wage of $7.25 per hour and another 1.7 million had wages below the federal minimum, and together these workers made up 3.9 percent (5 percent) of all hourly paid workers.[481]

Many states beginning in 2016 announced much higher minimum wages than that set by the Federal government. The District of Columbia has the highest minimum wage rate at $10.50 per hour and both California and Massachusetts mandate a $10-per-hour rate. Only fourteen states abide by the federal minimum wage rate ($7.25) with a couple caveats (such as Minnesota, which allows small companies to pay the minimum) and but one state (Wyoming) has a rate lower ($5.15).[482]

Who are these minimum- or below-wage earners? They tend to be young and unmarried with no dependents. Specifically, workers under age twenty-five make-up nearly half of those paid the federal minimum wage or less. Employed teenagers (ages sixteen to nineteen) who were paid by the hour are about 15 percent of those who earned the minimum wage or less.[483] Further, most (62 percent) of the minimum-wage earners are female and they are overwhelmingly (77 percent) white. The majority (72 percent) have at least a high school degree.[484] Also, most (64 percent) are part-timers who work less than thirty-five hours a week, according to the Bureau of Labor Statistics. Not surprisingly, the largest group (55 percent) of part-timers work in the leisure and hospitality industry, where tips are a common income supplement. So, many of the minimum-wage part-timers actually earn more than the minimum wage, although it likely comes in as tips, which may or may not be reported, meaning government wage statistics fail to represent the whole truth.

Obviously, the minimum-wage issue affects a very small part of the full-time work force, and of those in that category, only a few are supporting a family on those wages alone, however.

Some economic liberals call for raising the minimum wage because they argue that minimum-wage earners with families can't afford to live without government subsidies. That's true if we are to believe the analysis of groups like Opendatanation Inc. at the Massachusetts Institute of Technology in Boston.

Opendatanation Inc. poses the question: Can an individual or a family live on the minimum wage? Of course, this is almost a moot point because, as demonstrated above, there are very few typical families (defined as two married adults with two children) that fit this category. However, for hypothetical reasons we will pursue that thesis.

Their analysis shows the typical family needs both adults to have two full-time minimum wage jobs each (a seventy-seven-hour work week per adult) to earn a "living wage." Not surprisingly, a single parent with two children can't survive without outside assistance because he/she would have to work the equivalent of three and one-half full-time jobs (139 hours per week) to earn a "living wage" on a minimum wage income.[485]

The study also found that even the two working adult family making minimum wages could only cover 63.7 percent of the "living wage" before taxes in Washington State and 40.2 percent at worst in Hawaii. This means, given these two venues, the minimum wage family falls between the poverty threshold ($24,037 for two working adults and two children in 2014) and the "living wage" ($61,336 before taxes). Therefore, in this hypothetical situation the family's income can't cover basic needs without further assistance.

Further, when both parents are working outside the home, childcare is necessary and expensive (21 percent of their total wages) and housing is the second most expensive item at 20 percent of their wages. That leaves 59 percent for everything else, which falls well short of the "living wage" depending on where the family lives. Of course, the "living wage" varies depending on the part of the country—the north ($66,047) and west ($62,506) are higher than the south ($59,687), and midwest ($58,871).

The Opendatanation Inc. found that more than one-third of families (37.3 percent), nearly nineteen million families across America, earn less than the "living wage," compared to 20.3 percent that are below the poverty line in 2014. Further, over 8.6 million families earn above the poverty line ($24,037) but less than the "living wage," which leaves them ineligible for most public assistance. Opendatanation also logically concludes that families with children and single parents are the most vulnerable.

These facts explain why there is a political effort to radically increase the minimum wage to help the working families that can't make a "living wage" but aren't eligible for assistance. Supporters of raising the minimum wage argue that giving these families more purchasing power will help the economy and help them perhaps reach a living income.

Those opposed to raising the minimum wage argue that doing so will likely keep low-skilled workers like part-time working teenagers out of the labor market, and employers would likely hire fewer workers because of the higher wage costs. Who is right?

The 2014 Congressional Budget Office (CBO) studied the issue and found the effects of raising the federal minimum wage to $10.10 an hour. That study drew two conclusions to support those who oppose the wage hike: raising the minimum wage to $10.10 would reduce employment by about five hundred thousand jobs. That increase results in low-skilled workers not finding jobs because it is too expensive for employers to hire them and those same merchants say if they did hire workers at the higher wage they would have to pass the added costs on to customers thus impacting their bottom line.[486]

The CBO found an upside to raising the minimum wage, however. It would lift nine hundred thousand people out of poverty (still not necessarily a "living wage"), and families with income up to six times the poverty threshold would see increases in real income. The CBO estimates some twenty-five million Americans would benefit from a minimum wage increase.

There is another aspect of the "living wage" issue that must be explored before answering the question: Does every worker deserve a "living wage"? The missing piece to the puzzle is how many Americans receive government assistance, a supplement to their incomes. The U.S. Census Bureau reports that 52.2 million people in the U.S. participate in major means-tested government assistance programs such as Medicaid and the Supplemental Nutrition Assistance Program (the so-called food stamps program).[487]

"Participation in government programs is dynamic," said Shelley Irving, an analyst with the Census Bureau's Social, Economic and Housing Statistics Division. "The Survey of Income and Program Participation show how individuals move in and out of government programs and how long they participate in them." The largest share of participants (43.0 percent) in any of the public assistance programs stayed in the programs between 37 and 48 months.[488]

The demographics of those participating in government assistance programs are revealing. Children are twice as likely as adults to receive benefits. Further, female-householder families had the highest rates (50 percent) of participation in major means-tested programs. Finally, the unemployed accounted for one-third who received means-tested benefits in an average month of 2012 while 17.6 percent of part-time workers and 6.7 percent of full-time workers participated in means-tested programs.

So what should the future U.S. government do regarding wages and helping families to meet their basic needs?

There is a small part of the population that works hard but earns insufficient money and needs outside help. However, there is a much larger part of the population that doesn't work or only works part-time and relies on public assistance for their needs.

Clearly, this is a sticky issue that must be dealt with compassion but also tough love. Like the Scripture says, "The one who is unwilling to work shall not eat" (2 Thessalonians 3:10, NIV).

First, let the free market dictate the wages. Salaries will rise to attract the workers and government mandating a minimum wage is an artificiality that runs contrary to the free enterprise system.

Second, there is a part of our population that will work hard but can't earn a "living wage," and these people warrant assistance. However, they shouldn't become permanent wards of the state but for a to-be-determined period they are granted help from a compassionate citizenry. Job training may be helpful as well as employment search advice.

Finally, most (75 percent) Americans self-identify as Christian. The local church ought to do more to help the truly needy among us. No one in America ought to go to bed hungry and adults able to work must provide for their families.

#3: Our tax structure must encourage individual and corporate investment, and America must remain competitive in the globalized marketplace and bring home those jobs sent overseas because of high taxes.

This is a call for a FairTax system and a corporate tax that keeps jobs here in America. Justice seems to be the scriptural principle that applies to taxes. Most Americans who file personal income taxes believe our tax system is intrusive, complex, and difficult to understand. Specifically, the Internal Revenue Service (IRS) says it takes the average personal tax filer 26.5 hours to complete his or her annual taxes based on the person's understanding of our three-million-word tax code. Further, Americans believe the complexities of the code with six federal income tax brackets ranging from 10 to 35 percent punishes the most productive members of society. Besides, our tax code is riddled with loopholes and biases that favor special interest groups while the rest of us are stuck paying most of the government's bills. Also, there is widespread suspicion that the IRS is politicized, which explains why many informed citizens are more than ready to replace the current federal income tax system with a simple, hassle-free system. A 2015 Rasmussen survey found that 50 percent of likely U.S. voters don't trust the IRS to fairly enforce tax laws.[489]

The variously recommended FairTax plan is one approach that seems to treat everyone justly and one this author supports. The concept is simple: Government generates tax revenue by instituting a national sales tax on most purchases. Businesses would collect the tax at the point of sale and surrender that revenue to the government.[490]

This plan would make the IRS obsolete and your entire paycheck would be tax-free: no payroll taxes, Medicare taxes, and Social Security taxes. State and local governments would continue to collect taxes as before on income, property, and sales, however.

The tax shock occurs when the consumer buys something, because automatically everything one previously bought jumps in price because the tax is now included. A $100 item that really costs only $77 beforehand must now include the $23 tax. There is a silver lining here, however.

Currently, businesses pay sales tax on raw materials they use to create the goods sold, and then those finished goods are taxed a second time. One approach to a FairTax avoids this double taxation because it brings the wholesale cost of the purchase down by not taxing the raw materials used in production. That's a great approach.

The FairTax approach also makes allowance for those not making a living wage by offsetting some or all of their sales tax expenditures based on poverty-level guidelines. That's like raising their salary up to 23 percent, which can eliminate some or all government dependency.

The FairTax plan has numerous advantages and disadvantages. The high income earners are advantaged because only the amount of their income they actually spend gets taxed and the rest can be saved. It also helps investments, because there would be no capital gains tax, and for those saving for retirement, there would be no limits or penalties for early or late withdrawals. Businesses benefit as well because double taxation on raw material expenses is gone.

There are obvious disadvantages to a FairTax system as well. Lower-income Americans would bear the brunt of the tax burden if they live paycheck to paycheck. Of course, many low income persons will spend a larger percentage of their income on necessities than the wealthier, but that's a choice everyone makes, poor or rich.

Another disadvantage is the increased likelihood of tax evasion. Tax evasion will be a problem because there will inevitably be tax-free bartering and trading without taxes both here at home and with merchants in other countries. This system may also decrease overall spending, which will lower the taxes taken in by the government.

The FairTax also comes with a warning. Before enacting such a system, we must repeal the 16th Amendment (the income tax amendment to the U.S. Constitution); otherwise, given Washington's lust for money, we might end up with both an income tax and a national sales tax. This has already happened in Europe, where some governments have a national sales tax on top of their federal income tax.[491]

Let's agree with former British Prime Minister Winston Churchill:

"There is no such thing as a good tax."[492] But for the Christian, Jesus said in Mark 12:17 (NIV), "Give back to Caesar what is Caesar's and to God what is God's." The apostle Paul said in Romans 13:7 (NIV), "Give to everyone what you owe them: if you owe taxes, pay taxes."

The Scripture is clear: We pay our taxes. However, it doesn't say we have to agree with the current tax code, which is backwards and punishes hard work. It needs a total overhaul to make it fair.

America also needs a fair corporate tax rate that encourages businesses to keep jobs here at home. Unfortunately, the U.S. has the highest corporate tax rate (39.1 percent) in the developed world, which costs American workers jobs and higher wages.

The Organization for Economic Cooperation and Development (OECD) indicates that America's 39.1 percent corporate tax rate is significantly higher than the other thirty-three OECD countries, which average 24.8 percent. The OECD is a group of countries that work together to stimulate economic progress and world trade.[493]

The Business Roundtable, a conservative group of chief executive officers of major U.S. corporations, agrees with the OECD. It indicates the average effective corporate tax rate in the U.S. is actually much higher than elsewhere overseas. The Business Roundtable warns that our higher tax rate makes America less competitive for investment, which causes lower job creation and slower wage growth.[494]

The obvious result of high tax rates is that businesses leave for lower rates. They often accomplish "leaving" by merging with a foreign company and then move their headquarters abroad—a concept known as "inversion." Thus any income earned abroad by the new "foreign company" will not be taxed at the higher U.S. rate, which translates into a better corporate bottom line.

A good example of inversion is the American drug company Pfizer Inc., which plans to merge with an Irish company called Allergen PLC and then set up its new office overseas as an Irish company. Why? It will get a new deduction opportunity and won't have to pay hefty U.S. taxes on its foreign-earned money shipped back to the states.[495]

We could keep companies like Pfizer here with lower tax rates, and then profits would go up, wages would increase, and more cash would flow to the government coffers.

America needs a FairTax system and significantly lower corporate taxes.

#4: We will do more than just bring home jobs.

We will take the government off the backs of the American entrepreneur who creates jobs when markets are truly free. America will once again become the land of opportunity.

The Bible speaks about the importance of work and for whom we work. King Solomon wrote in Ecclesiastes 2:24 (NIV): "A person can do nothing better than to eat and drink and find satisfaction in their own toil." Evidently work should bring us enjoyment, and a byproduct of that toil is the fruits of labor which we need to live—food and drink.

There is a more important motivation for our work than just to eat and drink. Colossians 3:23 (NASB) states, "Whatever you do, do your work heartily, as for the Lord rather than for men." Proverbs goes even further to associate working with God's provision: "Commit your work to the Lord, and your plans will be established." The Pulpit Commentary explains this verse: "The plans and deliberations out of which the 'works' sprang shall meet with a happy fulfilment, because they are undertaken according to the will of God, and directed to the end by His guidance (Proverbs 19:21; Psalm 90:17; 1 Corinthians 3:9)."[496]

So, the Christian works first to please God and then to provide for his physical needs. Unfortunately, our economy is starved for jobs even for those who really want to work.

Evidently, the only person in America today who seems happy with our economy is President Obama. However, the rest of us look at the slow growth, flat wages, and a poor job market, and we know something isn't right.

The so-called Great Recession allegedly ended in June 2009, but

10.5 million Americans are still unemployed, and the labor force participation rate remains near a thirty-five-year low.[497] Why is our labor force participation so anemic, and what can we do?

The conservative Heritage Foundation offers a common-sense solution to the jobs crisis: Reduce labor restrictions, cut costly regulations, unleash American resources, and change our economic strategy.

President Obama believes that raising the minimum wage for federal contractors sets an example for private employers. No, Mr. Obama, private employers must compete in an ever-increasingly competitive global marketplace. Alternatively, according to the Heritage Foundation, we ought to help reduce firms' employment costs. Heritage wisely calls for the repeal of the Davis-Bacon Act, which mandates the government to pay construction workers a much higher wage (20 percent higher) than the marketplace demands. Repealing Davis-Bacon "would free up federal resources that could be used to employ more workers, build more infrastructure, and reduce the deficit."[498]

Second, Heritage calls for cutting job-killing regulations. Since Obama came to office, his administration has added 157 major new regulatory burdens such as Obamacare, which cost Americans another $73 billion annually. The nation's job creators, the small business community, cite government regulation as the single most important problem.

Heritage rightly argues that our job problem is not "that too many people are being laid off but that too few people are being hired." That's because our market is plagued with "a lack of new job creation" incentives. Heritage explains, "In the three years prior to the recession, the rate of new hires was 19 percent higher than it has been over the past three years." Reduce the regulatory burden and watch the number of jobs grow.

Unnecessary environmental restrictions restrict job creation as well. The U.S. is wealthy in natural resources that are off-limits thanks to government restrictions. Free up those restrictions and more jobs will flow.

Finally, we need a new job creation strategy. Our anemic labor market recovery can in part be blamed on the explosion of growth in federal spending, debt, regulations, and natural resource restrictions. "Government," as former President Ronald Reagan said, "is the problem."[499]

#5: America will become energy independent.

For decades we have depended on the flow of energy from some of the most unstable regions, and as a result, crises have whipsawed our economy. This will stop, and America will do what is necessary to exploit domestic energy resources to untether our economy from foreign manipulation.

Energy independence is a worthy and attainable goal that will stabilize our economy. For many years, our dependence on foreign oil and especially petroleum from the troubled Mideast has hurt our economy and drawn us into many conflicts. But that is changing.

In 2005, less than 70 percent of the oil consumed by Americans was produced here, but today, thanks to hydraulic fracturing technology, 89 percent of the oil we use comes from American soil. Not surprisingly, in 2005, we imported 60 percent of our fuel, but today it is only 25 percent and rapidly declining.[500]

The U.S. Energy Information Administration (EIA), a U.S. government agency that collects, analyzes, and disseminates energy information, reports that in March 2015, U.S. soil produced 9.69 million barrels per day, which at the time exceeded Saudi Arabia's 9.51 million barrels. Further, the EIA predicts that "the United States becomes a net exporter of energy in 2019."[501]

U.S. oil and gas production may soon exceed our domestic requirements, according to Citi Group's global head of commodities research, Edward Morse. He said in a report that America's energy production is reshaping the global energy industry because our role is morphing from being a big importer to an exporter of energy. This radical change limits the revenues of the producing nations of Organization of Petroleum Exporting Countries (OPEC), the Russian Federation, and West African countries like Nigeria. Not only are the U.S. and Canada increasing output, but so are Iraq and Iran, which contribute to the cheap oil flooding the global market.

The Citi report continues, "OPEC will find it challenging to survive another 60 years, let alone another decade.... But not all the consequences are positive, for when it comes to the geopolitics of energy, the likely outcomes are asymmetric, with clear cut winners and losers" and of course, the U.S. is a big winner.[502]

The new energy rush has very positive outcomes for America. Specifically, oil production is booming in places like Texas and North Dakota, where unemployment is at just 3 percent and holding.

The oil bomb that leads to cheaper power here at home will give the U.S. a new era of industrialization, according to Citi analysts. Expect to see new industries such as auto, chemicals, and steel.

Domestic production means wealth stays here and America's fuel supplies aren't interrupted by weather, political, environmental, and price factors as much. It also shouts "new jobs" for many thousands of Americans.

America's good news has negative consequences for many economies built on energy revenues, however. Some countries will have to stop pumping oil because as the price per barrel declines they can't afford to pump it out of the ground. For example, the break-even levels are $71 for Saudi Arabia and $44 per barrel for Kuwait, but for others like Russia, the break-even level is $117 per barrel of Urals crude. No wonder the Russian gross national product is shrinking with the global decline in oil prices.

The U.S. must take advantage of this energy boom. First, it must build sufficient pipelines to move the oil rather than rely on more long-term and expensive alternatives such as railroad tanker cars and heavy duty trucks. This is especially critical as we wean ourselves from overseas crude shipments. However, energy exports will keep our ports and refineries very busy, which is more good news.

Second, natural gas production is rising as well, and that resource warrants an immediate relook at our infrastructure to support use and export.

It is important to provide a caution here. Hydraulic fracturing becomes unprofitable at about $40 per barrel because the maintenance cost on the wells does not decline with oil prices. Canada's tar sands and United Kingdom's North Sea oil fields also become unprofitable around $30 per barrel and $50 per barrel respectively. Therefore, realization of energy independence depends in part on a global market price of at least $40 per barrel.[503]

#6: America believes in fair competition.

Countries that want to trade with the United States will provide open tax-free markets to American goods.

Earlier in this chapter, free enterprise was defined as "an economic system where there are few restrictions on business where competition is based on the factors of price, quality, and service." Unfortunately, the global marketplace is marred by unfair competition, and the U.S. should do its best to level the economic playing field for the American worker.

China is a major unfair trade partner, according to the European Union (EU). Evidently, Chinese steel companies sell their excess product in Europe for less than it costs to produce and export, thus suffocating rivals, and compelled the EU to slap tariffs on Chinese steel. China produces half of the world's steel, more than the U.S., the EU, Russia, and Japan combined.[504]

Why is that? China has a lot of cheap steel because of its slowing construction boom. In 2015, China's demand for steel dropped 3.5 percent and, while its plants continue to produce steel, China will continue to dump that excess product on the global market at uncompetitive low prices according to the World Steel Association. The impact is that steelmakers around the world are shuttering plants, and according to European officialism as many as forty thousand steel jobs have been lost across the EU.

The U.S. at this writing is considering tacking a 236 percent tariff on corrosive-resistant Chinese steel, according to the U.S. Department of Commerce. Evidently, as the EU learned, five Chinese exporters got subsidies of 236 percent, which explains the odd tariff figure.[505]

Chinese dumping is a pervasive problem. In early 2016 and for the second time, Titan Tire International and the United Steelworkers Union sought relief from heavily subsidized, unfairly underpriced off-the-road tire (OTR) imports from China, India, and Sri Lanka in accordance with Sections 701 and 731 of the Trade Act.

"The first time we were successful, but Chinese producers and importers appear to be gaming the system to avoid the duties they owe," said United Steelworkers International President Leo Gerard. "And now, subsidized producers in India and Sri Lanka have stepped in to get their own piece of the U.S. market."

"Too many domestic industries have been overwhelmed with unfair trade practices that capture sales of U.S. companies," said Maurice Taylor Jr., Titan chairman and CEO. "Titan has been fighting for the last eight years to safeguard the rights of U.S. producers of certain OTR tires and their workers. This case represents a significant effort by our company to restore conditions of fair trade to the U.S. market for OTR tires."[506]

China's enforcement of unfair trade is poor as well, according to a report by the U.S.-China Business Council. That organization found that 86 percent of firms surveyed expressed concern about competition enforcement activities in China.[507]

Suspicion about Chinese trade prompted then-presidential candidate and real-estate mogul Donald Trump to tell the *New York Times* he favors putting a massive 45 percent tariff on Chinese exports to the United States. Trump explained, "I would do a tax. And the tax let me tell you what the tax should be...the tax should be 45 percent." Many economists believe such a stiff tariff would contribute to what is called a "deadweight loss," or inefficiency built into the economy.[508]

Trump counters such arguments by saying that "the only power that we have with China is massive trade." Threatening that trade is our best leverage to change Chinese behavior.

Nelson Hultberg, author and director of Americans for a Free Republic, wrote an article, "Free Trade vs. Fair Trade," that agrees with

Mr. Trump and makes the case for allowing all people "to trade freely among themselves for goods and services without government intervention and restrictions." Unfortunately, as Hultberg points out, "'free trade' is the ideal only if it is practiced fairly and equally by all participant countries," and China and a few other countries like India in the OTR example above don't practice free trade.[509]

Our government shares some of the blame for stifling American trade competition as well. Specifically, America has a very anti-business environment: Our corporate taxes are among the highest; there are too many business regulations; our environmental restrictions make some operations uncompetitive; and we have "coercive support of labor union monopolies" that force up wages, pricing our manufacturing sector out of the world marketplace.[510]

I endorse Hultberg's relatively simple, common-sense trade solution. First, we need to play consistent hardball with unfair traders like China by imposing tariffs on their products until they lift their own restrictions on our exports. Second, we need, as mentioned earlier, to reduce corporate tax rates, cut the overregulation of businesses, and work with labor unions to set realistic wages.[511]

The U.S. must aggressively enforce fair competition, and it's clear that countries like China are either promoting unfair practices or turning a blind eye to such practices. In either case, unfair competition hurts America and her workers, and it must stop. Our government must be aggressive at making a trading marketplace level.

#7: Today America is living beyond its means, as evidenced by an overwhelming $19+ trillion debt.

We have mortgaged our grandchildren's future and this will stop. We will tighten our belts and give our economy permission to expand. Righting our trade balance and living within our means is our number-one priority. We will balance our budget while creating a FairTax system.

For years Congress has used a patch to raise the federal borrowing limit or face a government default. This phenomenon is known as raising the debt limit/ceiling, which means the government must increase the borrowing cap to unleash more spending. Of course, the U.S. has never defaulted on its debts. But such irresponsible spending has dug America into a deep debt, which former Admiral Mike Mullen, then-chairman of the Joint Chiefs of Staff, said is the biggest threat to our national security. He continued, "Obviously it's complex, but the way I looked at it, if we didn't get control of our debt, there would be continued loss of confidence in America." [512]

A bit of background is important to establish the debt framework before recommending a solution.

Recent governments with the consent of Congress spent to the limit and then kept asking to raise the cap in order to satisfy their unmet thirst for more cash. Of course, the Treasury needs Congress' okay to issue bonds and treasury notes to raise cash to meet the government's obligations. If the government runs out of money, then there will be delays in Medicare payments and Social Security benefits as well as paying interest on U.S. Treasuries (bonds), military and civilian federal salaries and many more obligations. [513]

Default occurs when the government is late in making payments on U.S. Treasury securities or paying other bills. What might happen if the government defaulted? Perhaps financial markets would implode or default would adversely affect the capital markets.

Ultimately, our debt does affect our budget. Government racks up debt when it spends more than it takes in and, as a result to continue spending, it must borrow more money, which is not free. It must pay interest on that loan, which is expected to be 15 percent of our total budget by year 2024.

Creditors have considered U.S. debt a secure investment, but as our fiscal stability comes into question, the Treasury will have to raise interest rates on U.S. bonds to attract enough funds. That in turn will consume a larger share of the annual budget to pay interest, leaving ever less for discretionary spending like defense.

Right now, the long-term debt is rising at unprecedented levels. At the current rate, public debt is projected to grow to about 100 percent of the economy by 2035 and nearly 150 percent by 2050. Debt at these levels threatens growth and our standard of living. It is not sustainable, which explains why drastic action is needed and soon.[514]

At the end of fiscal year 2016, the federal budget debt is expected to contribute $544 billion to our deficit, a rise of 6 percent in 2016 or 21.2 percent of the country's GDP. In the same year, the government will collect $3.4 trillion in taxes, or 18.3 percent of GDP.[515]

America's public debt, excluding Social Security and Medicare trust funds, accounts for 73.6 percent of current GDP. Not surprising and thanks to Obamacare, federal health programs will increase 11 percent in 2016. The CBO projects that our debt will reach 85 percent by 2026.

So, America, how much do you really owe? As of December 30, 2015, our official debt was $18.8 trillion ($18,825,061,664,536.00). That translates to $58,361 for every person in the U.S. or $151,361 for every household and 104 percent of the U.S. gross domestic product.[516] But that figure gets much higher when you consider obligations for current Medicare participants ($28.2 trillion).

The solution to our out-of-control spending and deficit is a giant bitter pill. We must stabilize the debt and then move to reduce it in the long term. But how might this be done before our economy reaches a point of no return?

Ohio Governor John Kasich and a 2016 Republican presidential candidate ran the House Ways and Means the last time we had a balanced budget. He promised during his failed campaign that "we will not only cut taxes, but we will balance our budget in eight years—both of them...because to balance a budget, you need economic growth while at the same time you manage your spending. Both of them lead us to a balanced budget, which ultimately can lead us to a place where we can begin to pay down our national debt." He called for cutting the top individual income tax rate from 39.6 percent to 28 percent, eliminating the estate tax, increasing the earned income tax credit by 10 percent, and cutting the top business tax rate from 35 percent to 25 percent.[517]

How might he accomplish this transformation? We may never know, because he's not going to be elected president. However, there are a number of possible places to begin the transformation and none will be very popular.

First, we can stop the out-of-control spending by passing a balanced budget amendment. This idea first resonated with founder Thomas Jefferson, who opined:

> I wish it were possible to obtain a single amendment to our Constitution; I would be willing to depend on that alone for the reduction of the administration of our government to the genuine principles of its constitution. I mean an additional article taking from the federal government the power of borrowing.[518]

President Ronald Reagan evidently saw the wisdom in Jefferson's idea because he openly advocated for a balanced budget amendment, but the proposal died because it would limit emergency spending say during a war for national survival.[519]

Rep. Paul Gosar (R-AZ) has a similar idea that he proposed in the Congress. He explains that Article V of the Constitution allows states to pass an amendment when three-fourths of them agree. Gosar's legislation would facilitate the convention of the states, House Congressional Resolution 26, Effectuating the Compact for a Balanced Budget, which treats the nation's out-of-control spending problem as a constitutional crisis.[520]

Second, sequestration is an option and currently used to force the federal government to slow its growth; we could do another sequestration. The Budget Control Act of 2011 (sequestration) is a federal statute that required equal cuts—national security and domestic programs—that was intended to force a balanced budget. It worked in part by reducing some expenditures, yet our debt keeps growing, and meanwhile, we cut our defenses. The problem is we live in a dangerous world. Fifty years ago, defense cost us 7.2 percent of our GDP and today it's only 3.5 percent. The balance of the sequestration cuts went to social-welfare programs that now account for 14.3 percent of GDP.

A third option is to impose painful spending cuts such as the following.[521]

Cut all domestic discretionary spending: Of course there are always caveats, exceptions to "everything is on the table" such as for Republicans, defense, and for Democrats, welfare. Unfortunately, the items left are fairly insignificant when the big ticket items aren't even considered for cuts. One recommendation is the so-called Green Scissors Plan, which cuts $70 billion in discretionary spending with a big slice from energy programs ($27 billion for subsidies like nuclear reactors and renewable energy).

Cut mandatory spending: Of course this one is very politically risky for risk-averse politicians. One way is to make cuts for "future seniors" in order to manage rising entitlement costs. This approach is used by past companies that exchanged annuities with 401(k) programs whereby the company puts far less into those programs. Alternatively, we can always increase contributions or even provide opt-outs whereby the individual covers his/her costs—which of course is risky, because a certain unknown percentage will fail to provide for the future expecting the government to provide a safety net.

Cut national defense: National security is considered a cash cow by many liberals who cringe at spending 20 percent of all federal discretionary money on defense. Of course, there are always options within the Pentagon's budget to find economies and savings such as uniform service health care—civilizing all but the combat-zone services. You can always redefine the Pentagon's missions, which translate into fewer personnel and equipment as well as expensive forward stationing. This is done via the Quadrennial Defense Review and the National Defense Strategy, which together, as outlined in Chapter 11 on national security, guide the budget process. Unfortunately, given the very dangerous state of the world today, we are gambling the nation's survival with the wrong and too-deep cuts in defense spending.

Raise taxes and/or cut deductions: Our tax code is larded with loopholes that save special interests while costing the rest of us more money. The Center on Budget and Policy Priorities found that $1.1 trillion in tax breaks fuel the wealthy. Eliminate these breaks and radically reduce government spending. Then the sacred cows of tax deductions like mortgage-interest deductions would make a significant dent in tax revenues. After all, why give a rich man an interest write-off on a second home at the beach?

Cut interest on our loans: This would put future loans at risk but tell those holding our securities we are only going to pay 85 cents on the dollar, thus saving 15 cents in interest payments. Of course, this puts a serious damper on future bond sales.

Hold federal spending to 2 percent per year: This would lead to a balanced budget within six years or limit growth to 3 percent and then the budget balances in nine years, according to Representative Gosar. Remember, federal spending exploded over the last two decades (a whopping 63 percent), and mandatory social spending doubled. Somehow, other developed nations like Germany managed to grow its economy while shrinking its debt. Perhaps we can take a lesson from the Swiss people who approved a constitutional initiative that required spending to follow revenue and it worked. We can do the same.[522]

Conclusion

It is past time America elects a president and a Congress that respect our Constitution and deliver sound fiscal policy. A big part of that outcome must be to live within our means—balance our budgets and pay off our debts. Why should an arguably hostile trading partner like China hold so much of our debt? It shouldn't, and that's why it would be prudent to limit the percentage of our debt any one country can buy so to minimize their leverage over our future.

The aforementioned recommendations are consistent with a biblical economic worldview. Christian America must insist our government become fiscally responsible.

Chapter 15

Strategy to Right the Spinning-Out-of-Control Culture

THE WORLD MAY seem to be spinning out of control because the American culture seems to have turned the country upside down. Even non-Christians are sensing that something is wrong and change is needed. But we cannot lead societal change unless we start with facts about the current situation and clear trend lines. The first question has to be: "Is there any evidence that our culture is spinning out of control?"

This chapter will answer that question. Yes, the evidence is overwhelming that our culture is rapidly slipping away from its once-Christian roots.

The obvious second question: "Is there anything we can do about it?" Many Christians believe that it is too late; America is too far gone to recover, and Christians just need to hunker down and ride out the storm. The true Christian view must be that a simple "yes" or "no" answer is irrelevant. Rather, we must be about our Lord's work until He returns.

The final question then, is: "What is the Lord's work that we should be doing?" This chapter presents answers in a five-part vision to restore

America's culture: (1) turn back to God, (2) elect godly leaders, (3) rebuild the American family, (4) make our communities and children safe, and (5) demand a morally responsible media.

Yes, America is worth fighting for, and it will require the weapons that God has provided to regain the high ground. Failure is certain doom for America.

The first part of restoring our culture is to recognize where it is today and from where it came.

#1: Reclaim Our National Christian Heritage

America's culture is morally corrupt and dysfunctional. Too many of us don't know right from wrong. We've become a nation of egotists having lost the ethic that made this nation great. We need to turn back to God.

America needs to turn back to God to have any hope of restoring its culture. We have a rich Christian heritage, but long ago, we lost our way and, as a result, our culture rotted, especially over the past two decades.

America's moral compass has radically shifted since the early 2000s, according to a 2015 Gallup poll of Americans' views of the moral acceptability of various behaviors. Specifically, majorities of Americans now believe the following are morally acceptable: homosexual relations (from 40 to 63 percent), having a baby outside of marriage (from 45 to 61 percent), and sex between an unmarried man and woman (from 53 to 68 percent).[523] It wasn't too many years ago when these figures were in the single digits. Obviously, something radically bad happened in America.

These shifts reflect the fact that Americans are becoming far more libertarian on social issues, which influences every aspect of their lives. The shift is further distressing because it is reflected across the age spectrum, with those sixty-five years and older shifting left just as fast as have the eighteen to twenty-nine year olds.

This shift is the result of a complex set of factors and will undoubtedly have a significant effect on virtually every aspect of our private and public life. Today it seems as if Americans can't abandon traditional, Christian values fast enough.

The largest cultural shift is regarding the acceptability of homosexuality. This shift to being mostly pro-homosexuality, which defies God's Word, has certainly been encouraged by the morally bankrupt entertainment industry that features gay and lesbian characters in mainstream programs, making them appear like the everyday modern person. Certainly President Obama helped the immoral shift by lifting the military's homosexual ban and promoting homosexual "marriage," not to mention his appointment of numerous openly gay persons to his administration and the judiciary.

These liberalizing shifts on moral issues are accompanied by an equally sobering but contrary phenomena. Most (72 percent) Americans told Gallup the state of moral values in America is getting worse rather than better. The obvious interpretation is that most Americans know we are sliding into the immoral abyss, but evidently they won't or don't know how to stop themselves. It reminds me of the familiar frog-in-the-pan illustration. The frog is in a pot of water being slowly brought to a boil, but the frog never jumps to safety. Evidently, Americans sense things are getting very bad morally, but they choose to be blind to the cause and won't or can't jump to safety.[524]

A few years ago, Gallup asked Americans what was wrong with our moral values. They listed a litany of behaviors common to our present declining culture: lack of consideration of others, deficits in the public's compassion, a general lack of personal accountability, respect, and tolerance; greed, selfishness, dishonesty, changes in family structure, lack of religion, and a general lack of morals. The evidence is undisputed, yet as the temperature rises, we play dumb.

This list of disturbing behaviors shouldn't surprise Christians who embrace a biblical worldview. These behaviors reflect biblical descriptions of past fallen people whether in Sodom and Gomorrah (Genesis 19), the pre-exile Israel and Judah (book of Isaiah), ancient Corinth

(1 Corinthians 5–6) or first-century Rome (Romans 1:18–32). Obviously, we aren't any better than those ancient people, and our demise is at hand if things don't begin to turn around.

The evidence of our pending collapse suggests that the further we as a people move away from God, the worse our lives become, and like the frog in the soon-to-boil water, we know what we are doing is wrong and where to go to get help, but we won't jump to safety. We refuse to seek the help that is within our reach—from God and His Word.

The American Bible Society conducts an annual "state of the Bible" survey to detail Americans' beliefs about the Bible, its role in society, its presence in our homes, and more. That survey found a significant disconnect in belief versus behavior, as evidenced by the previously mentioned Gallup poll showing broad-based acceptance of behaviors named by the Bible as sin—homosexual behavior, abortion, and sex outside of marriage.[525]

The survey found that two-thirds of Americans agree the Bible contains everything a person needs to know in order to live a meaningful life, but 58 percent say they do not personally want wisdom and advice from the Bible, and not surprisingly, about the same number (57 percent) read the Bible fewer than five times a year.

It is noteworthy that those least likely to read the Bible say they really want input and wisdom regarding tough life issues that are amply addressed in God's word: parenting (42 percent),[526] family conflict (40 percent),[527] dating and relationships (35 percent),[528] and romance and sexuality (30 percent).[529]

So we've come full circle. "Americans overwhelmingly recognize the decline of morality in our nation," said Doug Birdsall, president of the American Bible Society. They need to turn back to God and His Word, and in part, that's because they have bought into the left's distortion of truth about God, the Bible, and America.

Not long ago America was a blessed Christian nation that relied on God's Word for guidance and wisdom for every aspect of life. America has a long history of many faults, but until the last few decades, most

Americans had a biblical foundation thanks to mostly Christian homes, schools where God's Word was taught, and communities filled with God-fearing churches.

Consider a bit of American history to appreciate just how far we have fallen as a people.

Over the past half century, America shoved Christianity aside to embrace a host of "isms"—humanism, atheism, social liberalism, pluralism, and multiculturalism. Our once-solid moral and spiritual foundation has disintegrated, and there is no more obvious arena to see that corrosion than across our culture.

America has fought and appears to be losing a decades'-long spiritually driven cultural war. Dr. Dave Miller, the executive director of Apologetics Press, makes a compelling case that God is the central issue in our crumbling cultural war. "America is in the throes of a life-and-death struggle over whether the God of the Bible will continue to be acknowledged as the one true God, and Christianity as the one true religion."[530]

All the modern isms are openly hostile toward God, which is why Christians are facing the most perilous times ever. Admit it: Our culture is expunging every trace of Christianity and God—government, schools, parks, and the media—and political correctness is the left's tool to marginalize Christianity at every turn.

This massive assault on traditional biblical morality is cleansing our culture of everything Christian under the disingenuous argument that our founders advocated "separation of church and state." That is a clever ruse that has ushered American culture into a secularized, pluralistic state but light years from the truth from where it was for most of the nation's young history.

Dr. Miller wrote a three-part article, "America, Christianity, and the Culture War," which thoroughly documents America's rich Christian heritage. Dr. Miller's article is the source for the following brief account of America's Christianity, which should equip Christians engaged in America's ongoing anti-Christian cultural war.

Prior to the 1960s, a biblical worldview was embraced widely in

America, and anti-Christian forces had only a small cultural presence. Unfortunately, over the intervening decades, the Christian majority naively gave in to that small chorus that called the Christian majority "intolerant." Ever so slowly, those voices of evil suppressed biblical views and used the banner of "free speech" to censor Christianity in every corner of America. Today they are rewriting history by contending America was never a Christian nation, and they have rabidly gone about removing every vestige of that heritage.[531]

Our founders believed the republic's dependence on the Bible permeated every aspect of American society. The evidence of the Bible's influence on the young American society is absolutely overwhelming, but it is seldom celebrated in today's politically correct society. Rather, today's progressive movement (read "liberal, anti-Christian") in the name of "separation of church and state" uses intimidation and our misguided judicial system to remove God's Ten Commandments from the public square and replace it with monuments celebrating sinful behaviors like homosexuality and promiscuous sex such as the marker at the Sacramento, California, Veterans Memorial that reads: "In Honor of Gay, Lesbian, Bisexual, and Transgender Veterans Killed in Action."[532]

Anyone who reads the U.S. Constitution won't find the phrase "separation of church and state," and in fact, you would have to search long to find that the phrase only appears in a private letter written by Thomas Jefferson regarding assurances that the government would not interfere with Baptists' free exercise of religion. Even the late U.S. Supreme Court Chief Justice William Rehnquist called out the anti-Christian distortion by labeling the phrase a "misguided analytical concept," and saying there is an "absence of a historical basis for this theory of rigid separation."[533]

A copy of the U.S. Constitution is included at appendix A to this volume.

Our founders put Christianity in all founding documents and consistently and frequently demonstrated their dependence upon the Almighty in their public and private communications.

The American Declaration of Independence declares the founders' intent to separate from the British crown. That short document (1,137 words) clearly aligns the founders' intentions with their Christian faith in a number of phrases: "nature's God," "all men are created equal and endowed by their Creator," "appealing to the Supreme Judge of the world," and "with a firm reliance on the protection of divine Providence." Obviously, the founders put no "separation between church and state" in this document.[534]

A copy of the Declaration of Independence is found at appendix B to this volume.

The U.S. Constitution recognizes issues regarding the free exercise of the Christian faith ought to be a matter for the states. However, the founders align themselves with the Christian faith in the document as well. Specifically, the First Amendment to the Constitution states, "Congress shall make no law respecting an establishment of religion, or prohibiting the free exercise thereof." Evidently, the records from the Constitutional Convention delegate discussions clarify the intent of the amendment's wording. The founders use the term "religion" to distinguish among Protestant denominations to prevent any single denomination from being made the state religion like what happened in Great Britain with the Anglican Church. There is zero evidence that the founders ever considered other religions like Islam or Buddhism.

Our Constitution even assumes the government will not be open on Sundays in deference to the Christian day of worship. Article 1, Section 7 states: "If any bill shall not be returned by the president within ten days (Sundays excepted)." Obviously, the founders demonstrated their Christian orientation with the "Sundays excepted" clause—an accommodation to their widespread Christian faith.

Finally, the Constitution closes with a reference to "in the year of our Lord." The Christian world dates human history from the birth of Christ: "B.C." means "before Christ," and "A.D." is the abbreviation for the Latin *anno domini*, meaning "year of our Lord." This is a clear reference to Jesus Christ.

A copy of America's Bill of Rights includes the collection of twenty-seven amendments to the U.S. Constitution, which is found at appendix C to this volume.

The state constitutions are just as clearly aligned with the Christian faith. In fact, forty-five of the current state preambles to their constitutions make explicit appeals to the God of the Bible. The body of many of those documents contains numerous specific references to God as well.

The presidential oath of office comes from Article 2, Section 1 of the Constitution. That oath closes with "So help me God." This phrase affirms the official's belief in the God of the Bible. Even the oath taken by the vice president and all federal government employees, which is mandated by statute, includes the phrase "So help me God." And the tradition of putting one's hand on the Bible continues to this day.

Not surprisingly, every president of the U.S. has mentioned the Christian God in their inaugural address. George Washington set the tone with his first address that included the following phrases: "It would be peculiarly improper to omit in this first official act my fervent supplications to that Almighty being who rules over the universe," "His benediction may consecrate to the liberties and happiness of the people of the United States," "the invisible hand [God's hand] which conducts the affairs of men," "the propitious smiles of heaven," "since He has been pleased to favor the American people," and "so His divine blessing."

Other American presidents have invoked their dependence on God as well. John Adams, our second president, in his inaugural used phrases like "under an overruling Providence," "and may that Being who is supreme over all," "the protector in all ages," and "His Providence." Even the so-called deist president, Thomas Jefferson, said in his inaugural address: "an overruling Providence" and "that infinite power which rules the destinies of the universe."

Abraham Lincoln sought the God of the Bible as a new president facing the imminent Civil War. Lincoln said in his address: "Intelligence, patriotism, Christianity, and a firm reliance on Him who has never yet forsaken this favored land are still competent to adjust in the best way all our present difficulty."

President Obama tries to marginalize our Christian heritage in his inaugural address with the statement: "For we know that our patchwork heritage is a strength, not a weakness. We are a nation of Christians and Muslims, Jews and Hindus, and non-believers," Obama said. He ends his address claiming our eyes are fixed on "God's grace" and then, like those who came before him he ends with "God bless you. And God bless the United States of America."[535]

America's early judiciary avowed our affiliation with the God of the Bible. An opinion by Chief Justice James Kent rebuts a defendant's assertion that Christianity was not a part of the laws of the state. Kent wrote:

> [W]hatever strikes at the root of Christianity tends manifestly to the dissolution of civil government.... The people of this State, in common with the people of this country, profess the general doctrines of Christianity, as the rule of their faith and practice; and to scandalize the author of these doctrines is not only, in a religious point of view, extremely impious, but, even in respect to the obligations due to society, is a gross violation of decency and good order... [T]o revile, with malicious and blasphemous contempt, the religion professed by almost the whole community, is an abuse of that right. Nor are we bound, by any expressions in the constitution, as some have strangely supposed, either not to punish at all, or to punish indiscriminately the like attacks upon the religion of Mahomet or of the Grand Lama; and for this plain reason, that the case assumes that we are a Christian people, and the morality of the country is deeply **ingrafted upon Christianity**, and not upon the doctrines or worship of those imposters.

Our currency reflects America's past deep devotion to God. Consider President Abraham Lincoln's directive to the director of the mint in Philadelphia: "No nation can be strong except in the strength of God or safe except in his defense. The trust of our people in God should be declared on our national coins." Then on April 22, 1864, Congress approved the motto, "In God We Trust," for placement on American currency.

Not surprisingly, our national symbols reflect America's religious heritage. The Liberty Bell in Philadelphia has an engraved quote from Leviticus 25:10 (KJV): "Proclaim liberty throughout all the land, unto all the inhabitants thereof."

The French government's gift of the Statue of Liberty stands on Bedloe's Island in New York Harbor. That statue includes a plaque with the verse Leviticus 25:10 as well.

The American national seal reflects the beliefs of the founding fathers. That seal was approved on June 20, 1782, and both sides are printed on the back of our dollar bill. One side includes a pyramid with the eye of Providence (God) and the motto annuit coepis, "He (God) favors our undertakings." Above the "ONE" on the same side of the dollar bill are the words "In God We Trust."[536]

Government buildings across America reflect our biblical faith. Consider this short list.

- The Supreme Court building openly displays the Ten Commandants (Exodus 20) directly above the bench where the justices sit.
- Next door to the Supreme Court, the Library of Congress hosts eight marble statues above columns that represent the categories of knowledge.
  - One of those statues represents "history" and includes an inscription of words from Lord Tennyson's *In Memoriam*: "One God, one law, one element, and one far-off divine event, to which the whole creation moves."
  - Above the "religion" figure atop another column of knowledge are the words of Micah 6:8, "What doth the Lord require of thee, but to do justly, and to love mercy, and to walk humbly with thy God."
- The library also hosts sixteen bronze statues that represent renowned men such as the apostle Paul and Moses.

- The murals in the library's dome include the words: "Thou shalt love thy neighbor as thyself" (Leviticus 19:18).[537]
- The front entrance to the CIA headquarters displays a large engraved quote: "You shall know the truth and the truth shall make you free" (John 8:32, ASV).

There are Bible quotes and images on virtually every government building to include the White House, the U.S. Capitol complex, the House Chamber, the House Rotunda, and the Lincoln Memorial. The Washington Monument, which towers 555 feet above the capitol's mall, includes numerous Bible verses and religious acknowledgements: "'Holiness to the Lord' (Exodus 28:26;30:30; Isaiah 23:18, Zechariah 14:20), 'Search the Scriptures' (John 5:39), 'The memory of the just is blessed' (Proverbs 10:7), 'May Heaven to this Union continue its beneficence,' and 'In God We Trust', but the Latin inscription *Laus Deo*—'Praise be to God'—is engraved on the monument's capstone."[538]

America's national songs and hymns include a strong faith component. "God of Our Fathers" was the official hymn of the centennial observance of the U.S. Constitution. That hymn is full of references to God, such as "Thy word," "Thy paths," "Thy strong arm," and "Thy true religion." The writer's "Thy true religion" was Christianity.

Francis Scott Key wrote "The Star-Spangled Banner," America's national anthem, during the defense of Fort McHenry in the War of 1812. The fourth verse of that anthem affirms the historic attitude toward God: "Oh! Thus be it ever, when freemen shall stand / Between their loved homes and the war's desolation! / Blest with victory and peace, May the heaven-rescued land / Praise the power that hath made and preserved us a nation / Then conquer we must, when our cause it is just, / And this be our motto: 'In God Is Our Trust.' / And the star-spangled banner in triumph shall wave / O'er the land of the free and the home of the brave!"

Our national cemeteries evidence our Christian heritage with so many white crosses and quotes from the Bible. Even American cemeteries in foreign lands use those same white marble crosses for the fallen

Christian service members and not a few Stars of David for their Jewish brothers.

There is no doubt America's founders embraced the importance of God, front and center in our nation's founding. Long ago George Washington warned America not to abandon the Lord. On March 11, 1792 he wrote:[539]

I am sure there never was a people who had more reason to acknowledge a Divine interposition in their affairs than those of the United States; and I should be pained to believe that they have forgotten that Agency which was so often manifested during our revolution, or that they failed to consider the omnipotence of that God who is alone able to protect them.

In 1852, American statesman Daniel Webster prophetically warned America:

[I]f we and our posterity reject religious instruction and authority, violate the rules of eternal justice, trifle with the injunctions of morality, and recklessly destroy the political constitution which holds us together, **no man can tell how sudden a catastrophe may overwhelm us that shall bury all our glory in profound obscurity.** (emphasis added)

Appendix D to this volume includes a listing of quotations from our founders regarding their beliefs in Jesus, Christianity, and the Bible.

America was launched on a good path by men who honored God, but she has lost her way because, in part, we have men and women at the top of government and our culture who no longer lead with a heart devoted to God's direction.

We must turn back to God and His Word—the Bible—in our journey to restore American culture.

The second part of our vision to reform American culture addresses the need for godly leadership at the helm of our country.

#2: Moral Top Leadership

Putting America back on a morally correct course requires action by the nation's top leader, the president. Our president—and by association his administration—will govern with biblical values and principles, always being above reproach. Those who disappoint us will be removed and replaced by people with unquestioned scruples.

Presidents use their bully pulpit to make things happen mostly in terms of government policies. That's important because policies impact culture's real center of gravity, the home, and local communities of America. For example, life is complex and it is hard to predict how inevitably a new federal government policy might impact culture such as changes to laws that address family structure, relationships, jobs, and life decisions such as birth or abortion.

President Obama may be the historical exception regarding cultural influence, or it could be that he became commander-in-chief at the time America experienced a shift to a more immoral culture. Certainly for the past eight years, America has had a man at the helm who has a broken moral compass, at least from a biblical perspective, and arguably Obama and his policies have negatively influenced American culture. In fact, President Obama will likely go down in history as our most immoral and godless president to date. Many of his policies and decisions must be reversed to help this nation recover—that is, if it isn't already too late.

Some of Obama's social policies are an absolute abomination. Specifically, Obama pushed gay rights at home and abroad. For example, while visiting Kenya, he pressed leaders to accept the homosexual lifestyle and same-sex "marriage." That mostly Christian nation's president, Uhuru Kenyetta, called gay rights a "non-issue" for his country. "It's very difficult for us to be able to impose on people that which they themselves do not accept. This is why I repeatedly say that for Kenyans today, the issue of gay rights is really a non-issue." Obama made a similar appeal in 2013 during a visit to Senegal.[540]

Dr. Richard Land, president of Southern Evangelical Seminary, said Obama's attempts to promote homosexuality abroad only furthers a growing perception overseas that Americans are "morally degenerate." Land continued, "I think it just reinforces in the minds of many more traditional-minded folks in Africa that the United States has gone politically correct to the point of having lost our moral compass and lost our moral way…. And it just sort of feeds into the perception that we are morally degenerate."[541]

"In my opinion, President Obama being president of the United States is in itself a judgment of God on the United States—that a God who disposes and proposes has allowed him to become president because of our moral profligacy," the seminary president said.

Mr. Land's assessment that Mr. Obama's presidency is God's judgment on America certainly fits the behavior seen on the very evening of the Supreme Court's tragic and misguided ruling (5-4) that homosexual "marriage" was constitutional. On that evening Obama allowed the homosexual movement's rainbow colors to illuminate the White House, the so-called People's House.[542]

The Reverend Franklin Graham with the Billy Graham Evangelistic Association responded to Obama's White House pro-homosexual marriage rainbow light show. "The rainbow came after the iniquity of man had become such a stench in God's nostrils that God acted to cleanse the earth, sparing only Noah and his family. At Christ's return, the Earth, the heavens and mankind will come under the intense judgment of God."[543]

President Obama's promotion of homosexuality opened the "floodgates of acceptance" of "sexual immorality," said Reverend Graham. "Never, though, would I have dreamed that our president would light up the White House with the vivid, bright rainbow colors of the gay movement," said Graham. "It was a deliberate, measured step by the president to flaunt and celebrate a disastrous [Supreme Court] ruling that blatantly defies God's moral law."[544]

Obama also makes clear his morally degenerate views when he speaks to Planned Parenthood groups. In 2013, Obama said to a Planned Parenthood conference, "So long as we've got to fight to make sure women have access to quality, affordable health care, and as long as we've got to fight to protect a woman's right to make her own choices about her own health, I want you to know that you've also got a president who's going to be right there with you, fighting every step of the way." He closed with "Thank you, Planned Parenthood. God bless you." What a mockery of God!

Planned Parenthood is the country's largest abortion business. In 2013, it performed 327,653 abortions and received over a half billion dollars in taxpayer funding for those dark "services."[545]

Obama should also go down in history as our most hypocritical president when we examine his claim to be for racial equality. Doesn't he understand that a disproportionate number of all babies killed in American abortion clinics are black? The Alan Guttmacher Institute reports black women are more than five times more likely than white women to have an abortion, and on average, 1,876 black babies are aborted every day in America. Obviously, Mr. Obama doesn't favor racial equality for unborn black babies. That's an issue for the so-called "Black Lives Matter" movement![546]

Obama also embraces a savage view of abortion. In 2015, he demonstrated an extreme view by threatening to veto the Pain-Capable Unborn Child Protection Act, which would protect the unborn after twenty weeks from abortion based on the concept that they can feel pain.[547]

Mr. Obama's immoral behavior extends to his foreign policies regarding minority people in the Middle East. He favors the immigration of Muslims at the expense of Christians and other minorities in the Middle East who are facing genocide at the hands of ISIS.

Obama's pro-Islamic views extend to his support for the radical Muslim Brotherhood's takeover of Egypt, which resulted in tremendous suffering among Coptic Christians. Obama even welcomed Brotherhood members to the White House and gave them his moral support. Certainly Mr. Obama knew from intelligence reports that the Muslim Brotherhood espouses the "common goal of dismantling American institutions and turning the U.S. into a Muslim nation."

Further, the Brotherhood openly spoke of the extermination of our good ally Israel.[548]

Obama's disdain for Israel and Jews is evidenced in his appointment of Hannah Rosenthal as the State Department's special envoy to monitor anti-Semitism. During a 2012 visit to Sweden to investigate Muslim attacks on Jews, Rosenthal made clear she holds the "insidious belief that Jews are to blame for anti-Semitism because of their support for Israel." Certainly Obama knew when he appointed Rosenthal that she had served with the group Americans for Peace Now, which is part of the boycott, divestment, and sanctions movement, a movement that seeks to delegitimize Israel through any means possible.[549]

Samantha Power, Obama's United Nation's ambassador and another top Obama official that is biased against Israel, served as the special assistant to an Atrocities Prevention Board. Power is the author of *A Problem from Hell: America and the Age of Genocide,* but when she was asked what she would recommend with respect to the Palestine-Israel situation if one party appeared to be moving toward genocide, Ms. Power came down firmly with the Palestinian camp.[550]

Obama's bankrupt views about the Islamic Republic of Iran's frequent calls for Israel's annihilation are especially distressing. An Obama administration official told the *Washington Post* that the administration "is trying to make the [Israeli] decision to attack [Iran] as hard as possible for Israel." That prompted columnist Charles Krauthammer to write that this is "revealing and shocking. The world's greatest exporter of terror (according to the State Department), the systematic killer of Americans in Iraq and Afghanistan, the self-declared enemy that invented 'death to America day' is approaching nuclear capability—and the focus of U.S. policy is to prevent a democratic ally [Israel] threatened with annihilation from preempting the threat?"[551]

Then consider that Obama many times approved killing Islamists to include an American living abroad with drone-launched missiles, but wants similar killers captured years ago on the battlefield brought to our shores and given access to our court system to win their release.[552]

Yes, Obama argues to shut down our prison at Guantanamo Bay, which holds Islamist killers. It is very hard to believe that Mr. Obama takes his job of protecting Americans seriously when he advocates for bringing Islamic killers taken off the battlefields to our shores.

Obama's evil decisions and policies have perhaps forever transformed America. What he has done to our military is unforgiveable as well. He made lifting the military's 230-year-old ban on open homosexuality a priority and in late June 2016, Obama's secretary of defense lifted the Pentagon's ban on transgendered personnel serving openly after making the delusional statement that our military needs their services, which ignores the fact these people suffer from more psychiatric pathologies compared to the general population, and he promised the taxpayer will now pay for doctor-recommended sex changes.[553] Further, the Obama administration has worked overtime to open the armed forces to virtually any abhorrent group even though his social experiments without exception undermine our military's readiness, and that's especially true about his agenda to push women into formerly closed, direct-combat roles.[554]

Obama disregards military readiness, proven biological science, and thousands of years of combat history to push women into the most physically grueling ground combat situations known to man. Obama's decision may appease radical feminists, but it is morally unforgiveable and something that will deliver long-term harm to our armed forces and women. Finally, his anti-Christian crusade within the military's ranks is well documented by the Family Research Council and reprehensible, while Obama's politicos push a very sympathetic and pro-Islamic view within the Pentagon's ranks.[555]

Mr. Obama's other immoral actions include his well-publicized surrender after announcing a "red line" regarding chemical weapons use against Syrian civilians. He abandoned our people at the consulate in Benghazi, Libya. He prematurely withdrew our forces from Iraq in 2011, which contributed to the rise of ISIS and a renewed sectarian civil war. He pushed for the Patient Protection and Affordable Care Act, the so-called Obamacare, which resulted in untold additional national debt, seriously damaged America's healthcare system, and resulted in very few previously uncovered people getting healthcare. He even fanned the flames of racism, which resulted in riots in places like Ferguson, Missouri, and Baltimore, Maryland, and did untold harm to our police forces. He failed to contain our out-of-control federal spending to the tune of $19+ trillion, and he personally wasted taxpayer money through his lavish lifestyle and much, much more.

God will judge Obama for his evil ways. However, now America must elect and put in place leaders who have a biblical moral compass. Those leaders must reverse the damage of the past eight years and chart a moral path into the future. But, as said at the beginning of this section, presidents have only a secondary or tertiary impact on culture in spite of Obama's horrible immoral legacy.

David French, a *National Review* editor and lawyer, downplayed the president's influence on culture. He admits "even most famous actors, writers, and musicians—are culturally meaningless on the large scale but culturally vital in the small scale." French says the president's bully pulpit has little impact on culture and the real influence is "at home around the dinner table, or in the streets with friends, or in the hours of school with teachers who care (or don't)."[556]

Next we turn to America's "dinner table" and schoolhouses, whereas French writes, our culture is most influenced.

#3: Rebuild the American Family

Our children deserve the opportunity to learn right from wrong and to grow in stable homes, and they belong first to their parents and families, not to the local school—much less to government. Parents have the primary responsibility and authority over their child's education and life decisions.

Rebuilding the American family is critical because, as political scientist Charles Murray said, "America has far too many children born to men and women who do not provide safe, warm, and nurturing environments for their offspring—not because there's no money to be found for food, clothing, and shelter, but because they are not committed to fulfilling the obligations that child-bearing brings with it."[557]

Social science conclusively demonstrates that children do much better when they grow up in homes with their married biological parents. The problem today is that two significant trends collide: a declining marriage rate and an increasing divorce rate, which rock the foundation of our society, the family. The research is clear that children who grow up in mother-only homes suffer high rates of poverty, drug and alcohol abuse, danger to their physical and emotional health, lower levels of educational achievement, and crime, and they tend to become sexually active outside of marriage.[558]

The importance of a two-parent, biological, married family is a view shared by those on both the political right and left. A study by the left-leaning Child Trends concludes:[559]

> [I]t is not simply the presence of two parents…but the presence of two biological parents that seems to support children's development.…
>
> [R]esearch clearly demonstrates that family structure matters for children, and the family structure that helps children the most is a family headed by two biological parents in a low-conflict marriage. Children in single-parent families, children born to unmarried mothers, and children in stepfamilies or cohabiting relationships face higher risks of poor outcomes.… There is thus value for children in promoting strong, stable marriages between biological parents.

Government should embrace only policies that promote child-bearing within heterosexual marriage and encourage parents to remain together and responsibly share the childrearing tasks in order to produce well-adjusted future citizens.

Not surprisingly, children best learn about right and wrong at home. Becky Sweat, a freelance journalist, wrote for the United Church of God "10 Practical Ways to Teach Your Children Right Values." Mrs. Sweat begins with some common-sense advice for all parents: "Take the time to talk with your children. Talk about what they did right, what they did wrong, how to make better moral decisions, what character traits God wants to see in us, and why you've made certain choices in your own life."[560]

Mrs. Sweat warns parents that "teaching values takes time—a scarce commodity for many parents today." The alternatives to parental value-molding time are the potentially dangerous outside influencers like peers, the Internet, television, movies, and video games. Unfortunately, if the cultural shift is believable, as polls suggest, then evidently more children than ever before are learning their values from other than those who love them most.

A big part of the solution is easy to say but harder to do in this hectic world. "You need to make time to be with your kids and make the time you have with them really count," urges Dr. Gary Hill, director of Clinical Services at The Family Institute at Northwestern University. "Talk with them about what's right and wrong, and what constitutes good behavior and what doesn't."[561]

Mrs. Sweat provides practical suggestions to help parents teach their children right from wrong. Those suggestions are: model good values, apologize to your children when you make mistakes, use everyday experiences as a springboard for conversation, read the Bible with your children, share your personal experiences, hold your children accountable for their mistakes, don't let your children take the easy way out of challenges, involve your children in encouraging and helping others, monitor television viewing and Internet use, and applaud good behavior.

Government policy can't substitute for good parenting. However, where there aren't good parents like Becky Sweat, then the local church, neighbors, and local schools must step in to help.

A big aspect of being an effective parent is taking charge of our children's education. Parents, not a government school, should know best when it comes to their children's education. Unfortunately, too many parents have become so busy with their own lives that they have bought into the idea that the professional educators know best.

Chris Stewart, the former executive director of the African American Leadership Forum, disagrees. Mr. Stewart asked the Black Education Strategy Roundtable, "Who is responsible for the education of your child?" He answers his own question: "No one [is] but you [the parent] should be]."[562]

The general lack of accountability for educating our children is devastating. Stewart points out the following sobering facts about black children in America.

- Fourteen percent of African-American eighth graders score at or above the proficient level.
- Fifty-four percent of African Americans graduate from high school.
- On average, African-American twelfth-grade students read at the same level as white eighth-grade students.
- Four percent of black students finish high school college-ready in their core subjects.

Stewart concludes:

Our kids can't read, write and compute sufficiently to gain jobs and homes. That is a problem if we want strong families, strong communities, and a self-sufficient, free black race.

The only hope I have for freeing our children from broken schools is the emergence of parents, teachers and leaders into a movement for better schools. If we want the best for our kids, we must answer the question of responsibility for ourselves.

The lion's share of the responsibility for educating our children rests with the parent, not the child or the school. Higher standards, better curricula, and better teachers are all important but not nearly as important as an involved parent. If your child isn't performing well in school, then look in the mirror.

Most learning takes place outside the classroom—at home, while riding to school, on the playground, and elsewhere.

Schools in England are required to recruit parents to their child's education process. The schools work with parents to get them involved with the schools by attending teacher-parent meetings, helping around the school, and helping their children with homework. The results of parental involvement are very positive, according to a British study.[563]

That study also found teenagers seek moral support from their parents. High school students in particular want to know their parents care for them and that education is valued. One student explained, "If they don't care about it, why should you?"

Pope Francis, the leader of the Roman Catholic Church, is a cheerleader for parental involvement in educating their children. "If family education regains its prominence, many things will change for the better. It's time for fathers and mothers to return from their exile—they have exiled themselves from educating their children—and slowly reassume their educative role," the pope said.[564]

"On the one part, there are tensions and distrust between parents and educators; on the other part, there are more and more 'experts' who pretend to occupy the role of parents, who are relegated to second place," he said. He stressed that educating and raising children in the human values that form the "backbone" of a healthy society is a responsibility that each family has.

However, many difficulties often impede parents' ability to properly educate their children. Today parents are spending less and less time with their children, and meeting their needs after a long day of work can be exhausting, he noted.

There are no easy government tools to compel parents to become actively involved in their children's education. Shaming works for some, and that's why the pope's words are important. The British government's push to enlist parents will work with others. But, as many teachers will testify, some parents just don't care or say they are too busy.

Parents are critical to reversing our culture's decline into the abyss. Collectively, government, churches, employers, and local communities must embrace policies and activities that foster healthy, intact homes with involved parents if we have any chance of restoring the next generation. And it must be done from a biblical worldview perspective.

#4: Make Our Communities and Children Safe

At the same time as seeking family stability, we must provide effective law enforcement for our communities. Criminals must be deterred and those addicted to drugs and alcohol must be given hope. Crime is a problem, and too often the perpetrators come from broken homes—then these troubled people fill our jails to capacity. The evidence shows that kids who grow up in intact homes prosper, and that is our goal for all American children. We must scrub everything the government does to create environments that protect our children and give them the opportunity to succeed.

We make communities and children safer by having more unbroken homes, making a concerted effort to protect vulnerable children, and providing effective law enforcement.

The federal government ignores "the importance of the intact married family in shaping outcomes of its social policies," according to a Family Research Council study, "U.S. Social Policy: Dependence on the Family."[565] That study found that: "The state has hitherto ignored the importance of the intact married family in shaping the outcomes of its social policies. This neglect of marriage is an error of historical proportions." That is because strong, unbroken families vaccinate children from criminal behavior.

An avalanche of scientific studies done over many years demonstrates that broken homes significantly contribute to crime. The following are just a sample of the findings from published scientific studies on the issue.

- An analysis of fifty studies on the impact of broken homes found juvenile delinquency was 10–15 percent higher than in intact homes, and that's after controlling for race and gender.[566]
- Convicted adolescent murderers come from highly dysfunctional families: fewer than 20 percent were from married parent households, 42.9 percent of their parents had never married, 29.5 percent of their parents were divorced, and 8.9 percent of their parents were separated.[567]
- Fatherless boys are 3.061 times as likely to go to jail as peers from unbroken families.[568]
- Juvenile crime in rural areas is strongly correlated with broken homes, while poverty was not directly associated with violence.[569]
- Teens from intact families had a marked advantage over children from fractured families in terms of discipline and academic performance.[570]
- Divorce is a strong predictor for juvenile violent crime.[571]

The government should advance the benefits of the traditional marriage families to reduce the incidence of juvenile crime.

Family policy expert Chuck Donovan outlines a plan to rebuild marriage and intact families in an article for the Heritage Foundation, "A Marshall Plan for Marriage: Rebuilding Our Shattered Homes."[572] Mr. Donovan indicates "by maximizing the benefits of family life for the next generation of Americans, it could reduce the costs of family breakdown to taxpayers while fostering personal happiness, independence, and productivity."

Mr. Donovan's "Marshall Plan for Marriage" identifies four principles that offer a "framework of recommendations that, if implemented, would eliminate pervasive biases against marriage without placing significant new demands on the public purse."[573] Those principles are:

Principle #1: The decision to marry is inherently economically beneficial to couples and their children, if any. Any form of financial penalty in tax policy that masks or subverts this reality and deters marriage should be eliminated.

Principle #2: Policymakers and program managers should encourage pro-marriage messaging in existing government programs and other already available resources.

Principle #3: States should recognize that a significant percentage of divorcing couples, especially those with children, would respond to reconciliation efforts and restore their marriages. States should develop policies and programs that maximize the reconciliation option.

Principle #4: Policymakers should study, recognize, and reward success in marriage, recognizing the power of the bully pulpit, and civic leadership to shape consensus and define progress.

Obviously, intact families are often the best protection for children from a rampaging, crime-infested, and immoral culture. However, there are other measures communities should take to protect our children, such as better policing.

Community-oriented policing involves police and citizens working together in partnerships to address the causes of crime and to reduce fear by solving problems before they get out of hand.

Most (74 percent) local police departments provide crime prevention education to citizens.[574] Those activities include problem-solving partnerships between police, churches, and community/neighborhood groups that facilitate crime prevention. Those efforts should include many of the ideas promoted by the National Institute of Justice for adolescents, such as mentoring and school-based programs such as Big Brothers and Big Sisters. After all, there is no substitute for a responsible adult's attention whether a parent or a concerned adult in the life of a young person struggling with life.[575]

#5: Demand a Morally Responsible Media

Our culture is significantly impacted by various media. Although we respect our free speech liberty, some outlets abuse the privilege and must be restrained. This may be a fine line, but we will not tolerate the coarsening of our next generation.

Chuck Colson, the founder of Prison Fellowship and Colson Center for Christian Worldview, used the final speech of his life to survey America's cultural and political landscape: "What we are seeing now [March 2012] is the full fruits of 30 years of relativism, the death of truth, in the academy in particular, and in public discourse, and the coarsening of public discourse, [the] coarsening of politics."[576]

Colson, who was a significant figure in the 1970s Watergate scandal, the nation's worst political crisis, told the crowd at the Colson Center's Wilberforce Weekend:

> Everybody looks to the elections and thinks the elections will settle this problem or settle that problem. Elections are important. Whoever serves in office, it makes a difference what kind of person that is and what that person believes.
>
> But elections can't solve the problem we've got. The real problem that we've got is that our culture has been decaying from the inside for 30 or 40 years. And politics is nothing but an expression of culture. So…how do you fix the culture?
>
> So how do you fix the culture? Culture is actually formed by the belief system of the people, by the "cult," which is us, the church, [as it] has been historically. So if things are bad, don't think it's going to be solved by an election. It's going to be solved by us. You have a healthy cult, you have a healthy culture. You have a healthy culture, you have a healthy politics.
>
> So it comes right back to us. Look in the mirror, that's where the problem is. And if we can, through the church, renew the church to really bring a healthy cultural influence, then there's some hope that we can be changed.
>
> I think Eric [Metaxas, an American author best known for his book *Amazing Grace: William Wilberforce and the Heroic Campaign to End Slavery*] is right that this is a moment. This is a moment when the time is right for a movement of God's people under the power of the Holy Spirit to begin to impact the culture we live in. Desperately needed.

Colson was unable to finish his speech and he was helped off-stage by colleagues. He left us with an explanation of why our culture is so coarsened. Perhaps the rest of the speech would have exposed some of the evidence of the coarsening of American culture and then detailed just how "God's people under the power of the Holy Spirit" must "impact the culture."

Let me complete what Colson couldn't finish with a focus on the media's corrupting cultural influence.

We are constantly bombarded and surprised by media outlets spewing graphic images of sex, violence, hatred, misogyny, and out-of-bounds speech. The result is a very inhospitable society and a radical change from the last century.

Our cultural situation reminds me of the time after Joshua died and Israel took its eyes off God. Judges 17:6 (NASB) states: "In those days there was no king in Israel, but every man did that which was right in his own eyes."

Once Israel took its eyes off God, it lusted after material things, which *Matthew Henry's Concise Commentary* attributes to cultural "corruption."[577] *Adam Clarke's Commentary* goes further: "When a man's own will, passions, and caprice, are to be made the rule of law, society is in a most perilous and ruinous state."[578]

That's the condition of American culture today in the wake of abandoning God and the rule of law.

The twentieth-century's all-American respectable world of Potterville made famous in *It's a Wonderful Life* is virtually gone today and replaced by excessive sexual content thanks to the likes of so-called pop entertainers like Beyoncé, Madonna, Kanye West, Jay-Z, and Miley Cyrus. These cultural icons splash gratuitous violence, graphic sexual imagery, and vulgarity across our lives using media platforms like video games, television, Internet websites, cellphones, iPhones, and iPads. Even local radio stations play tunes that pound out their offensive lyrics that are no longer censored by a responsible government. Nothing shocks us anymore and we call it entertainment!

Mass media hasn't always been so irresponsible, however. John Seabrook, a writer for *The New Yorker* and author of *Nobrow: The Culture of Marketing—The Marketing of Culture*, explains in the early years, television "played a significant role in that standard-setting, enforcing certain decency among people. They took their role seriously, and the people behind the camera took their role seriously, too." That is no longer the case. Today, according to Seabrook, the media doesn't believe self-policing is important. Rather, that's the family's job.[579]

Even some so-called entertainers like Eminem, a coarse American rapper, believes he has no responsibility to self-police. Seabrook mimics Eminem's argument: "Well, it's up to the parents to see to that. If they don't want their children to watch me or listen to me, then they should not have the television set on." The rapper has a point; unfortunately, as outlined above, too few parents monitor their children's consumption of entertainment media.[580]

Clearly, when perpetrators of immorality like Eminem absolve themselves of direct responsibility for the consequences of their "entertainment," they create a perfect recipe for chaos.

That's an admonition for parents from an unlikely source, and one that is supported by Scripture. The *Matthew Henry Commentary* translates King Solomon's guidance in Proverbs 22:6 (NASB) as a charge for parents to mind every aspect of their children's upbringing. "Train up a child in the way he should go, even when he is old he will not depart from it."[581]

There's another aspect of the new coarser media that parents of young children must understand. Mark Crispin, a media critic and author of *Boxed In: The Culture of Television*, speaks of media's irresponsible marketing to young children.

Crispin said, "If you watch Saturday morning kids' television, you can see it in programming that is unrelievedly frantic, hyped-up, hysterical, and in its own way, quite violent and pervasively commercial." And that's the point, explains Crispin. "It's all about selling…the primary reason why there is something of a crisis nowadays, a cultural crisis involving children." Those who control the media are governed by "commercial logic and the commercial imperative overall…. They're trying to sell as much junk as they can by appealing to the worst in all of us, but they do it [by] some extremely civilized means."[582]

Parents must come to terms with the obvious: Television is addictive, while at the same time it's a great teacher of the things that aren't true. Today's commercialized television promotes bad values—concerning money, personal appearance, and fame—and meanwhile, it makes our kids into intellectual zombies by stifling creativity, lowering self-esteem, and promoting violence.[583]

What are our kids receiving from those electronic media devices? Abrasive, coarse language and horrible images are blasted out 24/7 by Twitter, Snapchat, Instagram, Facebook, U-Tube, and television in embarrassing repetition. Those often offensive words and images ripple through our handheld and other electronic devices every second of the day and even interrupt our sleep. Why? Those devices have addicted us to an unending blast of immoral trash that fills our minds corrupting our values.

The family is often attacked by these same technology-delivered messages, which challenge even the most involved parents who desperately try to protect their children. For example, many parents are unaware of the dangers associated with the likes of "applications" (apps) which their children often download to their handheld devices. These apps potentially put children at risk to sexual predators, cyber-bullying, and much more. Consider some of the most disturbing apps.[584]

The app known as Whisper allows the user to post secrets anonymously for strangers to read and to chat with those strangers.

Yikyak is a very dangerous app because it "can turn a school into a virtual chat room where everyone can post his or her comments, anonymously." That's an invitation to cyber-bullying because students use the app to slander teachers and other children.

Snapchat allows the user to make an image or video available to another user for a limited time and then the picture/video disappears. Evidently this app makes children who send sexually inappropriate photos (known as "sexting") feel better, because the image self-destructs automatically.

Poof hides other apps on the iPhone, denying responsible parents from finding the likes of Snapchat and Yikyak on their child's device.

Internet safety is getting more complex for concerned parents. The best approach is to teach children about the dangers, but, like former President Ronald Reagan said, "Trust but verify." Parents can verify their children's use of apps with the help of guides like the Net Smartz Workshop, netsmartz.org, a public service provided by the National Center for Missing & Exploited Children.

It seems as if the media purveyors of filth don't know right from wrong and certainly they have no respect for Christian values. No, it's about money and sex, and violence sells. So what to do?

Once again, intact involved families are the first line of defense. Parents need to set an example of responsible media consumption, and they need to monitor their children's use. They need to discuss the temptations and explain the damage done by media's coarsening of society.

Parents can always turn off the screens or block access to certain media sites. You don't have to give your child an iPhone until he or she is ready, and then monitor that use as well.

Clearly, Americans can use their buying power to send a clear message to media filth purveyors through advertisers to clean up their act. Meanwhile, government has a role as well. We need to call for standard setting and enforcement, which is called for in the Scriptures.

Psalm 82:3–4 (ESV) states: "Give justice to the weak and the fatherless; maintain the right of the afflicted and the destitute. Rescue the weak and the needy; deliver them from the hand of the wicked."

The government of the people has a moral obligation to protect the "weak and fatherless" from the "hand of the wicked" media.

Conclusion

America's culture is spinning out of control, as evidenced by an abandonment of God and His Word, an embracing of immoral leaders like

Obama, the disintegration of the intact family, and an unrestrained corrupting media.

This chapter explains the sad state of our culture and outlines how moral restoration is possible assuming the Lord tarries. We must believe America is worth fighting for or otherwise our culture is doomed.

Conclusion

Spinning-Out-of-Control America

FUTURE WAR is about our country's last chance to turn back to God before the end times. No one knows for certain whether the end times are right at our doorstep, but many prophetic signs seem to be lighting up the world. The end-times clock is known only to God (Matthew 24:36), which leaves us with but two choices: obediently carry on His work (Matthew 28:20) or follow the ways of the world so evident today in our crumbling culture.

America does not appear to be a factor in the Bible's end-times prophecies. From beginning to end, the Bible is focused on one piece of real estate on the planet: Israel. Other nations mentioned in this volume appear to have critical roles such as Iran, Jordan, and Egypt. Arguments are made that Russia and China are alluded to in prophecy, and maybe the European Union. But the United States is missing, which leaves Americans with a pregnant question: What happens to America?

I'm not a prophet, but I can speculate given the mounting evidence. One possibility is that America, as powerful as she is today, is deci-mated—no longer exists at the end times. Clearly the threats outlined in section II of this volume are frightening and could very well explain scenarios in which the U.S. is destroyed.

A second alternative is America becomes irrelevant. That scenario can possibly be understood from the emergence of the nonpolar reality outlined in chapter 4; many powers emerge and coalesce to dominate every aspect of life—security, economic—thus making the once-great America irrelevant, a pathetic giant on a chaotic world stage.

There is a hopeful alternative, and that is the outcome this author espouses in section III: America turns back to God. Stranger things have happened in history whereby entire nations and people switch direction—and that's what contemporary America needs today.

Our future as outlined in section I is pretty bleak. What's needed is a revival of the so-called Christian majority in America—a turning back to the God of our salvation through His son Jesus Christ. That's the seed that spawned the biblical worldview chapters that put legs on the vision in chapter 10 of this volume.

America needs godly leaders and people who capture a biblical vision and then carry it forth to execution through a future government. Those biblical-worldview leaders will consider the five domains of government and society (chapters 11–15) that translate godly, biblical values and principles into meaningful policies for our future government.

However, government alone is not the answer. In fact, it is often the root of the problem, as former President Reagan once said. It is doubtful that any president will ever hold the power and personality to single-handedly turn America back toward God. Our hope in government can go no further than putting the right godly men and women in positions of leadership to ease the way for a return to God throughout the nation. No law or executive order can make a return happen. The power lies within the individual Christian joining with others and being involved in the politics and service to others in their local arena; that's where the real work needs to done.

And just where does this rejuvenation of our nation begin? The church!

Believers must step out in faith and obediently embrace their biblical mandate. Like the writer of 2 Chronicles 7:14 advocates:

If my people, who are called by my name, will humble themselves and pray and seek my face and turn from their wicked
ways, then I will hear from heaven, and I will forgive their sin
and will heal their land.

America's founders came to this land on bent knee and with hum-
bled hearts seeking the freedom to worship and liberty. Today, godly
men across America must evidence a similar faith through their words
and actions if our land is to be healed. That's needed in the present
day, and if America does obediently surrender to God's love, there is
hope—and where there is hope, radical change as outlined in section III
is possible.

It just might be that America as we know it at present, which is
morally imploding, does turn back to God and recapture our founders'
vision for this great land—liberty, "In God We Trust" and *e pluribus
unum*. That possibility excites those of us who daily pray for revival in
America and believe this once-great nation isn't to be seen in the end-
times Scriptures because Christ returned in the air as He promised in
1 Thessalonians 4:15–16 and takes us to heaven before He initiates the
final judgments on an unbelieving and degenerate world.

That's my hope for America today. I don't want to believe that this
once-great land continues to morally decline into the abyss. No, it is my
prayer that today's Christians will answer Christ's call to obedience and
share the only hope this world has—the salvation of Jesus Christ. Then
we collectively can use the strategy outlined in section III of this volume
as a reference point to help guide the revived nation of Christian believers to harness the Word of God to serve Him until that day when we
join Him as the body of Christ in heaven.

Amen!

Appendix A

U.S. Constitution

The Constitution of the United States: A Transcription[585]

Note: The following text is a transcription of the Constitution as it was inscribed by Jacob Shallus on parchment (the document on display in the Rotunda at the National Archives Museum.) Items that are hyperlinked have since been amended or superseded. The authenticated text of the Constitution can be found on the website of the Government Printing Office.

WE THE PEOPLE of the United States, in Order to form a more perfect Union, establish Justice, insure domestic Tranquility, provide for the common defence, promote the general Welfare, and secure the Blessings of Liberty to ourselves and our Posterity, do ordain and establish this Constitution for the United States of America.

Article. I.

Section. 1.

All legislative Powers herein granted shall be vested in a Congress of the United States, which shall consist of a Senate and House of Representatives.

Section. 2.

The House of Representatives shall be composed of Members chosen every second Year by the People of the several States, and the Electors in each State shall have the Qualifications requisite for Electors of the most numerous Branch of the State Legislature.

No Person shall be a Representative who shall not have attained to the Age of twenty five Years, and been seven Years a Citizen of the United States, and who shall not, when elected, be an Inhabitant of that State in which he shall be chosen.

Representatives and direct Taxes shall be apportioned among the several States which may be included within this Union, according to their respective Numbers, which shall be determined by adding to the whole Number of free Persons, including those bound to Service for a Term of Years, and excluding Indians not taxed, three fifths of all other Persons. The actual Enumeration shall be made within three Years after the first Meeting of the Congress of the United States, and within every subsequent Term of ten Years, in such Manner as they shall by Law direct. The Number of Representatives shall not exceed one for every thirty Thousand, but each State shall have at Least one Representative; and until such enumeration shall be made, the State of New Hampshire shall be entitled to chuse three, Massachusetts eight, Rhode-Island and Providence Plantations one, Connecticut five, New-York six, New Jersey four, Pennsylvania eight, Delaware one, Maryland six, Virginia ten, North Carolina five, South Carolina five, and Georgia three.

When vacancies happen in the Representation from any State, the Executive Authority thereof shall issue Writs of Election to fill such Vacancies.

The House of Representatives shall chose their Speaker and other Officers; and shall have the sole Power of Impeachment.

Section. 3.

The Senate of the United States shall be composed of two Senators from each State, chosen by the Legislature thereof, for six Years; and each Senator shall have one Vote.

Immediately after they shall be assembled in Consequence of the first Election, they shall be divided as equally as may be into three Classes. The Seats of the Senators of the first Class shall be vacated at the Expiration of the second Year, of the second Class at the Expiration of the fourth Year, and of the third Class at the Expiration of the sixth Year, so that one third may be chosen every second Year; and if Vacancies happen by Resignation, or otherwise, during the Recess of the Legislature of any State, the Executive thereof may make temporary Appointments until the next Meeting of the Legislature, which shall then fill such Vacancies.

No Person shall be a Senator who shall not have attained to the Age of thirty Years, and been nine Years a Citizen of the United States, and who shall not, when elected, be an Inhabitant of that State for which he shall be chosen.

The Vice President of the United States shall be President of the Senate, but shall have no Vote, unless they be equally divided.

The Senate shall chuse their other Officers, and also a President pro tempore, in the Absence of the Vice President, or when he shall exercise the Office of President of the United States.

The Senate shall have the sole Power to try all Impeachments. When sitting for that Purpose, they shall be on Oath or Affirmation. When the President of the United States is tried, the Chief Justice shall preside: And no Person shall be convicted without the Concurrence of two thirds of the Members present.

Judgment in Cases of Impeachment shall not extend further than to removal from Office, and disqualification to hold and enjoy any Office of honor, Trust or Profit under the United States: but the Party convicted shall nevertheless be liable and subject to Indictment, Trial, Judgment and Punishment, according to Law.

Section. 4.

The Times, Places and Manner of holding Elections for Senators and Representatives, shall be prescribed in each State by the Legislature thereof; but the Congress may at any time by Law make or alter such Regulations, except as to the Places of chusing Senators.

The Congress shall assemble at least once in every Year, and such Meeting shall be on the first Monday in December, unless they shall by Law appoint a different Day.

Section. 5.

Each House shall be the Judge of the Elections, Returns and Qualifications of its own Members, and a Majority of each shall constitute a Quorum to do Business; but a smaller Number may adjourn from day to day, and may be authorized to compel the Attendance of absent Members, in such Manner, and under such Penalties as each House may provide.

Each House may determine the Rules of its Proceedings, punish its Members for disorderly Behaviour, and, with the Concurrence of two thirds, expel a Member.

Each House shall keep a Journal of its Proceedings, and from time to time publish the same, excepting such Parts as may in their Judgment require Secrecy; and the Yeas and Nays of the Members of either House on any question shall, at the Desire of one fifth of those Present, be entered on the Journal.

Neither House, during the Session of Congress, shall, without the Consent of the other, adjourn for more than three days, nor to any other Place than that in which the two Houses shall be sitting.

Section. 6.

The Senators and Representatives shall receive a Compensation for their Services, to be ascertained by Law, and paid out of the Treasury of the

United States. They shall in all Cases, except Treason, Felony and Breach of the Peace, be privileged from Arrest during their Attendance at the Session of their respective Houses, and in going to and returning from the same; and for any Speech or Debate in either House, they shall not be questioned in any other Place.

No Senator or Representative shall, during the Time for which he was elected, be appointed to any civil Office under the Authority of the United States, which shall have been created, or the Emoluments whereof shall have been encreased during such time; and no Person holding any Office under the United States, shall be a Member of either House during his Continuance in Office.

Section. 7.

All Bills for raising Revenue shall originate in the House of Representatives; but the Senate may propose or concur with Amendments as on other Bills.

Every Bill which shall have passed the House of Representatives and the Senate, shall, before it become a Law, be presented to the President of the United States; If he approve he shall sign it, but if not he shall return it, with his Objections to that House in which it shall have originated, who shall enter the Objections at large on their Journal, and proceed to reconsider it. If after such Reconsideration two thirds of that House shall agree to pass the Bill, it shall be sent, together with the Objections, to the other House, by which it shall likewise be reconsidered, and if approved by two thirds of that House, it shall become a Law. But in all such Cases the Votes of both Houses shall be determined by yeas and Nays, and the Names of the Persons voting for and against the Bill shall be entered on the Journal of each House respectively. If any Bill shall not be returned by the President within ten Days (Sundays excepted) after it shall have been presented to him, the Same shall be a Law, in like Manner as if he had signed it, unless the Congress by their Adjournment prevent its Return, in which Case it shall not be a Law.

Every Order, Resolution, or Vote to which the Concurrence of the Senate and House of Representatives may be necessary (except on a question of Adjournment) shall be presented to the President of the United States; and before the Same shall take Effect, shall be approved by him, or being disapproved by him, shall be repassed by two thirds of the Senate and House of Representatives, according to the Rules and Limitations prescribed in the Case of a Bill.

Section. 8.

The Congress shall have Power To lay and collect Taxes, Duties, Imposts and Excises, to pay the Debts and provide for the common Defence and general Welfare of the United States; but all Duties, Imposts and Excises shall be uniform throughout the United States;

To borrow Money on the credit of the United States;

To regulate Commerce with foreign Nations, and among the several States, and with the Indian Tribes;

To establish an uniform Rule of Naturalization, and uniform Laws on the subject of Bankruptcies throughout the United States;

To coin Money, regulate the Value thereof, and of foreign Coin, and fix the Standard of Weights and Measures;

To provide for the Punishment of counterfeiting the Securities and current Coin of the United States;

To establish Post Offices and post Roads;

To promote the Progress of Science and useful Arts, by securing for limited Times to Authors and Inventors the exclusive Right to their respective Writings and Discoveries;

To constitute Tribunals inferior to the supreme Court;

To define and punish Piracies and Felonies committed on the high Seas, and Offences against the Law of Nations;

To declare War, grant Letters of Marque and Reprisal, and make Rules concerning Captures on Land and Water;

To raise and support Armies, but no Appropriation of Money to that Use shall be for a longer Term than two Years;

To provide and maintain a Navy;

To make Rules for the Government and Regulation of the land and naval Forces;

To provide for calling forth the Militia to execute the Laws of the Union, suppress Insurrections and repel Invasions;

To provide for organizing, arming, and disciplining, the Militia, and for governing such Part of them as may be employed in the Service of the United States, reserving to the States respectively, the Appointment of the Officers, and the Authority of training the Militia according to the discipline prescribed by Congress;

To exercise exclusive Legislation in all Cases whatsoever, over such District (not exceeding ten Miles square) as may, by Cession of particular States, and the Acceptance of Congress, become the Seat of the Government of the United States, and to exercise like Authority over all Places purchased by the Consent of the Legislature of the State in which the Same shall be, for the Erection of Forts, Magazines, Arsenals, dock-Yards, and other needful Buildings;—And

To make all Laws which shall be necessary and proper for carrying into Execution the foregoing Powers, and all other Powers vested by this Constitution in the Government of the United States, or in any Department or Officer thereof.

Section. 9.

The Migration or Importation of such Persons as any of the States now existing shall think proper to admit, shall not be prohibited by the Congress prior to the Year one thousand eight hundred and eight, but a Tax or duty may be imposed on such Importation, not exceeding ten dollars for each Person.

The Privilege of the Writ of Habeas Corpus shall not be suspended, unless when in Cases of Rebellion or Invasion the public Safety may require it.

No Bill of Attainder or ex post facto Law shall be passed.

No Capitation, or other direct, Tax shall be laid, unless in

Proportion to the Census or enumeration herein before directed to be taken.

No Tax or Duty shall be laid on Articles exported from any State.

No Preference shall be given by any Regulation of Commerce or Revenue to the Ports of one State over those of another: nor shall Vessels bound to, or from, one State, be obliged to enter, clear, or pay Duties in another.

No Money shall be drawn from the Treasury, but in Consequence of Appropriations made by Law; and a regular Statement and Account of the Receipts and Expenditures of all public Money shall be published from time to time.

No Title of Nobility shall be granted by the United States: And no Person holding any Office of Profit or Trust under them, shall, without the Consent of the Congress, accept of any present, Emolument, Office, or Title, of any kind whatever, from any King, Prince, or foreign State.

Section. 10.

No State shall enter into any Treaty, Alliance, or Confederation; grant Letters of Marque and Reprisal; coin Money; emit Bills of Credit; make any Thing but gold and silver Coin a Tender in Payment of Debts; pass any Bill of Attainder, ex post facto Law, or Law impairing the Obligation of Contracts, or grant any Title of Nobility.

No State shall, without the Consent of the Congress, lay any Imposts or Duties on Imports or Exports, except what may be absolutely neces-sary for executing it's inspection Laws: and the net Produce of all Duties and Imposts, laid by any State on Imports or Exports, shall be for the Use of the Treasury of the United States; and all such Laws shall be sub-ject to the Revision and Controul of the Congress.

No State shall, without the Consent of Congress, lay any Duty of Tonnage, keep Troops, or Ships of War in time of Peace, enter into any Agreement or Compact with another State, or with a foreign Power, or engage in War, unless actually invaded, or in such imminent Danger as will not admit of delay.

Article. II.

Section. 1.

The executive Power shall be vested in a President of the United States of America. He shall hold his Office during the Term of four Years, and, together with the Vice President, chosen for the same Term, be elected, as follows

Each State shall appoint, in such Manner as the Legislature thereof may direct, a Number of Electors, equal to the whole Number of Senators and Representatives to which the State may be entitled in the Congress: but no Senator or Representative, or Person holding an Office of Trust or Profit under the United States, shall be appointed an Elector.

The Electors shall meet in their respective States, and vote by Ballot for two Persons, of whom one at least shall not be an Inhabitant of the same State with themselves. And they shall make a List of all the Persons voted for, and of the Number of Votes for each; which List they shall sign and certify, and transmit sealed to the Seat of the Government of the United States, directed to the President of the Senate. The President of the Senate shall, in the Presence of the Senate and House of Representatives, open all the Certificates, and the Votes shall then be counted. The Person having the greatest Number of Votes shall be the President, if such Number be a Majority of the whole Number of Electors appointed; and if there be more than one who have such Majority, and have an equal Number of Votes, then the House of Representatives shall immediately chuse by Ballot one of them for President; and if no Person have a Majority, then from the five highest on the List the said House shall in like Manner chuse the President. But in chusing the President, the Votes shall be taken by States, the Representation from each State having one Vote; A quorum for this Purpose shall consist of a Member or Members from two thirds of the States, and a Majority of all the States shall be necessary to a Choice. In every Case, after the Choice of the President, the Person having the greatest Number of Votes of the Electors shall be the Vice President. But if there should remain two or more who have equal Votes, the Senate shall chuse from them by Ballot the Vice President.

The Congress may determine the Time of chusing the Electors, and the Day on which they shall give their Votes; which Day shall be the same throughout the United States.

No Person except a natural born Citizen, or a Citizen of the United States, at the time of the Adoption of this Constitution, shall be eligible to the Office of President; neither shall any Person be eligible to that Office who shall not have attained to the Age of thirty five Years, and been fourteen Years a Resident within the United States.

In Case of the Removal of the President from Office, or of his Death, Resignation, or Inability to discharge the Powers and Duties of the said Office, the Same shall devolve on the Vice President, and the Congress may by Law provide for the Case of Removal, Death, Resignation or Inability, both of the President and Vice President, declaring what Officer shall then act as President, and such Officer shall act accordingly, until the Disability be removed, or a President shall be elected.

The President shall, at stated Times, receive for his Services, a Compensation, which shall neither be encreased nor diminished during the Period for which he shall have been elected, and he shall not receive within that Period any other Emolument from the United States, or any of them.

Before he enter on the Execution of his Office, he shall take the following Oath or Affirmation:—"I do solemnly swear (or affirm) that I will faithfully execute the Office of President of the United States, and will to the best of my Ability, preserve, protect and defend the Constitution of the United States."

Section. 2.

The President shall be Commander in Chief of the Army and Navy of the United States, and of the Militia of the several States, when called into the actual Service of the United States; he may require the Opinion, in writing, of the principal Officer in each of the executive Departments, upon any Subject relating to the Duties of their respective Offices, and he shall have Power to grant Reprieves and Pardons for Offences against the United States, except in Cases of Impeachment.

He shall have Power, by and with the Advice and Consent of the Senate, to make Treaties, provided two thirds of the Senators present concur; and he shall nominate, and by and with the Advice and Consent of the Senate, shall appoint Ambassadors, other public Ministers and Consuls, Judges of the supreme Court, and all other Officers of the United States, whose Appointments are not herein otherwise provided for, and which shall be established by Law: but the Congress may by Law vest the Appointment of such inferior Officers, as they think proper, in the President alone, in the Courts of Law, or in the Heads of Departments.

The President shall have Power to fill up all Vacancies that may happen during the Recess of the Senate, by granting Commissions which shall expire at the End of their next Session.

Section. 3.

He shall from time to time give to the Congress Information of the State of the Union, and recommend to their Consideration such Measures as he shall judge necessary and expedient; he may, on extraordinary Occasions, convene both Houses, or either of them, and in Case of Disagreement between them, with Respect to the Time of Adjournment, he may adjourn them to such Time as he shall think proper; he shall receive Ambassadors and other public Ministers; he shall take Care that the Laws be faithfully executed, and shall Commission all the Officers of the United States.

Section. 4.

The President, Vice President and all civil Officers of the United States, shall be removed from Office on Impeachment for, and Conviction of, Treason, Bribery, or other high Crimes and Misdemeanors.

Article III.

Section. 1.

The judicial Power of the United States, shall be vested in one supreme Court, and in such inferior Courts as the Congress may from time to time ordain and establish. The Judges, both of the supreme and inferior Courts, shall hold their Offices during good Behaviour, and shall, at stated Times, receive for their Services, a Compensation, which shall not be diminished during their Continuance in Office.

Section. 2.

The judicial Power shall extend to all Cases, in Law and Equity, arising under this Constitution, the Laws of the United States, and Treaties made, or which shall be made, under their Authority;—to all Cases affecting Ambassadors, other public Ministers and Consuls;—to all Cases of admiralty and maritime Jurisdiction;—to Controversies to which the United States shall be a Party;—to Controversies between two or more States;—between a State and Citizens of another State,—between Citizens of different States,—between Citizens of the same State claiming Lands under Grants of different States, and between a State, or the Citizens thereof, and foreign States, Citizens or Subjects.

In all Cases affecting Ambassadors, other public Ministers and Consuls, and those in which a State shall be Party, the supreme Court shall have original Jurisdiction. In all the other Cases before mentioned, the supreme Court shall have appellate Jurisdiction, both as to Law and Fact, with such Exceptions, and under such Regulations as the Congress shall make.

The Trial of all Crimes, except in Cases of Impeachment, shall be by Jury; and such Trial shall be held in the State where the said Crimes shall have been committed; but when not committed within any State, the

Trial shall be at such Place or Places as the Congress may by Law have directed.

Section. 3.

Treason against the United States, shall consist only in levying War against them, or in adhering to their Enemies, giving them Aid and Comfort. No Person shall be convicted of Treason unless on the Testimony of two Witnesses to the same overt Act, or on Confession in open Court.

The Congress shall have Power to declare the Punishment of Treason, but no Attainder of Treason shall work Corruption of Blood, or Forfeiture except during the Life of the Person attainted.

Article. IV.

Section. 1.

Full Faith and Credit shall be given in each State to the public Acts, Records, and judicial Proceedings of every other State. And the Congress may by general Laws prescribe the Manner in which such Acts, Records and Proceedings shall be proved, and the Effect thereof.

Section. 2.

The Citizens of each State shall be entitled to all Privileges and Immunities of Citizens in the several States.

A Person charged in any State with Treason, Felony, or other Crime, who shall flee from Justice, and be found in another State, shall on Demand of the executive Authority of the State from which he fled, be delivered up, to be removed to the State having Jurisdiction of the Crime.

No Person held to Service or Labour in one State, under the Laws thereof, escaping into another, shall, in Consequence of any Law or

Regulation therein, be discharged from such Service or Labour, but shall be delivered up on Claim of the Party to whom such Service or Labour may be due.

Section. 3.

New States may be admitted by the Congress into this Union; but no new State shall be formed or erected within the Jurisdiction of any other State; nor any State be formed by the Junction of two or more States, or Parts of States, without the Consent of the Legislatures of the States concerned as well as of the Congress.

The Congress shall have Power to dispose of and make all needful Rules and Regulations respecting the Territory or other Property belonging to the United States; and nothing in this Constitution shall be so construed as to Prejudice any Claims of the United States, or of any particular State.

Section. 4.

The United States shall guarantee to every State in this Union a Republican Form of Government, and shall protect each of them against Invasion; and on Application of the Legislature, or of the Executive (when the Legislature cannot be convened), against domestic Violence.

Article. V.

The Congress, whenever two thirds of both Houses shall deem it necessary, shall propose Amendments to this Constitution, or, on the Application of the Legislatures of two thirds of the several States, shall call a Convention for proposing Amendments, which, in either Case, shall be valid to all Intents and Purposes, as Part of this Constitution, when ratified by the Legislatures of three fourths of the several States, or by Conventions in three fourths thereof, as the one or the other Mode of Ratification may be proposed by the Congress; Provided that no Amendment which may be made prior to the Year One thousand eight hundred and eight shall in any Manner affect the first and fourth Clauses in the Ninth Section of the first Article; and that no State, without its Consent, shall be deprived of its equal Suffrage in the Senate.

Article. VI.

All Debts contracted and Engagements entered into, before the Adoption of this Constitution, shall be as valid against the United States under this Constitution, as under the Confederation.

This Constitution, and the Laws of the United States which shall be made in Pursuance thereof; and all Treaties made, or which shall be made, under the Authority of the United States, shall be the supreme Law of the Land; and the Judges in every State shall be bound thereby, any Thing in the Constitution or Laws of any State to the Contrary notwithstanding.

The Senators and Representatives before mentioned, and the Members of the several State Legislatures, and all executive and judicial Officers, both of the United States and of the several States, shall be bound by Oath or Affirmation, to support this Constitution; but no religious Test shall ever be required as a Qualification to any Office or public Trust under the United States.

Article. VII.

The Ratification of the Conventions of nine States, shall be sufficient for the Establishment of this Constitution between the States so ratifying the Same.

The Word, "the," being interlined between the seventh and eighth

Lines of the first Page, The Word "Thirty" being partly written on an Erazure in the fifteenth Line of the first Page, The Words "is tried" being interlined between the thirty second and thirty third Lines of the first Page and the Word "the" being interlined between the forty third and forty fourth Lines of the second Page.

Attest William Jackson Secretary

done in Convention by the Unanimous Consent of the States present the Seventeenth Day of September in the Year of our Lord one thousand seven hundred and Eighty seven and of the Independance of the United States of America the Twelfth In witness whereof We have hereunto subscribed our Names,

G°. Washington
Presidt and deputy from Virginia

Delaware
Geo: Read
Gunning Bedford jun
John Dickinson
Richard Bassett
Jaco: Broom

Maryland
James McHenry
Dan of St Thos. Jenifer
Danl. Carroll

Virginia
John Blair
James Madison Jr.

North Carolina
Wm. Blount
Richd. Dobbs Spaight
Hu Williamson

South Carolina
J. Rutledge
Charles Cotesworth Pinckney
Charles Pinckney
Pierce Butler

Georgia
William Few
Abr Baldwin

New Hampshire
John Langdon
Nicholas Gilman

Massachusetts
Nathaniel Gorham
Rufus King

Connecticut
Wm. Saml. Johnson
Roger Sherman

New York
Alexander Hamilton

New Jersey
Wil: Livingston
David Brearley
Wm. Paterson
Jona: Dayton

Pennsylvania
B Franklin
Thomas Mifflin
Robt. Morris
Geo. Clymer
Thos. FitzSimons
Jared Ingersoll
James Wilson
Gouv Morris

U.S. Declaration of Independence

The Declaration of Independence: A Transcription[586]

IN CONGRESS, July 4, 1776.

The unanimous Declaration of the thirteen [U]nited States of America,

When in the Course of human events, it becomes necessary for one people to dissolve the political bands which have connected them with another, and to assume among the powers of the earth, the separate and equal station to which the Laws of Nature and of Nature's God entitle them, a decent respect to the opinions of mankind requires that they should declare the causes which impel them to the separation.

We hold these truths to be self-evident, that all men are created equal, that they are endowed by their Creator with certain unalienable Rights, that among these are Life, Liberty and the pursuit of Happiness.—That to secure these rights, Governments are instituted among Men, deriving their just powers from the consent of the governed,—That whenever any Form of Government becomes destructive of these ends, it is the Right of the People to alter or to abolish it, and to institute new Government, laying its foundation on such principles and organizing its powers in such form, as to them shall seem most likely to effect their Safety and Happiness. Prudence, indeed, will dictate that

Governments long established should not be changed for light and transient causes; and accordingly all experience hath shewn, that mankind are more disposed to suffer, while evils are sufferable, than to right themselves by abolishing the forms to which they are accustomed. But when a long train of abuses and usurpations, pursuing invariably the same Object evinces a design to reduce them under absolute Despotism, it is their right, it is their duty, to throw off such Government, and to provide new Guards for their future security.—Such has been the patient sufferance of these Colonies; and such is now the necessity which constrains them to alter their former Systems of Government. The history of the present King of Great Britain is a history of repeated injuries and usurpations, all having in direct object the establishment of an absolute Tyranny over these States. To prove this, let Facts be submitted to a candid world.

He has refused his Assent to Laws, the most wholesome and necessary for the public good.

He has forbidden his Governors to pass Laws of immediate and pressing importance, unless suspended in their operation till his Assent should be obtained; and when so suspended, he has utterly neglected to attend to them.

He has refused to pass other Laws for the accommodation of large districts of people, unless those people would relinquish the right of Representation in the Legislature, a right inestimable to them and formidable to tyrants only.

He has called together legislative bodies at places unusual, uncomfortable, and distant from the depository of their public Records, for the sole purpose of fatiguing them into compliance with his measures.

He has dissolved Representative Houses repeatedly, for opposing with manly firmness his invasions on the rights of the people.

He has refused for a long time, after such dissolutions, to cause others to be elected; whereby the Legislative powers, incapable of Annihilation, have returned to the People at large for their exercise; the State remaining in the mean time exposed to all the dangers of invasion from without, and convulsions within.

He has endeavoured to prevent the population of these States; for that purpose obstructing the Laws for Naturalization of Foreigners; refusing to pass others to encourage their migrations hither, and raising the conditions of new Appropriations of Lands.

He has obstructed the Administration of Justice, by refusing his Assent to Laws for establishing Judiciary powers.

He has made Judges dependent on his Will alone, for the tenure of their offices, and the amount and payment of their salaries.

He has erected a multitude of New Offices, and sent hither swarms of Officers to harrass our people, and eat out their substance.

He has kept among us, in times of peace, Standing Armies without the Consent of our legislatures.

He has affected to render the Military independent of and superior to the Civil power.

He has combined with others to subject us to a jurisdiction foreign to our constitution, and unacknowledged by our laws; giving his Assent to their Acts of pretended Legislation:

For Quartering large bodies of armed troops among us:

For protecting them, by a mock Trial, from punishment for any Murders which they should commit on the Inhabitants of these States:

For cutting off our Trade with all parts of the world:

For imposing Taxes on us without our Consent:

For depriving us in many cases, of the benefits of Trial by Jury:

For transporting us beyond Seas to be tried for pretended offences

For abolishing the free System of English Laws in a neighbouring Province, establishing therein an Arbitrary government, and enlarging its Boundaries so as to render it at once an example and fit instrument for introducing the same absolute rule into these Colonies:

For taking away our Charters, abolishing our most valuable Laws, and altering fundamentally the Forms of our Governments:

For suspending our own Legislatures, and declaring themselves invested with power to legislate for us in all cases whatsoever.

He has abdicated Government here, by declaring us out of his Protection and waging War against us.

He has plundered our seas, ravaged our Coasts, burnt our towns, and destroyed the lives of our people.

He is at this time transporting large Armies of foreign Mercenaries to compleat the works of death, desolation and tyranny, already begun with circumstances of Cruelty & perfidy scarcely paralleled in the most barbarous ages, and totally unworthy the Head of a civilized nation.

He has constrained our fellow Citizens taken Captive on the high Seas to bear Arms against their Country, to become the executioners of their friends and Brethren, or to fall themselves by their Hands.

He has excited domestic insurrections amongst us, and has endeavoured to bring on the inhabitants of our frontiers, the merciless Indian Savages, whose known rule of warfare, is an undistinguished destruction of all ages, sexes and conditions.

In every stage of these Oppressions We have Petitioned for Redress in the most humble terms: Our repeated Petitions have been answered only by repeated injury. A Prince whose character is thus marked by every act which may define a Tyrant, is unfit to be the ruler of a free people.

Nor have We been wanting in attentions to our British brethren. We have warned them from time to time of attempts by their legislature to extend an unwarrantable jurisdiction over us. We have reminded them of the circumstances of our emigration and settlement here. We have appealed to their native justice and magnanimity, and we have conjured them by the ties of our common kindred to disavow these usurpations, which, would inevitably interrupt our connections and correspondence. They too have been deaf to the voice of justice and of consanguinity. We must, therefore, acquiesce in the necessity, which denounces our Separation, and hold them, as we hold the rest of mankind, Enemies in War, in Peace Friends.

We, therefore, the Representatives of the united States of America, in General Congress, Assembled, appealing to the Supreme Judge of the world for the rectitude of our intentions, do, in the Name, and by Authority of the good People of these Colonies, solemnly publish and declare, That these United Colonies are, and of Right ought to be Free and Independent States; that they are Absolved from all Allegiance to the British Crown, and that all political connection between them and the State of Great Britain, is and ought to be totally dissolved; and that as Free and Independent States, they have full Power to levy War, conclude Peace, contract Alliances, establish Commerce, and to do all other Acts and Things which Independent States may of right do. And for the support of this Declaration, with a firm reliance on the protection of divine Providence, we mutually pledge to each other our Lives, our Fortunes and our sacred Honor.

The 56 signatures on the Declaration appear in the positions indicated:

Column 1
Georgia:
Button Gwinnett
Lyman Hall
George Walton

Column 2
North Carolina:
William Hooper
Joseph Hewes
John Penn

South Carolina:
Edward Rutledge
Thomas Heyward, Jr.
Thomas Lynch, Jr.
Arthur Middleton

Column 3
Massachusetts:
John Hancock

Maryland:
Samuel Chase
William Paca
Thomas Stone
Charles Carroll of Carrollton

Virginia:
George Wythe
Richard Henry Lee
Thomas Jefferson
Benjamin Harrison
Thomas Nelson, Jr.
Francis Lightfoot Lee
Carter Braxton

Column 4
Pennsylvania:
Robert Morris
Benjamin Rush
Benjamin Franklin
John Morton

George Clymer
James Smith
George Taylor
James Wilson
George Ross

Delaware:
Caesar Rodney
George Read
Thomas McKean

Column 5

New York:
William Floyd
Philip Livingston
Francis Lewis
Lewis Morris

New Jersey:
Richard Stockton
John Witherspoon
Francis Hopkinson
John Hart
Abraham Clark

Column 6

New Hampshire:
Josiah Bartlett
William Whipple

Massachusetts:
Samuel Adams
John Adams

Robert Treat Paine
Elbridge Gerry
Rhode Island:
Stephen Hopkins
William Ellery

Connecticut:
Roger Sherman
Samuel Huntington
William Williams
Oliver Wolcott

New Hampshire:
Matthew Thornton

U.S. Bill of Rights

ON SEPTEMBER 25, 1789, the First Congress of the United States proposed twelve amendments to the Constitution. The 1789 Joint Resolution of Congress proposing the amendments is on display in the Rotunda in the National Archives Museum. Ten of the proposed twelve amendments were ratified by three-fourths of the state legislatures on December 15, 1791. The ratified Articles (Articles 3–12) constitute the first ten amendments of the Constitution, or the U.S. Bill of Rights. In 1992, 203 years after it was proposed, Article 2 was ratified as the 27th Amendment to the Constitution. Article 1 was never ratified.

Transcription of the 1789 Joint Resolution of Congress Proposing 12 Amendments to the U.S. Constitution

Congress of the United States begun and held at the City of New-York, on Wednesday the fourth of March, one thousand seven hundred and eighty nine.

THE Conventions of a number of the States, having at the time of their adopting the Constitution, expressed a desire, in order to prevent misconstruction or abuse of its powers, that further declaratory and restrictive clauses should be added: And as extending the ground of public confidence in the Government, will best ensure the beneficent ends of its institution.

RESOLVED by the Senate and House of Representatives of the United States of America, in Congress assembled, two thirds of both Houses concurring, that the following Articles be proposed to the Legislatures of the several States, as amendments to the Constitution of the United States, all, or any of which Articles, when ratified by three fourths of the said Legislatures, to be valid to all intents and purposes, as part of the said Constitution; viz.

ARTICLES in addition to, and Amendment of the Constitution of the United States of America, proposed by Congress, and ratified by the Legislatures of the several States, pursuant to the fifth Article of the original Constitution.

Article the first... After the first enumeration required by the first article of the Constitution, there shall be one Representative for every thirty thousand, until the number shall amount to one hundred, after which the proportion shall be so regulated by Congress, that there shall be not less than one hundred Representatives, nor less than one Representative for every forty thousand persons, until the number of Representatives shall amount to two hundred; after which the proportion shall be so regulated by Congress, that there shall not be less than two hundred Representatives, nor more than one Representative for every fifty thousand persons.

Article the second... No law, varying the compensation for the services of the Senators and Representatives, shall take effect, until an election of Representatives shall have intervened.

Article the third... Congress shall make no law respecting an establishment of religion, or prohibiting the free exercise thereof; or abridging the freedom of speech, or of the press; or the right of the people peaceably to assemble, and to petition the Government for a redress of grievances.

Article the fourth... A well regulated Militia, being necessary to the security of a free State, the right of the people to keep and bear Arms, shall not be infringed.

Article the fifth... No Soldier shall, in time of peace be quartered in any house, without the consent of the Owner, nor in time of war, but in a manner to be prescribed by law.

Article the sixth... The right of the people to be secure in their persons, houses, papers, and effects, against unreasonable searches and seizures, shall not be violated, and no Warrants shall issue, but upon probable cause, supported by Oath or affirmation, and particularly describing the place to be searched, and the persons or things to be seized.

Article the seventh... No person shall be held to answer for a capital, or otherwise infamous crime, unless on a presentment or indictment of a Grand Jury, except in cases arising in the land or naval forces, or in the Militia, when in actual service in time of War or public danger; nor shall any person be subject for the same offence to be twice put in jeopardy of life or limb; nor shall be compelled in any criminal case to be a witness against himself, nor be deprived of life, liberty, or property, without due process of law; nor shall private property be taken for public use, without just compensation.

Article the eighth... In all criminal prosecutions, the accused shall enjoy the right to a speedy and public trial, by an impartial jury of the State and district wherein the crime shall have been committed, which district shall have been previously ascertained by law, and to be informed of the nature and cause of the accusation; to be confronted with the witnesses against him; to have compulsory process for obtaining witnesses in his favor, and to have the Assistance of Counsel for his defence.

Article the ninth... In suits at common law, where the value in controversy shall exceed twenty dollars, the right of trial by jury shall be preserved, and no fact tried by a jury, shall be otherwise re-examined in any Court of the United States, than according to the rules of the common law.

Article the tenth... Excessive bail shall not be required, nor excessive fines imposed, nor cruel and unusual punishments inflicted.

Article the eleventh... The enumeration in the Constitution, of certain rights, shall not be construed to deny or disparage others retained by the people.

Article the twelfth... The powers not delegated to the United States by the Constitution, nor prohibited by it to the States, are reserved to the States respectively, or to the people.

ATTEST

Frederick Augustus Muhlenberg, Speaker of the House of Representatives

John Adams, Vice-President of the United States, and President of the Senate

John Beckley, Clerk of the House of Representatives.

Sam. A Otis Secretary of the Senate

Note: The capitalization and punctuation in this version is from the enrolled original of the Joint Resolution of Congress proposing the Bill of Rights, which is on permanent display in the Rotunda of the National Archives Building, Washington, D.C.

The U.S. Bill of Rights

The Preamble to The Bill of Rights

CONGRESS OF THE UNITED STATES begun and held at the City of New-York, on Wednesday the fourth of March, one thousand seven hundred and eighty nine.

THE Conventions of a number of the States, having at the time of their adopting the Constitution, expressed a desire, in order to prevent misconstruction or abuse of its powers, that further declaratory and restrictive clauses should be added: And as extending the ground of public confidence in the Government, will best ensure the beneficent ends of its institution.

RESOLVED by the Senate and House of Representatives of the United States of America, in Congress assembled, two thirds of both Houses concurring, that the following Articles be proposed to the Legislatures of the several States, as amendments to the Constitution of the United States, all, or any of which Articles, when ratified by three fourths of the said Legislatures, to be valid to all intents and purposes, as part of the said Constitution; viz.

ARTICLES in addition to, and Amendment of the Constitution of the United States of America, proposed by Congress, and ratified by the Legislatures of the several States, pursuant to the fifth Article of the original Constitution.

Note: The following text is a transcription of the first ten amendments to the Constitution in their original form. These amendments were ratified December 15, 1791, and form what is known as the ""Bill of Rights."

Amendment I

Congress shall make no law respecting an establishment of religion, or prohibiting the free exercise thereof; or abridging the freedom of speech, or of the press; or the right of the people peaceably to assemble, and to petition the Government for a redress of grievances.

Amendment II

A well regulated Militia, being necessary to the security of a free State, the right of the people to keep and bear Arms, shall not be infringed.

Amendment III

No Soldier shall, in time of peace be quartered in any house, without the consent of the Owner, nor in time of war, but in a manner to be prescribed by law.

Amendment IV

The right of the people to be secure in their persons, houses, papers, and effects, against unreasonable searches and seizures, shall not be violated, and no Warrants shall issue, but upon probable cause, supported by Oath or affirmation, and particularly describing the place to be searched, and the persons or things to be seized.

Amendment V

No person shall be held to answer for a capital, or otherwise infamous crime, unless on a presentment or indictment of a Grand Jury, except in cases arising in the land or naval forces, or in the Militia, when in actual service in time of War or public danger; nor shall any person be subject for the same offence to be twice put in jeopardy of life or limb; nor shall be compelled in any criminal case to be a witness against himself, nor be deprived of life, liberty, or property, without due process of law; nor shall private property be taken for public use, without just compensation.

Amendment VI

In all criminal prosecutions, the accused shall enjoy the right to a speedy and public trial, by an impartial jury of the State and district wherein the crime shall have been committed, which district shall have been previously ascertained by law, and to be informed of the nature and cause of the accusation; to be confronted with the witnesses against him; to have compulsory process for obtaining witnesses in his favor, and to have the Assistance of Counsel for his defence.

Amendment VII

In Suits at common law, where the value in controversy shall exceed twenty dollars, the right of trial by jury shall be preserved, and no fact tried by a jury, shall be otherwise re-examined in any Court of the United States, than according to the rules of the common law.

Amendment VIII

Excessive bail shall not be required, nor excessive fines imposed, nor cruel and unusual punishments inflicted.

Amendment IX

The enumeration in the Constitution, of certain rights, shall not be construed to deny or disparage others retained by the people.

Amendment X

The powers not delegated to the United States by the Constitution, nor prohibited by it to the States, are reserved to the States respectively, or to the people.

Amendment XI

Passed by Congress March 4, 1794. Ratified February 7, 1795.

Note: Article III, section 2, of the Constitution was modified by amendment 11.

The Judicial power of the United States shall not be construed to extend to any suit in law or equity, commenced or prosecuted against one of the United States by Citizens of another State, or by Citizens or Subjects of any Foreign State.

Amendment XII

Passed by Congress December 9, 1803. Ratified June 15, 1804.

Note: A portion of Article II, section 1 of the Constitution was superseded by the 12th amendment.

The Electors shall meet in their respective states and vote by ballot for President and Vice-President, one of whom, at least, shall not be an inhabitant of the same state with themselves; they shall name in their ballots the person voted for as President, and in distinct ballots the person voted for as Vice-President, and they shall make distinct lists of all persons voted for as President, and of all persons voted for as Vice-President, and of the number of votes for each, which lists they shall sign and certify, and transmit sealed to the seat of the government of the United States, directed to the President of the Senate; —the President of the Senate shall, in the presence of the Senate and House of Representatives, open all the certificates and the votes shall then be counted; —The person having the greatest number of votes for President, shall be the President, if such number be a majority of the whole number of Electors appointed; and if no person have such majority, then from the persons having the highest numbers not exceeding three on the list of those voted for as President, the House of Representatives shall choose immediately, by ballot, the President. But in choosing the President, the votes shall be taken by states, the representation from each state having one vote; a quorum for this purpose shall consist of a member or members from two-thirds of the states, and a majority of all the states shall be necessary to a choice. [And if the House of Representatives shall not choose a President whenever the right of choice shall devolve upon them, before the fourth day of March next following, then the Vice-President shall act as President, as in case of the death or other constitutional disability of the President.—]* The person having the greatest number of votes as Vice-President, shall be the Vice-President, if such number be a majority of the whole number of Electors appointed, and if no person have a majority, then from the two highest numbers on the list, the Senate shall choose the Vice-President; a quorum for the purpose shall consist of two-thirds of the whole number of Senators, and a majority of the whole number shall be necessary to a choice. But no person constitutionally ineligible to the office of President shall be eligible to that of Vice-President of the United States.

*Superseded by section 3 of the 20th amendment.

Amendment XIII

Passed by Congress January 31, 1865. Ratified December 6, 1865.

Note: A portion of Article IV, section 2, of the Constitution was superseded by the 13th amendment.

Section 1.

Neither slavery nor involuntary servitude, except as a punishment for crime whereof the party shall have been duly convicted, shall exist within the United States, or any place subject to their jurisdiction.

Section 2.

Congress shall have power to enforce this article by appropriate legislation.

Amendment XIV

Passed by Congress June 13, 1866. Ratified July 9, 1868.

Note: Article I, section 2, of the Constitution was modified by section 2 of the 14th amendment.

Section 1.

All persons born or naturalized in the United States, and subject to the jurisdiction thereof, are citizens of the United States and of the State wherein they reside. No State shall make or enforce any law which shall abridge the privileges or immunities of citizens of the United States; nor shall any State deprive any person of life, liberty, or property, without due process of law; nor deny to any person within its jurisdiction the equal protection of the laws.

Section 2.

Representatives shall be apportioned among the several States according to their respective numbers, counting the whole number of persons in each State, excluding Indians not taxed. But when the right to vote at any election for the choice of electors for President and Vice-President of the United States, Representatives in Congress, the Executive and Judicial officers of a State, or the members of the Legislature thereof, is denied to any of the male inhabitants of such State, being twenty-one years of age,* and citizens of the United States, or in any way abridged, except for participation in rebellion, or other crime, the basis of representation therein shall be reduced in the proportion which the number of such male citizens shall bear to the whole number of male citizens twenty-one years of age in such State.

Section 3.

No person shall be a Senator or Representative in Congress, or elector of President and Vice-President, or hold any office, civil or military, under the United States, or under any State, who, having previously taken an oath, as a member of Congress, or as an officer of the United States, or as a member of any State legislature, or as an executive or judicial officer of any State, to support the Constitution of the United States, shall have engaged in insurrection or rebellion against the same, or given aid or comfort to the enemies thereof. But Congress may by a vote of two-thirds of each House, remove such disability.

Section 4.

The validity of the public debt of the United States, authorized by law, including debts incurred for payment of pensions and bounties for services in suppressing insurrection or rebellion, shall not be questioned. But neither the United States nor any State shall assume or pay any debt or obligation incurred in aid of insurrection or rebellion against the United States, or any claim for the loss or emancipation of any slave; but all such debts, obligations and claims shall be held illegal and void.

Section 5.

The Congress shall have the power to enforce, by appropriate legislation, the provisions of this article.

*Changed by section 1 of the 26th amendment.

Amendment XV

Passed by Congress February 26, 1869. Ratified February 3, 1870.

Section 1.

The right of citizens of the United States to vote shall not be denied or abridged by the United States or by any State on account of race, color, or previous condition of servitude—

Section 2.

The Congress shall have the power to enforce this article by appropriate legislation.

Amendment XVI

Passed by Congress July 2, 1909. Ratified February 3, 1913.

Note: Article I, section 9, of the Constitution was modified by amendment 16.

The Congress shall have power to lay and collect taxes on incomes, from whatever source derived, without apportionment among the several States, and without regard to any census or enumeration.

Amendment XVII

Passed by Congress May 13, 1912. Ratified April 8, 1913.

Note: Article I, section 3, of the Constitution was modified by the 17th amendment.

The Senate of the United States shall be composed of two Senators from each State, elected by the people thereof, for six years; and each Senator shall have one vote. The electors in each State shall have the qualifications requisite for electors of the most numerous branch of the State legislatures.

When vacancies happen in the representation of any State in the Senate, the executive authority of such State shall issue writs of election to fill such vacancies: *Provided*, That the legislature of any State may empower the executive thereof to make temporary appointments until the people fill the vacancies by election as the legislature may direct.

This amendment shall not be so construed as to affect the election or term of any Senator chosen before it becomes valid as part of the Constitution.

Amendment XVIII

Passed by Congress December 18, 1917. Ratified January 16, 1919. Repealed by amendment 21.

Section 1.

After one year from the ratification of this article the manufacture, sale, or transportation of intoxicating liquors within, the importation thereof into, or the exportation thereof from the United States and all territory subject to the jurisdiction thereof for beverage purposes is hereby prohibited.

Section 2.

The Congress and the several States shall have concurrent power to enforce this article by appropriate legislation.

Section 3.

This article shall be inoperative unless it shall have been ratified as an amendment to the Constitution by the legislatures of the several States, as provided in the Constitution, within seven years from the date of the submission hereof to the States by the Congress.

Amendment XIX

Passed by Congress June 4, 1919. Ratified August 18, 1920.

The right of citizens of the United States to vote shall not be denied or abridged by the United States or by any State on account of sex.

Congress shall have power to enforce this article by appropriate legislation.

Amendment XX

Passed by Congress March 2, 1932. Ratified January 23, 1933.

Note: Article I, section 4, of the Constitution was modified by section 2 of this amendment. In addition, a portion of the 12th amendment was superseded by section 3.

Section 1.

The terms of the President and the Vice President shall end at noon on the 20th day of January, and the terms of Senators and Representatives at noon on the 3d day of January, of the years in which such terms would have ended if this article had not been ratified; and the terms of their successors shall then begin.

Section 2.

The Congress shall assemble at least once in every year, and such meeting shall begin at noon on the 3d day of January, unless they shall by law appoint a different day.

Section 3.

If, at the time fixed for the beginning of the term of the President, the President elect shall have died, the Vice President elect shall become President. If a President shall not have been chosen before the time fixed for the beginning of his term, or if the President elect shall have failed to qualify, then the Vice President elect shall act as President until a President shall have qualified; and the Congress may by law provide for the case wherein neither a President elect nor a Vice President elect shall have qualified, declaring who shall then act as President, or the manner in which one who is to act shall be selected, and such person shall act accordingly until a President or Vice President shall have qualified.

Section 4.

The Congress may by law provide for the case of the death of any of the persons from whom the House of Representatives may choose a President whenever the right of choice shall have devolved upon them, and for the case of the death of any of the persons from whom the Senate may choose a Vice President whenever the right of choice shall have devolved upon them.

Section 5.

Sections 1 and 2 shall take effect on the 15th day of October following the ratification of this article.

Section 6.

This article shall be inoperative unless it shall have been ratified as an amendment to the Constitution by the legislatures of three-fourths of the several States within seven years from the date of its submission.

Amendment XXI

Passed by Congress February 20, 1933. Ratified December 5, 1933.

Section 1.

The eighteenth article of amendment to the Constitution of the United States is hereby repealed.

Section 2.

The transportation or importation into any State, Territory, or possession of the United States for delivery or use therein of intoxicating liquors, in violation of the laws thereof, is hereby prohibited.

Section 3.

This article shall be inoperative unless it shall have been ratified as an amendment to the Constitution by conventions in the several States, as provided in the Constitution, within seven years from the date of the submission hereof to the States by the Congress.

Amendment XXII

Passed by Congress March 21, 1947. Ratified February 27, 1951.

Section 1.

No person shall be elected to the office of the President more than twice, and no person who has held the office of President, or acted as President, for more than two years of a term to which some other person was elected President shall be elected to the office of the President more than once. But this Article shall not apply to any person holding the office of President when this Article was proposed by the Congress, and shall not prevent any person who may be holding the office of President, or acting as President, during the term within which this Article becomes operative from holding the office of President or acting as President during the remainder of such term.

Section 2.

This article shall be inoperative unless it shall have been ratified as an amendment to the Constitution by the legislatures of three-fourths of the several States within seven years from the date of its submission to the States by the Congress.

Amendment XXIII

Passed by Congress June 16, 1960. Ratified March 29, 1961.

Section 1.

The District constituting the seat of Government of the United States shall appoint in such manner as the Congress may direct:

A number of electors of President and Vice President equal to the whole number of Senators and Representatives in Congress to which the District would be entitled if it were a State, but in no event more than the least populous State; they shall be in addition to those appointed by the States, but they shall be considered, for the purposes of the election of President and Vice President, to be electors appointed by a State; and they shall meet in the District and perform such duties as provided by the twelfth article of amendment.

Section 2.

The Congress shall have power to enforce this article by appropriate legislation.

Amendment XXIV

Passed by Congress August 27, 1962. Ratified January 23, 1964.

Section 1.

The right of citizens of the United States to vote in any primary or other election for President or Vice President, for electors for President or Vice President, or for Senator or Representative in Congress, shall not be denied or abridged by the United States or any State by reason of failure to pay any poll tax or other tax.

Section 2.

The Congress shall have power to enforce this article by appropriate legislation.

Amendment XXV

Passed by Congress July 6, 1965. Ratified February 10, 1967.

Note: Article II, section 1, of the Constitution was affected by the 25th amendment.

Section 1.

In case of the removal of the President from office or of his death or resignation, the Vice President shall become President.

Section 2.

Whenever there is a vacancy in the office of the Vice President, the President shall nominate a Vice President who shall take office upon confirmation by a majority vote of both Houses of Congress.

Section 3.

Whenever the President transmits to the President pro tempore of the Senate and the Speaker of the House of Representatives his written declaration that he is unable to discharge the powers and duties of his office, and until he transmits to them a written declaration to the contrary, such powers and duties shall be discharged by the Vice President as Acting President.

Section 4.

Whenever the Vice President and a majority of either the principal officers of the executive departments or of such other body as Congress may by law provide, transmit to the President pro tempore of the Senate and the Speaker of the House of Representatives their written declaration that the President is unable to discharge the powers and duties of his office, the Vice President shall immediately assume the powers and duties of the office as Acting President.

Thereafter, when the President transmits to the President pro tempore of the Senate and the Speaker of the House of Representatives his written declaration that no inability exists, he shall resume the powers and duties of his office unless the Vice President and a majority of either the principal officers of the executive department or of such other body as Congress may by law provide, transmit within four days to the President pro tempore of the Senate and the Speaker of the House of Representatives their written declaration that the President is unable to discharge the powers and duties of his office. Thereupon Congress shall decide the issue, assembling within forty-eight hours for that purpose if not in session. If the Congress, within twenty-one days after receipt of the latter written declaration, or, if Congress is not in session, within twenty-one days after Congress is required to assemble, determines by two-thirds vote of both Houses that the President is unable to discharge the powers and duties of his office, the Vice President shall continue to discharge the same as Acting President; otherwise, the President shall resume the powers and duties of his office.

Amendment XXVI

Passed by Congress March 23, 1971. Ratified July 1, 1971.

Note: Amendment 14, section 2, of the Constitution was modified by section 1 of the 26th amendment.

Section 1.

The right of citizens of the United States, who are eighteen years of age or older, to vote shall not be denied or abridged by the United States or by any State on account of age.

Section 2.

The Congress shall have power to enforce this article by appropriate legislation.

Amendment XXVII
Originally proposed Sept. 25, 1789. Ratified May 7, 1992.

No law, varying the compensation for the services of the Senators and Representatives, shall take effect, until an election of Representatives shall have intervened

Appendix D

Quotes from Christian Founders[587]

Wallbuilders: A Few Declarations of Founding Fathers and Early Statesmen on Jesus, Christianity, and the Bible

(Note: This appendix is taken from the Wallbuilders, PO Box, Aledo, TX 76008, and used with permission. This list is by no means exhaustive; many other founders could be included, and even with those who appear below, additional quotes could have been used.)

John Adams

Signer of the Declaration of Independence, judge, diplomat, one of two signers of the Bill of Rights, second president of the United States

The general principles on which the fathers achieved independence were the general principles of Christianity. I will avow that I then believed, and now believe, that those general principles of Christianity are as eternal and immutable as the existence and attributes of God.[588]

Without religion, this world would be something not fit to be mentioned in polite company: I mean hell.[589]

The Christian religion is, above all the religions that ever prevailed or existed in ancient or modern times, the religion of wisdom, virtue, equity and humanity.[590]

Suppose a nation in some distant region should take the Bible for their only law book and every member should regulate his conduct by the precepts there exhibited…. What a Eutopia—what a Paradise would this region be![591]

I have examined all religions, and the result is that the Bible is the best book in the world.[592]

John Quincy Adams

Sixth president of the United States, diplomat, secretary of state, U.S. senator, U.S. representative, "Old Man Eloquent," "Hell-Hound of Abolition"

My hopes of a future life are all founded upon the Gospel of Christ and I cannot cavil or quibble away [evade or object to]…. the whole tenor of His conduct by which He sometimes positively asserted and at others countenances [permits] His disciples in asserting that He was God.[593]

The hope of a Christian is inseparable from his faith. Whoever believes in the Divine inspiration of the Holy Scriptures must hope that the religion of Jesus shall prevail throughout the earth. Never since the foundation of the world have the prospects of mankind been more encouraging to that hope than they appear to be at the present time. And may the associated distribution of the Bible proceed and prosper till the Lord shall have made "bare His holy arm in the eyes of all the nations, and all the ends of the earth shall see the salvation of our God" [Isaiah 52:10].[594]

In the chain of human events, the birthday of the nation is indissolubly linked with the birthday of the Savior. The Declaration of Independence laid the cornerstone of human government upon the first precepts of Christianity.[595]

Samuel Adams

Signer of the Declaration of Independence, "Father of the American Revolution," ratifier of the U.S. Constitution, governor of Massachusetts

I…[rely] upon the merits of Jesus Christ for a pardon of all my sins.[596]

The name of the Lord (says the Scripture) is a strong tower; thither the righteous flee and are safe [Proverbs 18:10]. Let us secure His favor and He will lead us through the journey of this life and at length receive us to a better.[597]

I conceive we cannot better express ourselves than by humbly supplicating the Supreme Ruler of the world… that the confusions that are and have been among the nations may be overruled by the promoting and speedily bringing in the holy and happy period when the kingdoms of our Lord and Savior Jesus Christ may be everywhere established, and the people willingly bow to the scepter of Him who is the Prince of Peace.[598]

He also called on the State of Massachusetts to pray that…

- the peaceful and glorious reign of our Divine Redeemer may be known and enjoyed throughout the whole family of mankind.[599]
- we may with one heart and voice humbly implore His gracious

and free pardon through Jesus Christ, supplicating His Divine aid…[and] above all to cause the religion of Jesus Christ, in its true spirit, to spread far and wide till the whole earth shall be filled with His glory.[600]

- with true contrition of heart to confess their sins to God and implore forgiveness through the merits and mediation of Jesus Christ our Savior.[601]

Josiah Bartlett

Military office, signer of the Declaration of Independence,
judge, governor of New Hampshire

[He called on the people of New Hampshire]…to confess before God their aggravated transgressions and to implore His pardon and forgiveness through the merits and mediation of Jesus Christ…[t]hat the knowledge of the Gospel of Jesus Christ may be made known to all nations, pure and undefiled religion universally prevail, and the earth be fill with the glory of the Lord.[602]

Gunning Bedford

Military officer; member of the Continental Congress;
signer of the Constitution; federal judge

To the triune God—the Father, the Son, and the Holy Ghost—be ascribed all honor and dominion, forevermore—Amen.[603]

Elias Boudinot

President of Congress; signed the peace treaty to end the American
Revolution; first attorney admitted to the U.S. Supreme Court Bar;
framer of the Bill of Rights; director of the U.S. Mint

Let us enter on this important business under the idea that we are Christians on whom the eyes of the world are now turned... [L]et us earnestly call and beseech Him, for Christ's sake, to preside in our councils.... We can only depend on the all powerful influence of the Spirit of God, Whose Divine aid and assistance it becomes us as a Christian people most devoutly to implore. Therefore I move that some minister of the Gospel be requested to attend this Congress every morning...in order to open the meeting with prayer.[604]

[A letter to his daughter:]

You have been instructed from your childhood in the knowledge of your lost state by nature—the absolute necessity of a change of heart and an entire renovation of soul to the image of Jesus Christ—of salvation through His meritorious righteousness only—and the indispensable necessity of personal holiness without which no man shall see the Lord [Hebrews 12:14]. You are well acquainted that the most perfect and consummate doctrinal knowledge is of no avail without it operates on and sincerely affects the heart, changes the practice, and totally influences the will—and that without the almighty power of the Spirit of God enlightening your mind, subduing your will, and continually drawing you to Himself, you can do nothing.... And may the God of your parents (for many generations past) seal instruction to your soul and lead you to Himself through the blood of His too greatly despised Son, Who notwithstanding, is still reclaiming the world to God through that blood, not imputing to them their sins. To Him be glory forever![605]

For nearly half a century have I anxiously and critically studied that invaluable treasure [the Bible]; and I still scarcely ever take it up that I do not find something new—that I do not receive some valuable addition to my stock of knowledge or perceive some instructive fact never observed before. In short, were you to ask me to recommend the most valuable book in the world, I should fix on the Bible as the most instructive both to the wise and ignorant. Were you to ask me for one affording the most rational and pleasing entertainment to the inquiring mind, I should repeat, it is the Bible; and should you renew the inquiry for the best philosophy or the most interesting history, I should still urge you to look into your Bible. I would make it, in short, the Alpha and Omega of knowledge.[606]

Jacob Broom

Legislator; signer of the Constitution

[A letter to his son, James, attending Princeton University:]

I flatter myself you will be what I wish, but don't be so much flatterer as to relax of your application—don't forget to be a Christian. I have said much to you on this head, and I hope an indelible impression is made.[607]

Charles Carroll

Signer of the Declaration of Independence; selected as delegate to the Constitutional Convention; framer of the Bill of Rights; U.S. senator

On the mercy of my Redeemer I rely for salvation and on His merits, not on the works I have done in obedience to His precepts.[608]

Grateful to Almighty God for the blessings which, through Jesus Christ Our Lord, He had conferred on my beloved country in her emancipation and on myself in permitting me, under circumstances of mercy, to live to the age of 89 years, and to survive the fiftieth year of independence, adopted by Congress on the 4th of July 1776, which I originally subscribed on the 2d day of August of the same year and of which I am now the last surviving signer.[609]

I, Charles Carroll...give and bequeath my soul to God who gave it, my body to the earth, hoping that through and by the merits, sufferings, and mediation of my only Savior and Jesus Christ, I may be admitted into the Kingdom prepared by God for those who love, fear and truly serve Him.[610]

Congress, 1854

The great, vital, and conservative element in our system is the belief of our people in the pure doctrines and the divine truths of the Gospel of Jesus Christ.[611]

Congress, U. S. House Judiciary Committee, 1854

Had the people, during the Revolution, had a suspicion of any attempt to war against Christianity, that Revolution would have been strangled in its cradle.... In this age, there can be no substitute for Christianity.... That was the religion of the founders of the republic and they expected it to remain the religion of their descendants.[612]

John Dickinson

Signer of the Constitution; governor of Pennsylvania; governor of Delaware; general in the American Revolution

Rendering thanks to my Creator for my existence and station among His works, for my birth in a country enlightened by the Gospel and enjoying freedom, and for all His other kindnesses, to Him I resign myself, humbly confiding in His goodness and in His mercy through Jesus Christ for the events of eternity.[613]

[Governments] could not give the rights essential to happiness.... We claim them from a higher source: from the King of kings, and Lord of all the earth.[614]

Gabriel Duvall

Soldier; judge; selected as delegate to the Constitutional Convention; comptroller of the U.S. Treasury; U.S. Supreme Court justice

I resign my soul into the hands of the Almighty Who gave it, in humble hopes of His mercy through our Savior Jesus Christ.[615]

Benjamin Franklin

Signer of the Declaration; diplomat; printer; scientist; s igner of the Constitution; governor of Pennsylvania

As to Jesus of Nazareth, my opinion of whom you particularly desire, I think the system of morals and His religion as He left them to us, the best the world ever saw or is likely to see.[616]

The body of Benjamin Franklin, printer, like the cover of an old book, its contents torn out and stripped of its lettering and guilding, lies here, food for worms. Yet the work itself shall not be lost; for it will, as he believed, appear once more in a new and more beautiful edition, corrected and amended by the Author.[617] (Franklin's eulogy that he wrote for himself)

Elbridge Gerry

Signer of the Declaration of Independence; member of the Constitutional Convention; framer of the Bill of Rights, governor of Massachusetts, vice president of the United States

[He called on the State of Massachusetts to pray that:]

- with one heart and voice we may prostrate ourselves at the throne of heavenly grace and present to our Great Benefactor sincere and unfeigned thanks for His infinite goodness and mercy towards us from our birth to the present moment for having above all things illuminated us by the Gospel of Jesus Christ, presenting to our view the happy prospect of a blessed immortality.[618]

- And for our unparalleled ingratitude to that Adorable Being Who has seated us in a land irradiated by the cheering beams of the Gospel of Jesus Christ...let us fall prostrate before offended Deity, confess sincerely and penitently our manifold sins and our unworthiness of the least of His Divine favors, fervently implore His pardon through the merits of our mediator.[619]

- And deeply impressed with a scene of our unparalleled ingratitude, let us contemplate the blessings which have flowed from the unlimited grave and favor of offended Deity, that we are still permitted to enjoy the first of Heaven's blessings: the Gospel of Jesus Christ.[620]

Alexander Hamilton

Revolutionary general, signer of the Constitution,
author of the Federalist Papers, secretary of the Treasury

Following his duel with Aaron Burr, in those final twenty four hours while life still remained in him, Hamilton called for two ministers, the Rev. J. M. Mason and the Rev. Benjamin Moore, to pray with him and administer Communion to him. Each of those two ministers reported what transpired. The Rev. Mason recounted:

[General Hamilton said:]

"I went to the field determined not to take his life." He repeated his disavowal of all intention to hurt Mr. Burr; the anguish of his mind in recollecting what had passed; and his humble hope of forgiveness from his God. I recurred to the topic of the Divine compassion; the freedom of pardon in the Redeemer Jesus to perishing sinners. "That grace, my dear General, which brings salvation, is rich, rich"—"Yes," interrupted he, "it is rich grace." "And on that grace," continued I, "a sinner has the highest encouragement to repose his confidence, because it is tendered to him upon the surest foundation; the Scripture testifying that we have redemption through the blood of Jesus, the forgiveness of sins according to the richness of His grace." Here the General, letting go my hand, which he had held from the moment I sat down at his bed side, clasped his hands together, and, looking up towards Heaven, said, with emphasis, "I have a tender reliance on the mercy of the Almighty, through the merits of the Lord Jesus Christ."[621]

[The Rev. Benjamin Moore reported:]

[I]mmediately after he was brought from [the field]…a message was sent informing me of the sad event, accompanied by a request from General

Hamilton that I would come to him for the purpose of administering the Holy Communion. I went.... I proceeded to converse with him on the subject of his receiving the Communion; and told him that with respect to the qualifications of those who wished to become partakers of that holy ordinance, my inquires could not be made in language more expressive than that which was used by our [own] Church. —[I asked], "Do you sincerely repent of your sins past? Have you a lively faith in God's mercy through Christ, with a thankful remembrance of the death of Christ? And are you disposed to live in love and charity with all men?" He lifted up his hands and said, "With the utmost sincerity of heart I can answer those questions in the affirmative—I have no ill will against Col. Burr. I met him with a fixed resolution to do him no harm—I forgive all that happened." ...The Communion was then administered, which he received with great devotion, and his heart afterwards appeared to be perfectly at rest. I saw him again this morning, when, with his last faltering words, he expressed a strong confidence in the mercy of God through the intercession of the Redeemer. I remained with him until 2 o'clock this afternoon, when death closed the awful scene – he expired without a struggle, and almost without a groan. By reflecting on this melancholy event, let the humble believer be encouraged ever to hold fast that precious faith which is the only source of true consolation in the last extremity of nature. [And l]et the infidel be persuaded to abandon his opposition to that Gospel which the strong, inquisitive, and comprehensive mind of a Hamilton embraced.[622]

One other consequence of Hamilton's untimely death was that it permanently halted the formation of a religious society Hamilton had proposed. Hamilton suggested that it be named the Christian Constitutional Society, and listed two goals for its formation: first, the support of the Christian religion; and second, the support of the Constitution of the United States. This organization was to have numerous clubs throughout each state which would meet regularly and work to elect to office those who reflected the goals of the Christian Constitutional Society.[623]

John Hancock

Signer of the Declaration of Independence, president of Congress,
Revolutionary general, governor of Massachusetts

Sensible of the importance of Christian piety and virtue to the order
and happiness of a state, I cannot but earnestly commend to you every
measure for their support and encouragement.[624]

[He called on the entire state to pray]

"that universal happiness may be established in the world [and] that all
may bow to the scepter of our Lord Jesus Christ, and the whole earth be
filled with His glory."[625]

[He also called on the State of Massachusetts to pray:]

- that all nations may bow to the scepter of our Lord and Savior Jesus Christ and that the whole earth may be filled with his
  glory.[626]
- that the spiritual kingdom of our Lord and Savior Jesus Christ
  may be continually increasing until the whole earth shall be
  filled with His glory.[627]
- to confess their sins and to implore forgiveness of God through
  the merits of the Savior of the World.[628]
- to cause the benign religion of our Lord and Savior Jesus Christ
  to be known, understood, and practiced among all the inhabitants of the earth.[629]
- to confess their sins before God and implore His forgiveness
  through the merits and mediation of Jesus Christ, our Lord and
  Savior.[630]
- that He would finally overrule all events to the advancement
  of the Redeemer's kingdom and the establishment of universal
  peace and good will among men.[631]

- that the kingdom of our Lord and Savior Jesus Christ may be established in peace and righteousness among all the nations of the earth.[632]
- that with true contrition of heart we may confess our sins, resolve to forsake them, and implore the Divine forgiveness, through the merits and mediation of Jesus Christ, our Savior.... And finally to overrule all the commotions in the world to the spreading the true religion of our Lord Jesus Christ in its purity and power among all the people of the earth.[633]

John Hart

Judge; legislator; signer of the Declaration

[T]hanks be given unto Almighty God therefore, and knowing that it is appointed for all men once to die and after that the judgment [Hebrews 9:27]...principally, I give and recommend my soul into the hands of Almighty God who gave it and my body to the earth to be buried in a decent and Christian like manner...to receive the same again at the general resurrection by the mighty power of God.[634]

Patrick Henry

Revolutionary general; legislator; "the Voice of Liberty";
ratifier of the U.S. Constitution; governor of Virginia

Being a Christian...is a character which I prize far above all this world has or can boast.[635]

The Bible...is a book worth more than all the other books that were ever printed.[636]

Righteousness alone can exalt [America] as a nation.... Whoever thou art, remember this; and in thy sphere practice virtue thyself, and encourage it in others.[637]

The great pillars of all government and of social life [are] virtue, morality, and religion. This is the armor, my friend, and this alone, that renders us invincible.[638]

This is all the inheritance I can give to my dear family. The religion of Christ can give them one which will make them rich indeed.[639]

Samuel Huntington

Signer of the Declaration of Independence; president of Congress; judge; governor of Connecticut

It becomes a people publicly to acknowledge the over-ruling hand of Divine Providence and their dependence upon the Supreme Being as their Creator and Merciful Preserver...and with becoming humility and sincere repentance to supplicate the pardon that we may obtain forgiveness through the merits and mediation of our Lord and Savior Jesus Christ.[640]

James Iredell

Ratifier of the U.S. Constitution; attorney general of North Carolina; U.S. Supreme Court justice appointed by President George Washington

For my part, I am free and ready enough to declare that I think the Christian religion is a Divine institution; and I pray to God that I may never forget the precepts of His religion or suffer the appearance of an inconsistency in my principles and practice.[641]

John Jay

*President of Congress; diplomat; author of the Federalist Papers;
original chief justice of the U.S. Supreme Court; governor of New York*

Condescend, merciful Father! to grant as far as proper these imperfect petitions, to accept these inadequate thanksgivings, and to pardon whatever of sin hath mingled in them for the sake of Jesus Christ, our blessed Lord and Savior; unto Whom, with Thee, and the blessed Spirit, ever one God, be rendered all honor and glory, now and forever.[642]

Unto Him who is the author and giver of all good, I render sincere and humble thanks for His manifold and unmerited blessings, and especially for our redemption and salvation by His beloved Son.... Blessed be His holy name.[643]

Mercy and grace and favor did come by Jesus Christ, and also that truth which verified the promises and predictions concerning Him and which exposed and corrected the various errors which had been imbibed respecting the Supreme Being, His attributes, laws, and dispensations.[644]

By conveying the Bible to people...we certainly do them a most interesting act of kindness. We thereby enable them to learn that man was originally created and placed in a state of happiness, but, becoming disobedient, was subjected to the degradation and evils which he and his posterity have since experienced. The Bible will also inform them that our gracious Creator has provided for us a Redeemer in whom all the nations of the earth should be blessed – that this Redeemer has made atonement "for the sins of the whole world," and thereby reconciling the Divine justice with the Divine mercy, has opened a way for our redemption and salvation; and that these inestimable benefits are of the free gift and grace of God, not of our deserving, nor in our power to deserve. The Bible will also [encourage] them with many explicit and consoling assurances of the Divine mercy to our fallen race, and with repeated invitations to accept the offers of pardon and reconciliation.... They, therefore, who enlist in His service, have the highest encouragement to fulfill the du¬ties assigned to their respective stations; for most certain it is, that those of His followers who [participate in] His conquests will also participate in the transcendent glories and blessings of His Triumph.[645]

I recommend a general and public return of praise and thanksgiving to Him from whose goodness these blessings descend. The most effectual means of securing the continuance of our civil and religious liberties is always to remember with reverence and gratitude the source from which they flow.[646]

The Bible is the best of all books, for it is the word of God and teaches us the way to be happy in this world and in the next. Continue therefore to read it and to regulate your life by its precepts.[647]

[T]he evidence of the truth of Christianity requires only to be carefully examined to produce conviction in candid minds...they who undertake that task will derive advantages.[648]

Providence has given to our people the choice of their rulers, and it is the duty as well as the privilege and interest of our Christian nation, to select and prefer Christians for their rulers.[649]

Thomas Jefferson

*Signer of the Declaration of Independence; diplomat;
governor of Virginia; secretary of state;
third president of the United States*

The doctrines of Jesus are simple, and tend all to the happiness of man.[650]

The practice of morality being necessary for the well being of society, He [God] has taken care to impress its precepts so indelibly on our hearts that they shall not be effaced by the subtleties of our brain. We all agree in the obligation of the moral principles of Jesus and nowhere will they be found delivered in greater purity than in His discourses.[651]

I am a Christian in the only sense in which He wished anyone to be: sincerely attached to His doctrines in preference to all others.[652]

I am a real Christian—that is to say, a disciple of the doctrines of Jesus Christ.[653]

William Samuel Johnson

Judge; member of the Continental Congress; signer of the Constitution; framer of the Bill of Rights; president of Columbia College; U.S. senator

[I]…am endeavoring…to attend to my own duty only as a Christian… let us take care that our Christianity, though put to the test…be not shaken, and that our love for things really good wax not cold.[654]

In an address to graduates:

You this day...have, by the favor of Providence and the attention of friends, received a public education, the purpose whereof hath been to qualify you the better to serve your Creator and your country. You have this day invited this audience to witness the progress you have made.... Thus you assume the character of scholars, of men, and of citizens.... Go, then...and exercise them with diligence, fidelity, and zeal.... Your first great duties, you are sensible, are those you owe to Heaven, to your Creator and Redeemer. Let these be ever present to your minds, and exemplified in your lives and conduct. Imprint deep upon your minds the principles of piety towards God, and a reverence and fear of His holy name. The fear of God is the beginning of wisdom and its [practice] is everlasting [happiness].... Reflect deeply and often upon [your] relations [with God]. Remember that it is in God you live and move and have your being,—that, in the language of David, He is about your bed and about your path and spieth out all your ways—that there is not a thought in your hearts, nor a word upon your tongues, but lo! He knoweth them altogether, and that He will one day call you to a strict account for all your conduct in this mortal life. Remember, too, that you are the redeemed of the Lord, that you are bought with a price, even the inestimable price of the precious blood of the Son of God. Adore Jehovah, therefore, as your God and your Judge. Love, fear, and serve Him as your Creator, Redeemer, and Sanctifier. Acquaint yourselves with Him in His word and holy ordinances.... [G]o forth into the world firmly resolved neither to be allured by its vanities nor contaminated by its vices, but to run with patience and perseverance, with firmness and [cheerfulness], the glorious career of religion, honor, and virtue.... Finally... in the elegant and expressive language are honest, whatsoever things are just, whatsoever things are pure, whatsoever things are lovely, whatsoever things are of good report, if there be any virtue, and if there be any praise, think on these things"—and do them, and the God of peace shall be with you, to whose most gracious protection I now commend you, humbly imploring Almighty Goodness that He will be your guardian and your guide, your protector and the rock of your defense, your Savior and your God.[655]

James Kent

Judge; law professor; "Father of American Jurisprudence"

My children, I wish to talk to you. During my early and middle life I was, perhaps, rather skeptical with regard to some of the truths of Christianity. Not that I did not have the utmost respect for religion and always read my Bible, but the doctrine of the atonement was one I never could understand, and I felt inclined to consider as impossible to be received in the way Divines taught it. I believe I was rather inclined to Unitarianism; but of late years my views have altered. I believe in the doctrines of the prayer books as I understand them, and hope to be saved through the merits of Jesus Christ.... My object in telling you this is that if anything happens to me, you might know, and perhaps it would console you to remember, that on this point my mind is clear: I rest my hopes of salvation on the Lord Jesus Christ.[656]

Francis Scott Key

U.S. attorney for the District of Columbia;
author of the "Star Spangled Banner"

[M]ay I always hear that you are following the guidance of that blessed Spirit that will lead you into all truth, leaning on that Almighty arm that has been extended to deliver you, trusting only in the only Savior, and going on in your way to Him rejoicing.[657]

James Madison

Signer of the Constitution; author of the Federalist Papers; framer of the
Bill of Rights; secretary of state; fourth president of the United States

A watchful eye must be kept on ourselves lest, while we are building ideal monuments of renown and bliss here, we neglect to have our names enrolled in the Annals of Heaven.[658]

I have sometimes thought there could not be a stronger testimony in favor of religion or against temporal enjoyments, even the most rational and manly, than for men who occupy the most honorable and gainful departments and [who] are rising in reputation and wealth, publicly to declare their unsatisfactoriness by becoming fervent advocates in the cause of Christ; and I wish you may give in your evidence in this way.[659]

James Manning

Member of the Continental Congress; president of Brown University

I rejoice that the religion of Jesus prevails in your parts; I can tell you the same agreeable news from this quarter. Yesterday I returned from Piscataway in East Jersey, where was held a Baptist annual meeting (I think the largest I ever saw) but much more remarkable still for the Divine influences which God was pleased to grant. Fifteen were baptized; a number during the three days professed to experience a change of heart. Christians were remarkably quickened; multitudes appeared.[660]

Henry Marchant

Member of the Continental Congress; attorney general of Rhode Island; ratifier of the U.S. Constitution; federal judge appointed by president George Washington

And may God grant that His grace may really affect your heart with suitable impressions of His goodness. Remember that God made you, that God keeps you alive and preserves you from all harm, and gives you all the powers and the capacity whereby you are able to read of Him and of Jesus Christ, your Savior and Redeemer, and to do every other needful business of life. And while you look around you and see the great privileges and advantages you have above what other children have (of learning to read and write, of being taught the meaning of the great truths of the Bible), you must remember not to be proud on that account but to bless God and be thankful and endeavor in your turn to assist others with the knowledge you may gain.[661] (to his daughter)

George Mason

Delegate at the Constitutional Convention; "
Father of the Bill of Rights"

I give and bequeath my soul to Almighty God that gave it me, hoping that through the meritorious death and passion of our Savior and Redeemer Jesus Christ to receive absolution and remission for all my sins.[662]

My soul I resign into the hands of my Almighty Creator, Whose tender mercies are all over His works...humbly hoping from His unbounded mercy and benevolence, through the merits of my blessed Savior, a remission of my sins.[663]

James McHenry

Revolutionary officer; signer of the Constitution;
ratifier of the U.S. Constitution; secretary of war under presidents
George Washington and John Adams

[P]ublic utility pleads most forcibly for the general distribution of the Holy Scriptures. Without the Bible, in vain do we increase penal laws and draw entrenchments around our institutions.[664]

Bibles are strong protections. Where they abound, men cannot pursue wicked courses and at the same time enjoy quiet conscience.[665]

Thomas McKean

Signer of the Declaration of Independence; president of Congress;
ratifier of the U.S. Constitution; chief justice of the Supreme Court
of Pennsylvania; governor of Pennsylvania; governor of Delaware

In the case Respublica v. John Roberts,[666] John Roberts was sentenced
to death after a jury found him guilty of treason. Chief Justice McKean
then told him:

You will probably have but a short time to live. Before you launch into eternity, it behooves you to improve the time that may be allowed you in this
world: it behooves you most seriously to reflect upon your past conduct; to
repent of your evil deeds; to be incessant in prayers to the great and merciful God to forgive your manifold transgressions and sins; to teach you to
rely upon the merit and passion of a dear Redeemer, and thereby to avoid
those regions of sorrow—those doleful shades where peace and rest can
never dwell, where even hope cannot enter. It behooves you to seek the [fel-
lowship], advice, and prayers of pious and good men; to be [persistent] at
the Throne of Grace, and to learn the way that leadeth to happiness. May
you, reflecting upon these things, and pursuing the will of the great Father
of light and life, be received into [the] company and society of angels and
archangels and the spirits of just men made perfect; and may you be qualified to enter into the joys of Heaven—joys unspeakable and full of glory![667]

Gouverneur Morris

Revolutionary officer; member of the Continental Congress; signer of the
Constitution; "Penman of the Constitution"; diplomat; U.S. senator

There must be religion. When that ligament is torn, society is disjointed
and its members perish… [T]he most important of all lessons is the
denunciation of ruin to every state that rejects the precepts of religion.[668]

Your good morals in the army give me sincere pleasure as it hath long been my fixed opinion that virtue and religion are the great sources of human happiness. More especially is it necessary in your profession firmly to rely upon the God of Battles for His guardianship and protection in the dreadful hour of trial. But of all these things you will and I hope in the merciful Lord.[669]

Jedidiah Morse

Historian of the American Revolution; educator;
"Father of American Geography"; appointed by secretary
of state to document condition of Indian affairs

To the kindly influence of Christianity we owe that degree of civil freedom and political and social happiness which mankind now enjoys. All efforts made to destroy the foundations of our Holy Religion ultimately tend to the subversion also of our political freedom and happiness. In proportion as the genuine effects of Christianity are diminished in any nation…in the same proportion will the people of that nation recede from the blessings of genuine freedom… Whenever the pillars of Christianity shall be overthrown, our present republican forms of government—and all the blessings which flow from them—must fall with them.[670]

John Morton

Legislator; judge;
signer of the Declaration

With an awful reverence to the Great Almighty God, Creator of all mankind, being sick and weak in body but of sound mind and memory, thanks be given to Almighty God for the same.[671]

James Otis

Leader of the Sons of Liberty; attorney and jurist;
mentor of John Hancock and Samuel Adams

Has [government] any solid foundation? Any chief cornerstone?... I think it has an everlasting foundation in the unchangeable will of God.... The sum of my argument is that civil government is of God.[672]

Robert Treat Paine

Military chaplain; signer of the Declaration of Independence;
attorney general of Massachusetts; judge

I desire to bless and praise the name of God most high for appointing me my birth in a land of Gospel Light where the glorious tidings of a Savior and of pardon and salvation through Him have been continually sounding in mine ears.[673]

I am constrained to express my adoration of the Supreme Being, the Author of my existence, in full belief of His Providential goodness and His forgiving mercy revealed to the world through Jesus Christ, through whom I hope for never ending happiness in a future state.[674]

I believe the Bible to be the written word of God and to contain in it the whole rule of faith and manners.[675]

William Paterson

Attorney general of New Jersey; signer of the Constitution; U.S. senator;
governor of New Jersey; U.S. supreme court justice

When the righteous rule, the people rejoice; when the wicked rule, the people groan. [invoking Proverbs 29:2 to instruct a grand jury].[676]

Timothy Pickering

Revolutionary general; judge; ratifier of the U.S. Constitution;
postmaster general under president George Washington;
secretary of war under presidents George Washington and John Adams;
secretary of state under president John Adams

Pardon, we beseech Thee, all our offences of omission and commission; and grant that in all our thoughts, words, and actions, we may conform to Thy known will manifested in our consciences and in the revelations of Jesus Christ, our Savior.[677]

[W]e do not grieve as those who have no...resurrection to a life immortal. Here the believers in Christianity manifest their superior advantages, for life and immortality were brought to light by the gospel of Jesus Christ [II Timothy 1:10]. Prior to that revelation even the wisest and best of mankind were involved in doubt and they hoped, rather than believed, that the soul was immortal.[678]

Charles Cotesworth Pinckney

Revolutionary general; legislator;
signer of the Constitution; diplomat

To the eternal and only true God be all honor and glory, now and forever. Amen![679]

John Randolph of Roanoke

Congressman under presidents John Adams, Thomas Jefferson,
James Madison, James Monroe, John Quincy Adams,
Andrew Jackson; U.S. senator; diplomat

I have thrown myself, reeking with sin, on the mercy of God, through Jesus Christ His blessed Son and our (yes, my friend, our) precious Redeemer; and I have assurances as strong as that I now owe nothing to your rank that the debt is paid and now I love God—and with reason. I once hated him—and with reason, too, for I knew not Christ. The only cause why I should love God is His goodness and mercy to me through Christ.[680]

I am at last reconciled to my God and have assurance of His pardon through faith in Christ, against which the very gates of hell cannot prevail. Fear hath been driven out by perfect love.[681]

[I] have looked to the Lord Jesus Christ, and hope I have obtained pardon.[682]

[I] still cling to the cross of my Redeemer, and with God's aid firmly resolve to lead a life less unworthy of one who calls himself the humble follower of Jesus Christ.[683]

Benjamin Rush

Signer of the Declaration of Independence;
surgeon general of the Continental Army;
ratifier of the U.S. Constitution; "Father of American Medicine";
treasurer of the U.S. Mint; "Father of Public Schools
under the Constitution"

The Gospel of Jesus Christ prescribes the wisest rules for just conduct in every situation of life. Happy they who are enabled to obey them in all situations!... My only hope of salvation is in the infinite transcendent love of God manifested to the world by the death of His Son upon the Cross. Nothing but His blood will wash away my sins [Acts 22:16]. I rely exclusively upon it. Come, Lord Jesus! Come quickly! [Revelation 22:20][684]

I do not believe that the Constitution was the offspring of inspiration, but I am as satisfied that it is as much the work of a Divine Providence as any of the miracles recorded in the Old and New Testament.[685]

By renouncing the Bible, philosophers swing from their moorings upon all moral subjects... It is the only correct map of the human heart that ever has been published.[686]

[T]he greatest discoveries in science have been made by Christian philosophers and...there is the most knowledge in those countries where there is the most Christianity.[687]

[T]he only means of establishing and perpetuating our republican forms of government is the universal education of our youth in the principles of Christianity by means of the Bible.[688]

The great enemy of the salvation of man, in my opinion, never invented a more effective means of limiting Christianity from the world than by persuading mankind that it was improper to read the Bible at schools.[689]

[C]hristianity is the only true and perfect religion; and...in proportion as mankind adopt its principles and obey its precepts, they will be wise and happy.[690]

The Bible contains more knowledge necessary to man in his present state than any other book in the world.[691]

The Bible, when not read in schools, is seldom read in any subsequent period of life…[T]he Bible…should be read in our schools in preference to all other books because it contains the greatest portion of that kind of knowledge which is calculated to produce private and public happiness.[692]

Roger Sherman

Signer of the Declaration; signer of the Constitution;
"Master Builder of the Constitution"; judge;
framer of the Bill of Rights; U.S. senator

I believe that there is one only living and true God, existing in three persons, the Father, the Son, and the Holy Ghost, the same in substance, equal in power and glory. That the Scriptures of the Old and New Testaments are a revelation from God, and a complete rule to direct us how we may glorify and enjoy Him…. That He made man at first perfectly holy; that the first man sinned, and as he was the public head of his posterity, they all became sinners in consequence of his first transgression, are wholly indisposed to that which is good and inclined to evil, and on account of sin are liable to all the miseries of this life, to death, and to the pains of hell forever. I believe that God…did send His own Son to become man, die in the room and stead of sinners, and thus to lay a foundation for the offer of pardon and salvation to all mankind, so as all may be saved who are willing to accept the Gospel offer…. I believe a visible church to be a congregation of those who make a credible profession of their faith in Christ, and obedience to Him, joined by the bond of the covenant…. I believe that the sacraments of the New Testament are baptism and the Lord's Supper…. I believe that the souls of believers are at their death made perfectly holy, and immediately taken to glory: that at the end of this world there will be a resurrection of the dead, and a final judgment of all mankind, when the righteous shall be publicly acquitted by Christ the Judge and admitted to everlasting life and glory, and the wicked be sentenced to everlasting punishment.[693]

God commands all men everywhere to repent. He also commands them to believe on the Lord Jesus Christ, and has assured us that all who do repent and believe shall be saved…[G]od…has absolutely promised to bestow them on all these who are willing to accept them on the terms of the Gospel—that is, in a way of free grace through the atonement. "Ask and ye shall receive [John 16:24]. Whosoever will, let him come and take of the waters of life freely [Revelation 22:17]. Him that cometh unto me I will in no wise cast out" [John 6:37].[694]

[I]t is the duty of all to acknowledge that the Divine Law which requires us to love God with all our heart and our neighbor as ourselves, on pain of eternal damnation, is Holy, just, and good…. The revealed law of God is the rule of our duty.[695]

True Christians are assured that no temptation (or trial) shall happen to them but what they shall be enabled to bear; and that the grace of Christ shall be sufficient for them.[696]

"The volume which he consulted more than any other was the Bible. It was his custom, at the commencement of every session of Congress, to purchase a copy of the Scriptures, to peruse it daily, and to present it to one of his children on his return."[697]

Richard Stockton

Judge; signer of the Declaration of Independence

[A]s my children will have frequent occasion of perusing this instrument, and may probably be particularly impressed with the last words of their father, I think it proper here not only to subscribe to the entire belief of the great and leading doctrines of the Christian religion, such as the being of God; the universal defection and depravity of human nature; the Divinity of the person and the completeness of the redemption purchased by the blessed Savior; the necessity of the operations of the Divine Spirit; of Divine faith accompanied with an habitual virtuous life; and the universality of the Divine Providence: but also, in the bowels of a father's affection, to exhort and charge [my children] that the fear of God is the beginning of wisdom, that the way of life held up in the Christian system is calculated for the most complete happiness that can be enjoyed in this mortal state, [and] that all occasions of vice and immorality is injurious either immediately or consequentially—even in this life.[698]

Thomas Stone

Signer of the Declaration of Independence; selected as a delegate to the Constitutional Convention

Shun all giddy, loose, and wicked company; they will corrupt and lead you into vice and bring you to ruin. Seek the company of sober, virtuous and good people...which will lead [you] to solid happiness.[699]

Joseph Story

U.S. congressman; "Father of American Jurisprudence"; U.S. Supreme Court justice appointed by president James Madison

One of the beautiful boasts of our municipal jurisprudence is that Christianity is a part of the Common Law. There never has been a period in which the Common Law did not recognize Christianity as lying at its foundations.[700]

I verily believe that Christianity is necessary to support a civil society and shall ever attend to its institutions and acknowledge its precepts as the pure and natural sources of private and social happiness.[701]

Caleb Strong

Delegate at the Constitutional Convention to frame the U.S. Constitution; ratifier of the Constitution; U.S. senator; governor of Massachusetts

[He called on the State of Massachusetts to pray that:] ...all nations may know and be obedient to that grace and truth which came by Jesus Christ.[702]

Zephaniah Swift

U.S. congressman; diplomat; judge; author of America's first legal text (1795)

Jesus Christ has in the clearest manner inculcated those duties which are productive of the highest moral felicity and consistent with all the innocent enjoyments, to which we are impelled by the dictates of nature. Religion, when fairly considered in its genuine simplicity and uncorrupted state, is the source of endless rapture and delight.[703]

Charles Thomson

Secretary of the Continental Congress; designer of the Great Seal of the United States; along with John Hancock, Thomson was one of only two founders to sign the initial draft of the Declaration of Independence approved by Congress

I am a Christian. I believe only in the Scriptures, and in Jesus Christ my Savior.[704]

Jonathan Trumbull

Judge; legislator; governor of Connecticut; confidant of George Washington and called "Brother Jonathan" by him

The examples of holy men teach us that we should seek Him with fasting and prayer, with penitent confession of our sins, and hope in His mercy through Jesus Christ the Great Redeemer.[705]

Principally and first of all, I bequeath my soul to God the Creator and giver thereof, and my body to the earth to be buried in a decent Christian burial, in firm belief that I shall receive the same again at the general resurrection through the power of Almighty God, and hope of eternal life and happiness through the merits of my dear Redeemer Jesus Christ.[706]

[He called on the State of Connecticut to pray that:]

God would graciously pour out His Spirit upon us and make the blessed Gospel in His hand effectual to a thorough reformation and general revival of the holy and peaceful religion of Jesus Christ.[707]

George Washington

Judge; member of the Continental Congress; commander-in-chief of the Continental Army; president of the Constitutional Convention; first president of the United States; "Father of His Country"

You do well to wish to learn our arts and ways of life, and above all, the religion of Jesus Christ. These will make you a greater and happier people than you are.[708]

While we are zealously performing the duties of good citizens and soldiers, we certainly ought not to be inattentive to the higher duties of religion. To the distinguished character of Patriot, it should be our highest glory to add the more distinguished character of Christian.[709]

The blessing and protection of Heaven are at all times necessary but especially so in times of public distress and danger. The General hopes and trusts that every officer and man will endeavor to live and act as becomes a Christian soldier, defending the dearest rights and liberties of his country.[710]

I now make it my earnest prayer that God would…most graciously be pleased to dispose us all to do justice, to love mercy, and to demean ourselves with that charity, humility, and pacific temper of the mind which were the characteristics of the Divine Author of our blessed religion.[711]

Daniel Webster

U. S. senator; secretary of state, "Defender of the Constitution"

[T]he Christian religion—its general principles—must ever be regarded among us as the foundation of civil society.[712]

Whatever makes men good Christians, makes them good citizens.[713]

[T]o the free and universal reading of the Bible…men [are] much indebted for right views of civil liberty.[714]

The Bible is a book...which teaches man his own individual responsibility, his own dignity, and his equality with his fellow man.[715]

Noah Webster

Revolutionary soldier; judge; legislator; educator; "
Schoolmaster to America"

[T]he religion which has introduced civil liberty is the religion of Christ and His apostles.... This is genuine Christianity and to this we owe our free constitutions of government.[716]

The moral principles and precepts found in the Scriptures ought to form the basis of all our civil constitutions and laws.[717]

All the...evils which men suffer from vice, crime, ambition, injustice, oppression, slavery and war, proceed from their despising or neglecting the precepts contained in the Bible.[718]

[O]ur citizens should early understand that the genuine source of correct republican principles is the Bible, particularly the New Testament, or the Christian religion.[719]

[T]he Christian religion is the most important and one of the first things in which all children under a free government ought to be instructed. No truth is more evident than that the Christian religion must be the basis of any government intended to secure the rights and privileges of a free people.[720]

The Bible is the chief moral cause of all that is good and the best corrector of all that is evil in human society—the best book for regulating the temporal concerns of men.[721]

[T]he Christian religion…is the basis, or rather the source, of all genuine freedom in government…I am persuaded that no civil government of a republican form can exist and be durable in which the principles of Christianity have not a controlling influence.[722]

John Witherspoon

*Signer of the Declaration of Independence;
ratifier of the U.S. Constitution; president of Princeton*

[C]hrist Jesus—the promise of old made unto the fathers, the hope of Israel [Acts 28:20], the light of the world [John 8:12], and the end of the law for righteousness to every one that believeth [Romans 10:4]—is the only Savior of sinners, in opposition to all false religions and every uninstituted rite; as He Himself says (John 14:6): "I am the way, and the truth, and the life: no man cometh unto the Father but by Me."[723]

[N]o man, whatever be his character or whatever be his hope, shall enter into rest unless he be reconciled to God though Jesus Christ.[724]

[T]here is no salvation in any other than in Jesus Christ of Nazareth.[725]

I shall now conclude my discourse by preaching this Savior to all who hear me, and entreating you in the most earnest manner to believe in Jesus Christ; for "there is no salvation in any other" [Acts 4:12].[726]

It is very evident that both the prophets in the Old Testament and the apostles in the New are at great pains to give us a view of the glory and dignity of the person of Christ. With what magnificent titles is He adorned! What glorious attributes are ascribed to him!… All these conspire to teach us that He is truly and properly God—God over all, blessed forever![727]

[I]f you are not reconciled to God through Jesus Christ—if you are not clothed with the spotless robe of His righteousness—you must forever perish.[728]

[H]e is the best friend to American liberty who is the most sincere and active in promoting true and undefiled religion, and who sets himself with the greatest firmness to bear down profanity and immorality of every kind. Whoever is an avowed enemy of God, I scruple not to call him an enemy to his country.[729]

Oliver Wolcott

Signer of the Declaration of Independence;
military general; governor of Connecticut

Through various scenes of life, God has sustained me. May He ever be my unfailing friend; may His love cherish my soul; may my heart with gratitude acknowledge His goodness; and may my desires be to Him and to the remembrance of His name.... May we then turn our eyes to the bright objects above, and may God give us strength to travel the upward road. May the Divine Redeemer conduct us to that seat of bliss which He himself has prepared for His friends; at the approach of which every sorrow shall vanish from the human heart and endless scenes of glory open upon the enraptured eye. There our love to God and each other will grow stronger, and our pleasures never be dampened by the fear of future separation. How indifferent will it then be to us whether we obtained felicity by travailing the thorny or the agreeable paths of life—whether we arrived at our rest by passing through the envied and unfragrant road of greatness or sustained hardship and unmerited reproach in our journey. God's Providence and support through the perilous perplexing labyrinths of human life will then forever excite our astonishment and love. May a happiness be granted to those I most tenderly love, which shall continue and increase through an endless existence. Your cares and burdens must be many and great, but put your trust in that God Who has hitherto supported you and me; He will not fail to take care of those who put their trust in Him.... It is most evident that this land is under the protection of the Almighty, and that we shall be saved not by our wisdom nor by our might, but by the Lord of Host Who is wonderful in counsel and Almighty in all His operations.[730]

Notes

1. "Civil War Casualties," Civil War Trust, accessed March 4, 2016, http://www.civilwar.org/education/civil-war-casualties.html.
2. "Financial Cost of World War II," CaseAgainstBush.com, accessed March 4, 2016, http://caseagainstbush.blogspot.com/2005/04/financial-cost-of-world-war-ii1u.html#!/2005/04/financial-cost-of-world-war-ii1u.html.
3. Dennis Prager, CPAC 2016, American Conservative Union, accessed March 4, 2016, https://www.youtube.com/watch?v=vi6OeTkxt7w (accessed March 4, 2016).
4. Ibid.
5. Ibid.
6. Ibid.
7. Ibid.
8. Congressional Performance, Rasmussen Reports, February 22, 2016, http://www.rasmussenreports.com/public_content/politics/mood_of_america/congressional_performance.
9. Daniel Halper, "Obama: 'There's A Sense...The World Is Spinning So Fast and Nobody Is Able to Control It,' *The Weekly Standard*, October 7, 2014, http://www.weeklystandard.com/obama-theres-a-sense-...-the-world-is-spinning-so-fast-and-nobody-is-able-to-control-it/article/810784.
10. "Direction of the Country," RealClear Politics, accessed March 22, 2016, http://www.realclearpolitics.com/epolls/other/direction_of_country-902.html.

11. David Leonhart and Marjoi Connelly, "81 Percent in Poll Say Nation Is Headed on Wrong Track," *New York Times*, April 4, 2008, http://www.nytimes.com/2008/04/04/us/04poll.html?_r=0.

12. Edward Rogers, "The Insiders: Voters Blame Obama for Country Being Off-track," *The Washington Post*, September 5, 2014, https://www.washingtonpost.com/blogs/post-partisan/wp/2014/09/05/the-insiders-voters-blame-obama-for-country-being-off-track/.

13. "America's Top Fears 2015," Wilkinson College of Arts, Humanities, and Social Sciences, October 13, 2015, https://blogs.chapman.edu/wilkinson/2015/10/13/americas-top-fears-2015/.

14. Elizabeth Palermo, "What Do Americans Fear Most? Big Brother & Cybercrime," Live Science, October 15, 2015, http://www.livescience.com/52535-american-fear-survey-2015.html.

15. Mark Murray, "NBC/WSJ Poll: Terror Fears Reshape 2016 Landscape," NBC News, December 14, 2015, http://www.nbcnews.com/meet-the-press/nbc-wsj-poll-terror-fears-reshape-2016-landscape-n479831.

16. Margaret Griffis (ed.), "Casualties in Iraq," AntiWar.Com, accessed March 11, 2016, http://www.antiwar.com/casualties/.

17. Barbara Staff, "Obama approves Afghanistan troop increase," CNN, February 18, 2009, http://www.cnn.com/2009/POLITICS/02/17/obama.troops/index.html?eref.

18. Operation Enduring Freedom, ICasulties, accessed July 3, http://icasualties.org/oef/.

19. Nika Knight, "Deeper and Deeper into War: Obama Authorizes More Military Force in Afghanistan," CommonDreams, June 10, 2016, http://www.commondreams.org/news/2016/06/10/deeper-and-deeper-war-obama-authorizes-more-military-force-afghanistan.

20. Luis Martinez, "General Austin: Only '4 or 5' US-Trained Syrian Rebels Fighting ISIS," ABC News, September 15, 2015, http://abcnews.go.com/Politics/general-austin-us-trained-syrian-rebels-fighting-isis/story?id=33802596.

21. Daniel Goure, "Missing U.S. Leadership & A Fraying International Order," InFocus Winter 2016, Volume X: Number 1, The Jewish Policy Center, accessed March 11, 2016, http://www.jewishpolicycenter.org/5716/us-leadership-international-order.

22. Michaela DoDge, "Russian Intermediate-Range Nuclear Forces: What They Mean for the United States," Heritage Foundation, July 30, 2015, http://www.heritage.org/research/reports/2015/07/russian-intermediate-range-nuclear-forces-what-they-mean-for-the-united-states.

23. Patricia Zengerle, "U.S. Military Aid for Egypt Seen Continuing Despite Rights Concerns," *Reuters*, November 13, 2015, http://www.reuters.com/article/us-egypt-usa-aid-idUSKCN0T22E520151113.

24. Gabriel Bourouvich, "France Delivers First Rafales to Egypt," *Defense News*, July 20, 2015, http://www.defensenews.com/story/defense/air-space/air-force/2015/07/20/france-delivers-first-rafales-egypt/30419843/.

25. "Iraq, Russia Sign Military Deal," *Al-Monitor*, accessed March 11, 2016, http://www.al-monitor.com/pulse/tr/security/01/10/iraq-security-deal-russia.html#ixzz3wq1hJqWe.

26. Robert Kagan, "The Crisis of World Order," *Wall Street Journal*, November 20, 2015, http://commentators.com/the-crisis-of-world-order-wsj/.

27. Daniel Goure, "Missing U.S. Leadership & A Fraying International Order," InFocus Winter 2016, Volume X: Number 1, The Jewish Policy Center, accessed March 11, 2016, http://www.jewishpolicycenter.org/5716/us-leadership-international-order.

28. "Global Trends 2030: Alternative Worlds, National Intelligence Council, December 2012, http://www.dni.gov/files/documents/GlobalTrends_2030.pdf.

29. "The American Middle Class Is Losing Ground," PewResearchCenter, December 9, 2015, http://www.pewsocialtrends.org/2015/12/09/the-american-middle-class-is-losing-ground/.

30. http://www.bbc.com/news/magazine-30483762.

31. "Global Trends 2030: Alternative Worlds, National Intelligence Council, December 2012, p.18, http://www.dni.gov/files/documents/GlobalTrends_2030.pdf.

32. "Modern Immigration Wave Brings 59 Million to U.S., Driving Population Growth and Change Through 2065, PewResearch Center, September 28, 2015, http://www.pewhispanic.org/2015/09/28/modern-immigration-wave-brings-59-million-to-u-s-driving-population-growth-and-change-through-2065/.

33. "U.S. Residents Now Include 61 Million Immigrants and Their Minor Children," Limits to Growth, accessed March 11, 2016, http://www.limitstogrowth.org/.

34. "It's No Longer a 'Leave It to Beaver' World for American Families—But It Wasn't Back Then, Either," Pew Research Center, December 30, 2015, http://www.pewresearch.org/fact-tank/2015/12/30/its-no-longer-a-leave-it-to-beaver-world-for-american-families-but-it-wasnt-back-then-either/.

35. "Four-in-Ten Couples Are Saying 'I Do,' Again," PewResearchCenter,

November 14, 2014, http://www.pewsocialtrends.org/2014/11/14/four-in-ten-couples-are-saying-i-do-again/.

36. "Growing Number of Dads Home with the Kids," PewResearchCenter, June 5, 2014, http://www.pewsocialtrends.org/2014/06/05/growing-number-of-dads-home-with-the-kids/.

37. "U.S. Birth Rate Falls to a Record Low; Decline Is Greatest Among Immigrants," PewResearchCenter, November 29, 2012, http://www.pewsocialtrends.org/2012/11/29/u-s-birth-rate-falls-to-a-record-low-decline-is-greatest-among-immigrants/.

38. Neil Howe, "U.S. Birthdate Falls—Again," *Forbes*, January 28, 2015, http://www.forbes.com/sites/neilhowe/2015/01/28u-s-birthrate-falls-again/2/#532044552d02.

39. Mark Mather, "Fact Sheet: The Decline in U.S. Fertility," Population Reference Bureau, July 2012, http://www.prb.org/publications/datasheets/2012/world-population-data-sheet/fact-sheet-us-population.aspx. *The total fertility rate estimates the number of births a woman is expected to have during her lifetime based on current age-specific fertility rates. Replacement level fertility is the level of fertility at which a couple has only enough children to replace themselves, or about 2.1 children per couple.*

40. Kevin Mathews, "Do You Trust the Government? 87 Percent of Americans Don't," Truth-Out.org, August 16, 2014, http://www.truth-out.org/news/item/25628-do-you-trust-the-government-87-of-americans-dont.

41. Demetrius Minor, "A Majority of Americans Believe That the US Is in a State of Moral Decline, Independent Journal," accessed March 12, 2016, http://www.ijreview.com/2015/06/338257-theres-collapse-morals-america-study-finds/.

42. "Americans Divided on the Importance of Church," Barna, March 24, 2014, https://www.barna.org/barna-update/culture/661-americans-divided-on-the-importance-of-church.

43. Ibid.

44. Ibid.

45. "Billy Graham: 'My Heart Aches for America,' New Letter Addresses Nation's Declining Morality and the Need for Spiritual Revival," Joel Rosenberg's Blog, July 24, 2012, https://flashtrafficblog.wordpress.com/2012/07/24/billy-graham-my-heart-aches-for-america-new-letter-addresses-nations-declining-morality-and-the-need-for-spiritual-revival.

46. David Rohde, "The Swelling Middle," *Reuters*, 2012, http://www.reuters.com/middle-class-infographic.

47. "Growing Urbanization," Citelum Group, accessed March 12, 2016, http://www.citelum.com/en/our-mission/modern-city/growing-urbanization.

48. David Bryce Haden, "The Future of Spirituality," The Huffpost Religion, May 2, 2015, http://www.huffingtonpost.com/david-bryce-yaden/the-future-of-spirituality_b_7235490.html.

49. Nassin Nicholas Talib, "The Black Swan: The Impact of the Highly Improbable," *New York Times*, April 22, 2007, http://www.nytimes.com/2007/04/22/books/chapters/0422-1st-tale.html?_r=0.

50. "Latest Ebola Outbreak Over in Liberia; West Africa Is at Zero, But New Flare-ups Are Likely to Occur," World Health Organization, January 14, 2016, http://www.who.int/mediacentre/news/releases/2016/ebola-zero-liberia/en/.

51. Andrew Mark Miller, "Debunking the "97 percent of Scientists Agree on Man-made 'Climate Change' Myth," *Communities Digital News*, July 23, 2015, http://www.commdiginews.com/politics-2/debunking-the-97-of-scientists-agree-on-man-made-climate-change-myth-45412/#FLH1G4ecimghILoD.99.

52. "Greece's Debt Crisis Explained," Editorial, *The New York Times*, November 9, 2015, http://www.nytimes.com/interactive/2015/business/international/greece-debt-crisis-euro.html?_r=0.

53. "Bible Signs of the End Times," Signs of the End Times, accessed March 12, 2016, http://www.signs-of-end-times.com/.

54. "Catastrophes: U.S.," Insurance Information Institute, accessed March 12, 2016, http://www.iii.org/fact-statistic/catastrophes-us.

55. "Tornadoes: A Rising Risk?," Lloyd's, accessed March 12, 2016, http://www.lloyds.com/~/media/Lloyds/Reports/Emergingpercent20Riskpercent20Reports/Tornadoespercent20finalpercent20report.pdf.

56. "Top 10 Most Expensive Disasters in U.S. History," Trusted Choice, December 17, 2013, https://www.trustedchoice.com/insurance-articles/weather-nature/most-expensive-disasters/.

57. "List of Ongoing Armed Conflicts," Wikipedia, accessed March 12, 2016, https://en.wikipedia.org/wiki/List_of_ongoing_armed_conflicts.

58. Jeremy Burns, "5 Signs of the End Times Unfolding Before Our Eyes," CharismaNews, May 8, 2015, http://www.charismanews.com/world/49561-5-signs-of-the-end-times-unfolding-before-our-eyes.

59. "William J. Clinton Quotes," Brainy Quotes, accessed March 12, 2016, http://www.brainyquote.com/quotes/authors/w/william_j_clinton.html.

60. "Obama: Police Who Arrested Professor 'Acted Stupidly,'" CNN, July 23, 2009, http://www.cnn.com/2009/US/07/22/harvard.gates.interview/.

61. Peter Labarbara, "Obama's Radical 'LGBT' Legacy— An Immoral 'Transformation' of America," *Barbwire*, January 15, 2016, http://barbwire.com/2016/01/15/ barack-obamas-radical-lgbt-legacy-an-immoral-transformation-of-america/.

62. Ali Meyer, "11,472,000 Americans Have Left Workforce Since Obama Took Office," CNS News, August 1, 2014, http://www.cnsnews.com/news/article/ ali-meyer/11472000-americans-have-left-workforce-obama-took-office.

63. Mortimer Zuckerman, "Mortimer Zuckerman: Those Jobless Numbers Are Even Worse Than They Look," *Wall Street Journal*, September 2, 2012, http://www.wsj.com/articles/SB100008723963904442737045776356812 06305056.

64. "Welfare Statistics," Statistic Brain Research Institute, accessed March 12, 2016, http://www.statisticbrain.com/welfare-statistics/.

65. Terrance Jeffrey, "10,996,447: Disability Beneficiaries Hit New Record," CNS News, May 20, 2014, http://cnsnews.com/news/article/ terence-p-jeffrey/10996447-disability-beneficiaries-hit-new-record.

66. "United States Loses Prized AAA Credit Rating from S&P," *Reuters*, August 7, 2011, http://www.reuters.com/article/ us-usa-debt-downgrade-idUSTRE7746VF20110807.

67. Kevin Robillard, "Brookings: Cash for Clunkers Failed," *Politico*, October 13, 2013, http://www.politico.com/story/2013/10/brookings-cash-for-clunkers-program-barack-obama-administration-cars-stimulus-099134#ixzz3xVxlUQgm.

68. Julia Ryan, "American Schools vs. the World: Expensive, Unequal, Bad at Math," *The Atlantic*, December 3, 2013, http://www.theatlantic.com/education/archive/2013/12/ american-schools-vs-the-world-expensive-unequal-bad-at-math/281983/.

69. "World Health Organization Ranking of Health Systems in 2000," Wikipedia, accessed March 12, 2016, https://en.wikipedia.org/wiki/ World_Health_Organization_ranking_of_health_systems_in_2000.

70. Christine Moyer, "U.S. Found to be Unhealthiest among 17 Affluent Countries," amednews.com, January 21, 2013, http://www.amednews. com/article/20130121/health/130129983/4/.

71. "Status of the Social Security and Medicare Programs," Social Security Administration, accessed March 12, 2016, https://www.ssa.gov/oact/ trsum/.

72. Ronald Reagan, "Freedom Is Never More than One Generation away from Extinction," YouTube, accessed March 12, 2016, https://www.youtube.com/watch?v=SDouNtnR_IA.

73. Richard Langworth, "Democracy Is the Worst Form of Government," Richard Langworth Blog, Hillsdale College, June 26, 2009, https://richardlangworth.com/worst-form-of-government.

74. Michael Joyce, "Renewing Our Experiment in Ordered Liberty," Acton Institute, accessed March 12, 2016, http://www.acton.org/pub/religion-liberty/volume-8-number-5/renewing-our-experiment-ordered-liberty.

75. Robert Kraynack, "The American Founders and Their Relevance Today," Intercollegiate Studies Institute, Winter 2015–Vol. 57, No. 1, https://home.isi.org/american-founders-and-their-relevance-today.

76. Ibid.

77. Ibid.

78. "The Laws of Nature," Founding.com, accessed March 12, 2016, http://www.founding.com/the_declaration_of_i/pageID.2415/default.asp.

79. Robert Kraynack, "The American Founders and Their Relevance Today," Intercollegiate Studies Institute, Winter 2015–Vol. 57, No. 1, https://home.isi.org/sites/default/files/MA_57.1_3Essays_TheAmericanFounders.pdf.

80. Ibid.

81. Ibid.

82. Ibid.

83. Danny M. Francis, "GOP Constantly Roots for Obama's Failure," Letters, *Watertown Daily Times* (New York), July 25, 2014.

84. John D. Gartner, "How to Pick a President," *Psychology Today*, January/February 2016, https://www.psychologytoday.com/articles/201601/how-pick-president.

85. Ibid.

86. Ibid.

87. Ibid.

88. Ibid.

89. Ibid.

90. Ibid.

91. Ibid.

92. Ibid.

93. Ibid.

94. Ibid.
95. Richard N. Haass, "The Age of Non-polarity," *Foreign Affairs*, May/June 2008, Vol. 87, Issue 3, p. 44–56, https://www.foreignaffairs.com/articles/united-states/2008-05-03/age-non-polarity.
96. Ibid.
97. Ibid.
98. Ibid.
99. Henry Kissinger, "Henry Kissinger on the Assembly of a New World Order," *Wall Street Journal*, August 29, 2014, http://www.wsj.com/articles/henry-kissinger-on-the-assembly-of-a-new-world-order-1409328075.
100. Ibid.
101. Ibid.
102. Ibid.
103. Ibid.
104. Andrea Edoardo, "Towards a Multi-Polar International System: Which Prospects for Global Peace?," E-International Relations Students, June 3, 2013, http://www.e-ir.info/2013/06/03/towards-a-multi-polar-international-system-which-prospects-for-global-peace/.
105. Ibid.
106. Wilfred Hahn, "The Role of Crisis: Greasing the Road to Multipolar Globalism," Watchman Newsletter, October 21, 2009, http://watchmannewsletter.typepad.com/news/2009/10/the-role-of-crisis-greasing-the-road-to-multipolar-globalism.html.
107. Ibid.
108. "Economy of China," Wikipedia, accessed March 12, 2016, https://en.wikipedia.org/wiki/Economy_of_China.
109. Alex Newman, "China Staking Claim in the New World order," TheNewAmerican.com, June 9, 2015, http://www.thenewamerican.com/world-news/asia/item/21025-china-staking-claim-in-the-new-world-order.
110. John McLaughlin, "Rethinking U.S. Strategy Toward China, the Cipher Brief, January 18, 2016, http://thecipherbrief.com/article/rethinking-us-strategy-toward-china.
111. Bill Gertz, "Chinese Militarization in S. China Sea Aimed at Rapid Power Projection," Free Beacon, March 10, 2016, http://freebeacon.com/national-security/dni-chinese-militarization-in-s-china-sea-aimed-at-rapid-power-projection/.
112. Ibid.
113. "History of the United Nations," United Nations, accessed March 12, 2016, http://www.un.org/en/sections/history/history-united-nations/index.html.

114. Ibid.

115. "Criticism of the United Nations," Wikipedia, accessed March 12, 2016, https://en.wikipedia.org/wiki/Criticism_of_the_United_Nations.

116. Caitlin Dickson, "Agenda 21: The UN Conspiracy that Just Won't Die," *The Daily Beast*, April 13, 2014, http://www.thedailybeast.com/articles/2014/04/13/agenda-21-the-un-conspiracy-that-just-won-t-die.html.

117. Rachel Alexander, "Agenda 21: Conspiracy Theory or Real Threat," Townhall.com, July 2, 2011, http://townhall.com/columnists/rachelalexander/2011/07/02/agenda_21_conspiracy_theory_or_real_threat/page/full.

118. Caitlin Dickson, "Agenda 21: The UN Conspiracy that Just Won't Die," *The Daily Beast*, April 13, 2014, http://www.thedailybeast.com/articles/2014/04/13/agenda-21-the-un-conspiracy-that-just-won-t-die.html.

119. Andrew Cohen, "Is the UN Using Bake Paths to Achieve World Domination?," *The Atlantic*, February 7, 2012, http://www.theatlantic.com/national/archive/2012/02/is-the-un-using-bike-paths-to-achieve-world-domination/252572/.

120. Rachel Alexander, "Agenda 21: Conspiracy Theory or Real Threat," Townhall.com, July 2, 2011, http://townhall.com/columnists/rachelalexander/2011/07/02/agenda_21_conspiracy_theory_or_real_threat/page/full.

121. Ibid.

122. "Number of Employees at Exon Mobile 2001-2014," Statista, accessed March 12, 2016, http://www.statista.com/statistics/264122/number-of-employees-at-exxon-mobil-since-2002/.

123. "Exxon Mobile Corporation Institutional Ownership," Nasdaq.com, accessed March 12, 2016, http://www.nasdaq.com/symbol/xom/institutional-holdings.

124. "World GDP Ranking 2015," Knoema.com, accessed March 12, 2016, http://knoema.com/nwnfkne/world-gdp-ranking-2015-data-and-charts.

125. "Exxon Mobile," OpenSecrets.org, accessed March 12, 2016, https://www.opensecrets.org/orgs/summary.php?id=D000000129.

126. "50 Reasons We Are Living in the End Times," Raptureforums.com, accessed March 12, 2016, http://www.raptureforums.com/Signs/50reasons.cfm.

127. Ibid.

128. Ibid.

129. Ibid.
130. Laura Beth Neilson, "What Is Terrorism?," *al Jazeera*, April 17, 2013, http://www.aljazeera.com/indepth/opinion/2013/04/20134179548891867.html.
131. "Definitions of Terrorism in the U.S. Code," Federal Bureau of Investigation, accessed March 12, 2016, https://www.fbi.gov/about-us/investigate/terrorism/terrorism-definition.
132. "What Is Terrorism?," Terrorism Research, accessed March 12, 2016, http://www.terrorism-research.com/.
133. Ibid.
134. Ibid.
135. "Frequently Asked Questions about Waco," FrontLine, *PBS*, accessed March 12, 2016, http://www.pbs.org/wgbh/pages/frontline/waco/topten.html.
136. "Yogi Berra Quotes," Goodreads.com, accessed March 12, 2016, http://www.goodreads.com/quotes/261863-it-s-tough-to-make-predictions-especially-about-the-future.
137. "Domestic Terrorism: In the Post 9/11 Era," Federal Bureau of Investigation, September 7, 2009, https://www.fbi.gov/news/stories/2009/september/domterror_090709.
138. "Country Reports on Terrorism 2014, Department of State, accessed March 12, 2016, http://www.state.gov/j/ct/rls/crt/2014/index.htm.
139. "Global Terrorism Database," Department of Homeland Security, accessed March 12, 2016, http://www.start.umd.edu/gtd/.
140. "Types of Terrorist Incidents," Terrorism Research, accessed March 12, 2016, http://www.terrorism-research.com/incidents/.
141. "Global Terrorism Database," Department of Homeland Security, accessed March 12, 2016, http://www.start.umd.edu/gtd/.
142. "Types of Terrorist Incidents," Terrorism Research, accessed March 12, 2016, http://www.terrorism-research.com/goals/.
143. "Statement by the President on ISIS," The White House, September 10, 2014, https://www.whitehouse.gov/the-press-office/2014/09/10/statement-president-isil-1.
144. "Salafi Movement," Wikipedia, accessed March 12, 2016, https://en.wikipedia.org/wiki/Salafi_movement.
145. "Iraqi Journalist Comes Out Against Claim That ISIS Has Nothing to Do with Islam," MEMRI, February 1, 2016, http://www.memri.org/report/en/0/0/0/0/0/0/0/8985.htm.

146. Robert L. Maginnis, *Never Submit: Will the Extermination of Christians Get Worse Before it Gets Better?* (Crane, MO: Defender, 2015).

147. Sahih Bukhari, Vol. 5, book 58, No. 148 (as cited in Bill Warner, *The Islamic Trilogy,* Vol, 2, *The Political Traditions of Mohammed, The Hadith for the Unbelievers*).

148. Ishaq, 125 (as quoted in Bill Warner, *The Islamic Doctrine of Christians and Jews,* (Center for the Study of Political Islam, 2010).

149. Koran 47:4 (as cited in Warner, *The Islamic Trilogy,* Vol, 2, *The Political Traditions of Mohammed, The Hadith for the Unbelievers*).

150. Maginnis, *Never Submit,* p. 205.

151. "Types of Terrorist Incidents," Terrorism Research, accessed March 12, 2016, http://www.terrorism-research.com/future/.

152. "Nasir al-Fahd," Wikipedia, accessed March 12, 2016, https://en.wikipedia.org/wiki/Nasir_al-Fahd.

153. Harold Doornbos and Jenan Moosa, "Found: The Islamic State's Terror Laptop of Doom," *Foreign Policy,* August 28, 2014.

154. Helene Cooper and Eric Schmidt, "ISIS Detainee's Information Led to 2 U.S. Airstrikes, Officials Say," The *New York Times,* March 9, 2016, http://www.nytimes.com/2016/03/10/world/middleeast/isis-detainee-mustard-gas.html?emc=edit_na_20160309&nlid=26545726&ref=cta&_r=0.

155. "161 Chemical Weapons Attacks in Syria's War, New Report Says," *ABC News,* March 14, 2016, http://abcnews.go.com/International/wireStory/161-chemical-weapons-attacks-syrias-war-report-37624560.

156. Joseph Votel, "Understanding Terrorism Today and Tomorrow," Combating Terrorism Center, July 28, 2015, https://www.ctc.usma.edu/posts/understanding-terrorism-today-and-tomorrow.

157. Ibid.

158. Ibid.

159. Nick Bostrom, "Transhumanist Values," Oxford University, accessed March 12, 2016, http://www.nickbostrom.com/ethics/values.html.

160. J. Hughes, "Problems with Transhumanism," IEET, January 6, 2010, http://ieet.org/index.php/IEET/more/hughes20100105/.

161. David Gelernter, "Machines That Will Think and Feel," *The Wall Street Journal,* March 18, 2016, http://www.wsj.com/articles/when-machines-think-and-feel-1458311760.

162. Ibid.

163. Ibid.

164. Ibid.

165. Annie Jacobsen, *The Pentagon's Brain*, Little, Brown and Company, New York, 2015.

166. Meghan Neal, "DARPAS New Biotech Unit Will Try to Create Artificial Life Forms," Motherboard, April 1, 2014, http://motherboard.vice.com/read/darpas-new-biotech-unit-will-try-to-create-artificial-life-forms.

167. John Stanton, "Zika, GMOs, the Pentagon and Wall Street," *Sri Lanka Guardian*, February 6, 2016, http://www.slguardian.org/2016/02/the-buzz-around-intrexon-corporation-zika-gmos-the-pentagon-and-wall-street/.

168. "Nazi Medical Experiments," Holocaust Encyclopedia, U.S. Holocaust Memorial Museum, accessed March 12, 2016, http://www.ushmm.org/wlc/en/article.php?ModuleId=10005168.

169. "The 30 Most Disturbing Human Experiments in History," Bestpsychologydegrees.com, accessed March 12, 2016, http://www.bestpsychologydegrees.com/30-most-disturbing-human-experiments-in-history/.

170. "Poison Laboratory of the Soviet secret services," Wikipedia, accessed March 12, 2016, https://en.wikipedia.org/wiki/Poison_laboratory_of_the_Soviet_secret_services.

171. "Human experimentation in North Korea," Wikipedia, accessed March 12, 2016, https://en.wikipedia.org/wiki/Human_experimentation_in_North_Korea.

172. "Unit 731," Wikipedia, accessed March 12, 2016, https://en.wikipedia.org/wiki/Unit_731.

173. "Project MKUltra," Wikipedia, accessed March 12, 2016, https://en.wikipedia.org/wiki/Project_MKUltra.

174. "Project ARTICHOKE," Wikipedia, accessed March 12, 2016, https://en.wikipedia.org/wiki/Project_ARTICHOKE.

175. "Operation Midnight Climax," Wikipedia, accessed March 12, 2016, https://en.wikipedia.org/wiki/Operation_Midnight_Climax.

176. "Project MKUltra," Wikipedia, accessed March 12, 2016, https://en.wikipedia.org/wiki/Project_MKUltra.

177. "Project 4.1," Wikipedia, accessed March 12, 2016, https://en.wikipedia.org/wiki/Project_4.1.

178. Rick Anderson, "Great Balls of Fire," Motherjones, January 4, 2000, http://www.motherjones.com/politics/2000/01/great-balls-fire.

179. "Unethical Human Experimentation in the United States," Wikipedia, accessed March 12, 2016, https://en.wikipedia.org/wiki/Unethical_human_experimentation_in_the_United_States.

180. Patrick Cochburn, "US Navy Tested Mustard Gas on its Own Sailors: In 1943 the Americans Used Humans in Secret Experiments. Patrick Cockburn in Washington Reports on the Survivors Who Bear the Scars," *Independent*, March 13, 1993, http://www.independent.co.uk/news/world/us-navy-tested-mustard-gas-on-its-own-sailors-in-1943-the-americans-used-humans-in-secret-1497508.html.

181. Robert Maginnis, review of Richard Gabriel, *No More Heroes: Madness and Psychiatry in War* (Hill and Wang, 1987), *Infantry*, Volume 77, Number 3, May-June 1987, http://www.benning.army.mil/infantry/magazine/issues/1987/MAY-JUN/pdfs/MAY-JUN1987.pdf.

182. Adam Estes, "The Enemy Can't Run from These Brain Wave-Powered Binoculars," Motherboard, September 19, 2012, http://motherboard.vice.com/blog/the-enemy-can-t-run-from-these-brain-wave-powered-binoculars.

183. Paul Philips, "DARPA: Genetically Modified Humans For A Super Soldier Army," Activist Post, October 11, 2015, http://www.activistpost.com/2015/10/darpa-genetically-modified-humans-for-a-super-soldier-army.html.

184. Ibid.

185. Meghan Neal, "DARPA Is Developing an Implant that Can Read Brain Signals in Real-Time," Motherboard, October 25, 2013, http://motherboard.vice.com/blog/darpa-is-developing-an-implant-that-can-read-brain-signals-in-real-time.

186. Jordan Pearson, "DARPA Is Developing Implants That Can Heal Soldiers' Bodies and Minds," Motherboard, August 27, 2014, http://motherboard.vice.com/read/darpa-is-developing-implants-that-can-heal-soldiers-bodies-and-minds.

187. Ryan Browne, "U.S. Military Spending Millions to Make Cyborgs a Reality," CNN, March 7, 2016, http://edition.cnn.com/2016/03/07/politics/pentagon-developing-brain-implants-cyborgs/index.html

188. Ibid.

189. Ibid.

190. The Pentagon's Brain, p. 425.

191. Paul Philips, "DARPA: Genetically Modified Humans For A Super Soldier Army," Activist Post, October 11, 2015, http://www.activistpost.com/2015/10/darpa-genetically-modified-humans-for-a-super-soldier-army.html.

192. Eileen Shim, "DARPA Has Genetically Engineered Super Blood to Protect Soldiers From a Major Threat," Mic.com, July 2, 2014, http://mic.com/

articles/92685/darpa-has-genetically-engineered-super-blood-to-protect-soldiers-from-a-major-threat.

193. Ibid.

194. Derek Mead, "New Radiation Treatment Will Still Save You a Day After the Bomb," Motherboard, January 11, 2012, http://motherboard.vice.com/blog/proteins-key-to-darpa-s-new-radiation-treatment.

195. Jason Koebler, "The Military Is Working on Making Humans as Disease-Resistant as Rats," Motherboard, July 2, 2015, http://motherboard.vice.com/read/the-military-is-working-on-making-humans-as-disease-resistant-as-rats.

196. Ibid.

197. Katie Drummond, "Pentagon Looks to Breed Immortal 'Synthetic Organisms,' Molecular Kill-Switch Included," Wired.com, February 5, 2010, http://www.wired.com/2010/02/pentagon-looks-to-breed-immortal-synthetic-organisms-molecular-kill-switch-included/.

198. The Pentagon's Brain, p. 435.

199. Ibid.

200. Loren Thompson, "Gene Wars: Targeted Mutations Will Spawn Unique Dangers, And Soon," Forbes, January 29, 2016, http://www.forbes.com/sites#/sites/lorenthompson/2016/01/29/gene-wars-targeted-mutations-will-spawn-unique-dangers-and-soon/#5b2c7cc841d0.

201. Ibid.

202. "Cyborg," Wikipedia, accessed March 12, 2016, https://en.wikipedia.org/wiki/Cyborg.

203. The Pentagon's Brain, p. 444.

204. George Devorsky, "It Could Be A War Crime To Use Biologically Enhanced Soldiers," io9, January 22, 2013, http://io9.gizmodo.com/5977986/would-it-be-a-war-crime-to-use-biologically-enhanced-soldiers.

205. Ibid.

206. Ibid.

207. Ibid.

208. Ibid.

209. Geneva Conventions, United Nations, accessed March 12, 2016, https://treaties.un.org/doc/Publication/UNTS/Volumepercent201125/volume-1125-I-17512-English.pdf.

210. "War Crimes," Wikipedia, accessed March 12, 2016, https://en.wikipedia.org/wiki/War_crime.

211. George Devorsky, "It Could Be A War Crime To Use Biologically Enhanced

Soldiers," io9, January 22, 2013, http://io9.gizmodo.com/5977986/
would-it-be-a-war-crime-to-use-biologically-enhanced-soldiers.

212. C. K. Quarterman, "Giants as described by Rabbinic Literature," Fallen
Angles, August 15, 2011, http://www.fallenangels-ckquarterman.com/
giants-as-described-by-rabbinic-literature/.

213. Keith Cowing, "Recent Space Poll: The Public is Not Always in Synch with
Space Advocates," NASA Watch, February 22, 2015, http://nasawatch.
com/archives/2015/02/recent-space-po.html.

214. "How Much Did It Cost the US Taxpayers to Put the First Man on the
Moon?," Quora, accessed March 16, 2016, https://www.quora.com/How-
much-did-it-cost-the-US-taxpayers-to-put-the-first-man-on-the-moon.

215. Cassandra Szklarski, "Stephen Hawking: Space Exploration Crucial
to Human Survival," *Huffington Post*, November 18, 2011, http://
www.huffingtonpost.ca/2011/11/18/stephen-hawking-space-
exploration_n_1101975.html.

216. Zaina Adamu, "Exploring Space: Why's It So Important?," CNN,
October 20, 2012, http://lightyears.blogs.cnn.com/2012/10/20/
exploring-space-whys-it-so-important/.

217. Eric Niiler, "3D-Printed Ceramics Could Build Next-Gen Spaceships,"
Space.com, January 5, 2016, http://www.space.com/31516-3d-printed-
ceramics-next-gen-spaceships.html.

218. Mike Wall, "Asteroid Metal 3D Printing Test Planetary Resources," Space.
com, January 8, 2016, http://www.space.com/31553-asteroid-metal-3d-
printing-test-planetary-resources.html.

219. Leonard David, "NASA Planetary Defense Office Launched to Protect
Earth from Asteroids," Space.com, January 8, 2016, http://www.space.
com/31551-nasa-planetary-defense-office-launched.html.

220. Peter de Selding, "European Governments Boost SATCOM
Spending," SpaceNews.com, January 19, 2016, http://spacenews.com/
european-governments-boost-satcom-spending/.

221. Walter Scott, "Growth of Satellite Technology," TheCipherBrief.
com, October 4, 2015, https://www.thecipherbrief.com/article/
growth-satellite-technology.

222. "NASA Spinoff 2015 Features Space Technology
Making Life Better on Earth," NASA, January 20,
2016, https://www.nasa.gov/press/2015./january/
nasa-spinoff-2015-features-space-technology-making-life-better-on-earth.

223. "What's next for NASA?," NASA, September 3, 2013, http://www.nasa.
gov/about/whats_next.html.

224. David Dickinson, "Robotic Flyers: Future of Space Exploration?," SkyandTelescope.com, August 18, 2015, http://www.skyandtelescope.com/astronomy-news/robotic-flyers-drones-the-future-of-space-exploration-0818201544/.

225. "The Battle to Militarize Space Has Begun," STRATFOR.com, November 11, 2015, https://www.stratfor.com/analysis/battle-militarize-space-has-begun.

226. Mike Gruss, "New U.S. Kill Vehicle Qill Dly in 2018, Rake on Its First Target in 2019," Spacenews.com, January 20, 2016, http://spacenews.com/new-u-s-kill-vehicle-will-fly-in-2018-take-on-its-first-target-in-2019/.

227. Michael Peck, "Get Ready America Here Comes China's Ballistic Missile," *National Interest*, October 25, 2015, http://nationalinterest.org/feature/get-ready-america-here-comes-chinas-ballistic-missile-14162.

228. John Costello, "China Finally Centralizes Its Space, Cyber, Information Forces," TheDiplomat.com, January 20, 2016, http://thediplomat.com/2016/01/china-finally-its-centralizes-space-cyber-information-forces/.

229. Joe MacReynolds, "China's Evolving Perspectives on Network Warfare: Lessons from the Science of Military Strategy," The Jamestown Foundation, April 20, 2015, http://www.jamestown.org/single/?tx_ttnewspercent5Btt_newspercent5D=43798&no_cache=1#.VqfVm__2a71.

230. Anatoly Zak, "Here is the Soviet Union's Secret Space Cannon," Popularmechanics.com, November 16, 2015, http://www.popularmechanics.com/military/weapons/a18187/here-is-the-soviet-unions-secret-space-cannon/.

231. J. Michael Cole, "Five Futuristic Weapons Could Change Warfare," NationalInterest.org, accessed March 12, 2016, http://nationalinterest.org/print/commentary/five-futuristic-weapons-could-change-warfare-9866.

232. Marcus Geisgerber, "Pentagon Eyes Laser Armed Drones to Shoot Down Ballistic Missiles, DefenseOne.com, January 19, 2016, http://www.defenseone.com/technology/2016/01/pentagon-laser-drones-ballistic-missiles/125232/.

233. Aaron Mehta, "Laser Weapons Ready for Use Today, Lockheed Executives Say," Defense News, March 15, 2016, http://www.defensenews.com/story/defense/innovation/2016/03/15/laser-weapons-directed-energy-lockheed-pewpew/81826876/.

234. Robert Beckhusen, "Russia Is Concerned about America's Far Off Space Weapons," Motherboard, September 18, 2015, http://motherboard.vice.com/read/russia-is-concerned-about-americas-far-off-space-weapons.

235. Ibid.
236. Marcia Smith, "SecDef Carter Worried about Russian Space Activities," spacepolicyonline.com, November 8, 2015, http://www.spacepolicyonline. com/news/secdef-carter-worried-about-russian-space-activities.
237. Ibid.
238. Franz-Stephan Gady, "Russia to Test Launch 16 Intercontinental Ballistic Missiles in 2016," thediplomat. com, January 12, 2016, http://thediplomat.com/2016/01/ russia-to-test-launch-16-intercontinental-ballistic-missiles-in-2016/.
239. James R. Clapper, Director of National Intelligence, "Worldwide Cyber Threats," House Permanent Select Committee on Intelligence, September 15, 2015, http://www.dni.gov/files/documents/HPSCIpercent2010 percent20Septpercent20Cyberpercent20Hearingpercent20SFR.pdf.
240. Amber Corrin, "Capitol Hill Testimonies Paint a Grim Cyber Picture," C4ISRnet.com, September 29, 2015, http://www.c4isrnet.com/story/ military-tech/cyber/2015/09/29/cybersecurity-testimonies/73044574/.
241. Ellen Nakashima and Steven Mufson, "Hackers Have Attacked Foreign Utilities, CIA Analyst Says," washingtonpost.com, January 19, 2008, http://www.washingtonpost.com/wp-dyn/content/article/2008/01/18/ AR2008011803277.html.
242. Sioghan Gorman, "Electricity Grid in U.S. Penetrated By Spies," *Wall Street Journal,* April 8, 2009, http://www.wsj.com/articles/ SB123914805204099085.
243. James R. Clapper, Director of National Intelligence, "Worldwide Cyber Threats," House Permanent Select Committee on Intelligence, September 15, 2015, http://www.dni.gov/files/documents/HPSCIpercent2010 percent20Septpercent20Cyberpercent20Hearingpercent20SFR.pdf.
244. "Cyberwarfare 101: The Internet Is Mightier Than the Sword," Stratfor.com, April 15, 2008, https://www.stratfor.com/analysis/ cyberwarfare-101-Internet-mightier-sword.
245. Ibid.
246. Oxford Dictionaries, accessed March 12, 2016, http://www. oxforddictionaries.com/definition/english/cyber.
247. "Cyberwarfare," Wikipedia.org, accessed March 12, 2016, https:// en.wikipedia.org/wiki/Cyberwarfare.
248. "Cyberwarfare 101: The Internet Is Mightier Than the Sword," Stratfor.com, April 15, 2008, https://www.stratfor.com/analysis/ cyberwarfare-101-Internet-mightier-sword.

249. Julie Hirschfeld Davis, "Hacking of Government Computers Exposed 21.5 Million People," *New York Times*, July 9, 2015, http://www.nytimes.com/2015/07/10/us/office-of-personnel-management-hackers-got-data-of-millions.html.

250. Jessica Silver-Greenberg, Matthew Goldstein and Nicole Perlroth, "JPMorgan Chase Hacking Affects 76 Million Households," New York Times, October 2, 2014, http://dealbook.nytimes.com/2014/10/02/jpmorgan-discovers-further-cyber-security-issues/.

251. Kevin Grandville, "9 Recent Cyberattacks Against Big Businesses," *New York Times*, February 5, 2015, http://www.nytimes.com/interactive/2015/02/05/technology/recent-cyberattacks.html.

252. Kara Scannell, "Typo Trips the Alarm in $101m Cyber Bank Heist," *Financial Times*, March 10, 2016, http://www.ft.com/intl/cms/s/0/357257c6-e6ee-11e5-bc31-138df2ae9ee6.html#axzz42tfoHY00.

253. Sean Lyngaas, "Rogers: Cyber Command Capabilities at 'Tipping Point,'" FCW, January 21, 2016, https://fcw.com/articles/2016/01/21/rogers-nsa-lyngaas.aspx.

254. James R. Clapper, Director of National Intelligence, "Worldwide Cyber Threats," House Permanent Select Committee on Intelligence, September 15, 2015, http://www.dni.gov/files/documents/HPSCIpercent2010 percent20Septpercent20Cyberpercent20Hearingpercent20SFR.pdf.

255. Written testimony of USCG Assistant Commandant for Prevention Policy RDML Paul Thomas for a House Committee on Homeland Security, Subcommittee on Border and Maritime Security hearing titled 'Protecting Maritime Facilities in the 21st Century: Are Our Nation's Ports at Risk for a Cyber-Attack?,' Homeland Security, October 8, 2015, https://www.dhs.gov/news/2015/10/08/written-testimony-uscg-house-homeland-security-subcommittee-border-and-maritime.

256. "2016 Emerging Cyber Threats Report," Georgia Tech Institute for Information Security and Privacy, accessed March 12, 2016, http://www.iisp.gatech.edu/2016-emerging-cyber-threats-report.

257. Holly Ellyatt, "Top 5 Cyber Security Risks for 2015," CNBC, January 5, 2015, http://www.cnbc.com/2014/12/19/top-5-cyber-security-risks-for-2015.html.

258. Jon Harper, "Defense Department Moving Slowly on Internet of Things," nationaldefensemagazine.org, February 2016, http://www.nationaldefensemagazine.org/archive/2016/February/Pages/DefenseDepartmentMovingSlowlyonInternetofThings.aspx?PF=.

259. Holly Ellyatt, "Top 5 Cyber Security Risks for 2015," CNBC, January

5, 2015, http://www.cnbc.com/2014/12/19/top-5-cyber-security-risks-for-2015.html.

260. "The Digital Talent Gap," Capgemini Consulting, accessed March 12, 2016, https://www.capgemini.com/resource-file-access/resource/pdf/the_digital_talent_gap27-09_0.pdf.

261. Nicole Perlroth, "Cyberattack that Hit Target a Widespread Threat to Consumers," *Boston Globe*, August 23, 2014, https://www.bostonglobe.com/business/2014/08/22/cyberattack-that-hit-target-affecting-businesses/AmsccErTlI4vLhQpUfSorL/story.html.

262. Steve Kroft, "The Attack on Sony," CBS News, April 12, 2015, http://www.cbsnews.com/news/north-korean-cyberattack-on-sony-60-minutes/.

263. Holly Ellyatt, "Top 5 Cyber Security Risks for 2015," *CNBC*, January 5, 2015, http://www.cnbc.com/2014/12/19/top-5-cyber-security-risks-for-2015.html.

264. Ibid.

265. Ibid.

266. Ibid.

267. James R. Clapper, Director of National Intelligence, "Worldwide Cyber Threats," House Permanent Select Committee on Intelligence, September 15, 2015, http://www.dni.gov/files/documents/HPSCIpercent2010percent20Septpercent20Cyberpercent20Hearingpercent20SFR.pdf.

268. Lisa Ferdinando, "Dempsey: Cyber Vulnerabilities Threaten National Security, DoD News, January 21, 2015, http://www.defense.gov/DesktopModules/ArticleCS/Print.aspx?PortalId=1&ModuleId=753&Article=603952.

269. Ibid.

270. Ibid.

271. Ibid.

272. James R. Clapper, Director of National Intelligence, "Worldwide Cyber Threats," House Permanent Select Committee on Intelligence, September 15, 2015, http://www.dni.gov/files/documents/HPSCIpercent2010percent20Septpercent20Cyberpercent20Hearingpercent20SFR.pdf.

273. Ibid.

274. Ibid.

275. Ibid.

276. Amber Corrin, "Capitol Hill Testimonies Paint a Grim Cyber Picture," C4ISRnet.com, September 29, 2015, http://www.c4isrnet.com/story/military-tech/cyber/2015/09/29/cybersecurity-testimonies/73044574/.

277. Ibid.

278. Ibid.
279. Ibid.
280. Bill Gertz, "China Continuing Cyber Attacks on U.S. Networks," *Free Beacon*, March 18, 2016, http://freebeacon.com/national-security/china-continuing-cyber-attacks-on-u-s-networks/.
281. Lincoln Davidson, "Xi Jinping's New Plans for China's Cyber Soldiers," nationalinterest.org, January 21, 2016, http://nationalinterest.org/blog/the-buzz/xi-jinpings-new-plans-chinas-cyber-soldiers-14978?page=show.
282. Ibid.
283. Ibid.
284. Ibid.
285. "Wave of Cyber Attacks on US Originating in Iran," *The Jerusalem Post*, May 25, 2013, http://www.jpost.com/Iranian-Threat/News/Report-Wave-of-cyber-attacks-on-US-originating-in-Iran-314308.
286. James R. Clapper, Director of National Intelligence, "Worldwide Cyber Threats," House Permanent Select Committee on Intelligence, September 15, 2015, http://www.dni.gov/files/documents/HPSCIpercent2010percent20Septpercent20Cyberpercent20Hearingpercent20SFR.pdf.
287. "Iran's Cyber Chief Killed in 'Internal Dispute,' Israel speculates," *The Times of Israel*, October 3, 2013, http://www.timesofisrael.com/irans-cyber-chief-killed-in-internal-dispute-israel-speculates/.
288. "South Korean Banks and Media Report Computer Network Crash, Causing Speculation of North Korea Cyberattack," Fox News, March 20, 2013, http://www.foxnews.com/world/2013/03/20/south-korean-banks-and-media-report-computer-network-crash.html.
289. James R. Clapper, Director of National Intelligence, "Worldwide Cyber Threats," House Permanent Select Committee on Intelligence, September 15, 2015, http://www.dni.gov/files/documents/HPSCIpercent2010percent20Septpercent20Cyberpercent20Hearingpercent20SFR.pdf.
290. Ibid.
291. Ibid.
292. Julian Barnes, "Pentagon Digs In on Cyberwar Front," *Wall Street Journal*, July 6, 2012, http://www.wsj.com/articles/SB10001424052702303684004577508850690121634.
293. Ibid.
294. "After Years of Secrecy and Denial, American Cyber Weapons Are Coming Out of the Closet," Motherboard, July 6, 2012, http://motherboard.vice.com/blog/slowly-but-surely-the-u-s-is-preparing-for-cyber-war.
295. Bill Gertz, "China Continuing Cyber Attacks on U.S. Networks," Free

Beacon, March 18, 2016, http://freebeacon.com/national-security/china-continuing-cyber-attacks-on-u-s-networks/.

296. Joe Gould, "Constructing a Cyber Superpower," Defense News, June 29, 2015, http://www.defensenews.com/story/defense/policy-budget/cyber/2015/06/27/us-cyber-command-budget-expand-fort-meade-offensive/28829321/.

297. Ibid.

298. "After Years of Secrecy and Denial, American Cyber Weapons Are Coming Out of the Closet," Motherboard, July 6, 2012, http://motherboard.vice.com/blog/slowly-but-surely-the-u-s-is-preparing-for-cyber-war.

299. "Remarks by President Barack Obama In Prague As Delivered," White House, April 5, 2009, https://www.whitehouse.gov/the-press-office/remarks-president-barack-obama-prague-delivered.

300. Ibid.

301. Lyman, Edwin. Chernobyl on the Hudson? The Health and Economic Impacts of a Terrorist Attack at the Indian Point Nuclear Plant. Union of Concerned Scientists, 2004, http://www.ucsusa.org/global_security/nuclear_terrorism/impacts-of-a-terrorist-attack-at-indian-point-nuclear-power-plant.html.

302. Edwin Lyman, Chernobyl on the Hudson? The Health and Economic Impacts of a Terrorist Attack at the Indian Point Nuclear Plant. Union of Concerned Scientists, 2004, http://www.ucsusa.org/global_security/nuclear_terrorism/impacts-of-a-terrorist-attack-at-indian-point-nuclear-power-plant.html.

303. "Chemical Warfare in The Iran-Iraq War 1980-1988, Iran Chamber Society, accessed March 12, 2016, http://www.iranchamber.com/history/articles/chemical_warfare_iran_iraq_war.php.

304. Robin Wright, "Iran Still Haunted and Influenced By Chemical Weapons Attacks," Time, January 20, 2014, http://world.time.com/2014/01/20/iran-still-haunted-and-influenced-by-chemical-weapons-attacks/.

305. "DNA," Wikipedia, accessed March 12, 2016, https://en.wikipedia.org/wiki/DNA.

306. Peter Rothman, "Biology is Technology—DARPA Is Back in the Game with A Big Vision and It Is H+," Humanity +, February 15, 2015, http://hplusmagazine.com/2015/02/15/biology-technology-darpa-back-game-big-vision-h/.

307. Jordan Pearson, "Programmable Bioweapons Could Be the Nuclear Bombs of Future Wars," Motherboard, February 25, 2015, http://motherboard.vice.com/read/programmable-bioweapons-could-be-the-nuclear-bombs-of-the-future.

308. Jordan Pearson, "Nanoscale Biosensors in Your Eyes and Brain Could Collect Data on Your Health," Motherboard, October 22,2014, http://motherboard.vice.com/read/nanoscale-biosensors-in-your-eyes-and-brain-could-collect-data-on-your-health.

309. Jordan Pearson, "Programmable Bioweapons Could Be the Nuclear Bombs of Future Wars," Motherboard, February 25, 2015, http://motherboard.vice.com/read/programmable-bioweapons-could-be-the-nuclear-bombs-of-the-future.

310. Jullian Hattam, "Nation 'Dangerously Vulnerable' to Biological Attack, Says Report," The Hill, November 3, 2015, http://thehill.com/policy/national-security/258973-nation-dangerously-vulnerable-to-bioterror-lawmakers-warned.

311. Ibid.

312. James Clapper, "Worldwide Threat Assessment," Senate Armed Services Committee, February 26, 2015, https://archive.org/stream/TheUSIntelligenceCommunity/Docpercent2001-Globalpercent20Threatpercent20Assesmentpercent20(2015)_djvu.txt.

313. "List of states with nuclear weapons," Wikipedia, accessed March 12, 2016, https://en.wikipedia.org/wiki/List_of_states_with_nuclear_weapons#cite_note-sipri.org-3.

314. Alex Lederman, "Republicans say White House dangerously unfocused on true nuclear threat," Medill News Service, March 17, 2016, http://www.militarytimes.com/story/military/capitol-hill/2016/03/17/republicans-say-white-house-dangerously-unfocused-true-nuclear-threat/81929980/.

315. Ibid.

316. "Russia," NTI, July 2015, http://www.nti.org/country-profiles/russia/.

317. Ibid.

318. Ibid.

319. Ibid.

320. Information Office, State Council of the People's Republic of China, " 中国武装力量的多样化运用 [The Diversified Employment of China's Armed Forces]" 16 April 2013, www.xinhuanet.com. For some of the debate surrounding the meaning of the 2013 white paper, see: Fravel M. Taylor, "China Has Not (Yet) Changed Its Position on Nuclear Weapons" The Diplomat, 22 April 2013, http://thediplomat.com; James Acton, "Debating China's No-First-Use Commitment: James Acton Responds," Carnegie Endowment for International Peace, 22 April 2012, http://carnegieendowment.org; Gregory Kulacki, "China Still Committed

to No First Use of Nuclear Weapons," *All Things Nuclear*, Union of Concerned Scientists, http://allthingsnuclear.org; Gregory Kulacki, "Reconceiving China's No First Use Policy," *All Things Nuclear*, Union of Concerned Scientists, http://allthingsnuclear.org; Ministry of National Defense, "China's Military Strategy," May 2015, www.eng.mod.gov.cn.

321. Con Coughlin, "The Saudis Are Ready to Go Nuclear," *Telegraph*, June 8, 2015, http://www.telegraph.co.uk/news/worldnews/middleeast/saudiarabia/11658338/The-Saudis-are-ready-to-go-nuclear.html.

322. "Chronology of U.S.-North Korean Nuclear and Missile Diplomacy," Arms Control Association, March 2016, https://www.armscontrol.org/factsheets/dprkchron.

323. Ankit Panda, "Pakistan Clarifies Conditions for Tactical Nuclear Weapon Use against India," thediplomat.com, October 20, 2015, http://thediplomat.com/2015/10/pakistan-clarifies-conditions-for-tactical-nuclear-weapon-use-against-india/.

324. "Father of Pakistani Bomb Sold Nuclear Secrets," Arms Control Association, accessed March 12, 2016, https://www.armscontrol.org/act/2004_03/Pakistan.

325. "Introduction to Chemical Weapons," Federal of American Scientists, accessed March 12, 2016, fas.org/cw/intro.htm.

326. "Pentagon Looks to Breed Immortal 'Synthetic Organisms,' Molecular Kill Switch Included," samadhisoft.com, February 11, 2010, http://samadhisoft.com/category/the-perfect-storm/nanotechnology-the-perfect-storm/.

327. "Former DHS Secretary Warns on Biodefense," Emergency Management, December 2, 2015, http://www.emergencymgmt.com/safety/-Former-DHS-Secretary-Warns-on-Biodefense.html?utm_medium=email&utm_source=Act-On+Software&utm_content=email&utm_campaign=Sanpercent20Bernardinopercent20Shooting:percent20Textbookpercent20Response&utm_term=Formerpercent20DHSpercent20Secretarypercent20Warnspercent20onpercent20Biodefense:

328. Ibid.

329. "Chemical and Biological Weapons Status at a Glance," Arms Control Association, February 2014, https://www.armscontrol.org/factsheets/cbwprolif.

330. "Russia," NTI, July 2015, http://www.nti.org/country-profiles/russia/.

331. "Chemical and Biological Weapons Status at a Glance," Arms Control Association, February 2014, https://www.armscontrol.org/factsheets/cbwprolif.

332. Ibid.

333. The Perfect Storm, *Dabiq*, p. 77. http://jihadology.net/2015/05/21/al-per centelpercentb8percenta5ayat-media-center-presents-a-new-issue-of-the-islamic-states-magazine-dabiq-9/.

334. Julian Hattem, "Nation 'Dangerously Vulnerable' to Biological Attack, Says Report," The Hill, November 3, 2015, http://thehill.com/policy/national-security/258973-nation-dangerously-vulnerable-to-bioterror-lawmakers-warned.

335. John Cohen, "Homegrown Terrorists," thecipherbrief.com, January 19, 2016, https://www.thecipherbrief.com/article/homegrown-terrorists.

336. Helen Kennedy, "Yemen Bomb Suspect, Al Qaeda Expert Ibrahim Hassan Tali al-Asiri, Blew Up His Own Brother," *New York Daily News*, October 31, 2010, http://www.nydailynews.com/news/world/yemen-bomb-suspect-al-qaeda-expert-ibrahim-hassan-tali-al-asiri-blew-brother-article-1.454883.

337. Cathy Burke, "Bomb Maker to U.S., Saudis: Will Hit You While There's a 'Pulsing Vein Within Us,'" *Newsmax*, January 11, 2016, http://www.newsmax.com/Newsfront/ibrahim-bin-hassan-al-asiri-bomb-maker-vow-attack/2016/01/11/id/709053/#ixzz43BKN1gtc .

338. Alyssa Pone, Alex Hosenball and Lee Ferran, "San Bernardino Shooter's Neighbor Enrique Marquez Pleads Not Guilty," ABC News, January 6, 2016, http://abcnews.go.com/Blotter/san-bernardino-shooters-neighbor-enrique-marquez-pleads-guilty/story?id=36117532

339. "Pestilence," Bible Hub, accessed March 12, 2016, http://Biblehub.com/topical/p/pestilence.htm.

340. "2015 Sees Sharp Rise in Post-Christian Population," Barna, August 12, 2015, https://barna.org/barna-update/culture/728-america-more-post-christian-than-two-years-ago#.VrZI4PkrK70.

341. Ed Stetzer, "Barna: How Many Have a Biblical Worldview?," *Christianity Today*, March 9, 2009, http://www.christianitytoday.com/edstetzer/2009/march/barna-how-many-have-Biblical-worldview.html.

342. Del Tackett, "What's a Christian Worldview?," Focus on the Family, accessed March 17, 2016, http://www.focusonthefamily.com/faith/christian-worldview/whats-a-christian-worldview/whats-a-worldview-anyway.

343. Bill Bailey, "Moral Ethics—Quotes from the Founding Fathers," The Federalist Papers Project, accessed March 12, 2016, https://www.thefederalistpapers.org/history/moral-ethics-quotes-from-the-founding-fathers.

344. Mark David Hall, "Did America Have a Christian Founding?," Heritage Foundation, June 7, 2011, http://www.heritage.org/research/lecture/2011/06/did-america-have-a-christian-founding.

345. Robert Kraynack, "The American Founders and Their Relevance Today," *Modern Age*, Winter 2015, https://home.isi.org/sites/default/files/MA_57.1_3Essays_TheAmericanFounders.pdf.

346. "The Founders' Vision & Today's Issues," The Jeffersonian Perspective, accessed March 12, 2016, http://eyler.freeservers.com/JeffPers/jefpco01.htm.

347. "Ronald Reagan," The American Presidency Project, accessed March 12, 2016, http://www.presidency.ucsb.edu/ws/index.php?pid=43130.

348. "John F. Kennedy," The American Presidency Project, accessed March 12, 2016, http://www.presidency.ucsb.edu/ws/index.php?pid=8032.

349. "President Barack Obama's Inaugural Address," The White House, January 21, 2009, https://www.whitehouse.gov/blog/2009/01/21/president-barack-obamas-inaugural-address.

350. "Fact Sheet: The 2015 National Security Strategy," The White House, February 6, 2015, https://www.whitehouse.gov/the-press-office/2015/02/06/fact-sheet-2015-national-security-strategy.

351. "China: The Power of Military Organization," Stratfor, January 25, 2016, http://www.realcleardefense.com/articles/2016/01/25/china_the_power_of_military_organization_108946.html.

352. Colin Clark, "McCain Launches Goldwater-Nichols Review; How Far Will He Go?," Breaking Defense, March 26, 2015, http://breakingdefense.com/2015/03/mccain-launches-goldwater-nichols-review-how-far-will-he-go/.

353. Luke Johnson and Ryan Grim, "GAO Cannot Audit Federal Government, Cites Department Of Defense Problems," Huffington Post, January 18, 2013, http://www.huffingtonpost.com/2013/01/18/gao-audit-federal-government-defense_n_2507097.html.

354. Lauren Chadwick, "You Could Buy an Australian Island for What the Pentagon Says It Would Cost to Take Inventory—of One Item," Public Integrity, March 15, 2016, http://www.publicintegrity.org/2016/03/15/19428/you-could-buy-australian-island-what-pentagon-says-it-would-cost-take-inventory-one.

355. 31 U.S. Code § 1502—Balances available, Legal Information Institute, Cornell University Law School, accessed July 3, 2016, https://www.law.cornell.edu/uscode/text/31/1502. (31 U.S. Code, Section 1502, states:

(a) The balance of an appropriation or fund limited for obligation to a definite period is available only for payment of expenses properly incurred during the period of availability or to complete contracts properly made within that period of availability and obligated consistent with section 1501 of this title. However, the appropriation or fund is not available for expenditure for a period beyond the period otherwise authorized by law. (b) A provision of law requiring that the balance of an appropriation or fund be returned to the general fund of the Treasury at the end of a definite period does not affect the status of lawsuits or rights of action involving the right to an amount payable from the balance. (Pub. L. 97–258, Sept. 13, 1982, 96 Stat. 928.).

356. Lora Lumpe and Jeremy Ravinsky, "The Pentagon's Secret Foreign Aid Budget," *Politico*, March 8, 2016, http://www.politico.com/agenda/story/2016/03/the-pentagons-foreign-aid-budget-needs-oversight-000060#ixzz42tmbQmHd.

357. "Overseas Contingency Operations: The Pentagon Slush Fund," National Priorities Project, accessed March 14, 2016, https://www.nationalpriorities.org/campaigns/overseas-contingency-operations/.

358. Chuck McCutcheon and David Mark, "'Elections Have Consequences': Does Obama Regret Saying That Now?," *Christian Science Monitor*, November 21, 2014, http://www.csmonitor.com/USA/Politics/Politics-Voices/2014/1121/Elections-have-consequences-Does-Obama-regret-saying-that-now.

359. Robert Maginnis, *Deadly Consequences: How Cowards Are Pushing Women into Combat*, Regnery Publishing Inc., 2013.

360. Colin Clark, "NSC Staff Too Big, Too Activist: Top Former Generals, Officials," BreakingDefense.com, March 14, 2016, http://breakingdefense.com/2016/03/nsc-staff-too-big-too-activist-top-former-generals-officials/.

361. Ibid.

362. Rick Maze, "Too Much World Not Enough Army," *Army*, December 14, 2015, WWW.ARMYMAGAZINE.ORG/2015/12/14/TOO-MUCH-WORLD-NOT-ENOUGH-ARMY.

363. Tara Copp, "Marine General to Congress: We Might Not Be Ready for Another War," Stars and Stripes, March 15, 2016, http://www.stripes.com/news/marine-general-to-congress-we-might-not-be-ready-for-another-war-1.399376.

364. Mark F. Cancian, "Is the Navy Too Small?," CSIS, September 29, 2015, http://csis.org/publication/navy-too-small.

365. Ibid.
366. Mark A. Welsh III, Chief of Staff U.S. Air Force, testimony, Senate Armed Services, Committee, January 28, 2015, http://www.af.mil/Portals/1/documents/Speeches/CSAF_SequestrationTranscript.pdf.
367. Ibid.
368. Bill Gertz, "China, Russia Planning Space Attacks on U.S. Satellites," Free Beacon, March 16, 2016, http://freebeacon.com/national-security/china-russia-planning-space-attacks-on-u-s-satellites/.
369. Ibid.
370. Ibid.
371. Ibid.
372. Robert Maginnis, "Badda Bing Badda Boom," *Human Events*, February 26, 2008, http://humanevents.com/2008/02/26/badda-bing-badda-boom/.
373. Jeffrey Lewis, "They Shoot Satellites, Don't They?," *Foreign Policy*, August 9, 2014, http://foreignpolicy.com/2014/08/09/they-shoot-satellites-dont-they/.
374. Steve Almasy and Euan McKirdy, "North Korea Claims to Have Nuclear Warheads That Can Fit on Missiles," CNN, March 8, 2016, http://www.cnn.com/2016/03/08/asia/north-korea-nuclear-warheads/.
375. Dustin Volz and Mark Hosenball, "Concerned by Cyber Threat, Obama Seeks Big Increase in Funding," Reuters, February 10, 2016, http://www.reuters.com/article/us-obama-budget-cyber-idUSKCN0VI0R1.
376. Caroline Glick, "An Urgent Memo to the Next Government," carolineglick.com, accessed March 12, 2016, http://carolineglick.com/an_urgent_memo_to_the_next_gov/.
377. "What the Next Arms Race Will Look Like," Stratfor, March 21, 2016, https://www.stratfor.com/analysis/what-next-arms-race-will-look.
378. Ibid.
379. Ibid.
380. Ibid.
381. Dustin Volz and Mark Hosenball, "Concerned by Cyber Threat, Obama Seeks Big Increase in Funding," Reuters, February 10, 2016, http://www.reuters.com/article/us-obama-budget-cyber-idUSKCN0VI0R1.
382. Barack Obama, "Protecting U.S. Innovation from Cyberthreats," February 9, 2016, http://www.wsj.com/articles/protecting-u-s-innovation-from-cyberthreats-1455012003.
383. "CYBERSECURITY: Challenges in Securing the Electricity Grid," GAO, July 17, 2012, http://gao.gov/products/GAO-12-926T.

384. "Satellite Images: An Iranian Military Complex," stratfor. com, February 8, 2016, https://www.stratfor.com/video/ satellite-images-iranian-military-complex.

385. Rebecca Hersman, "Nuclear Deterrence in a Disordered World," CSIS, November 16, 2015, http://csis.org/publication/ nuclear-deterrence-disordered-world.

386. "What Train: China Tests Rail-based Long-range Missile Capable of Hitting U.S.," Sputnik News, December 22, 2015, http://sputniknews. com/asia/20151222/1032117958/china-rail-based-missile.html.

387. Brendon Thomas-Noone, "Russia's Nuclear Doctrine Takes an Aarming Step Backwards," The Interpreter, July 3, 2015, http://www. lowyinterpreter.org/post/2015/07/03/Russias-nuclear-doctrine-takes-an-alarming-step-backwards.aspx?p=true.

388. "Top U.S. Intelligence Official Calls Gene Editing a WMD Threat," Geek Journal, accessed March 12, 2016, http://www.geekjournal.net/ articles/2016/02/top-u-s-intelligence-official-calls-gene-editing-a-wmd-threat-57354.html.

389. "A Clear and Present Danger: The Threat to Religious Liberty in the Military," Family Research Council, accessed March 12, 2016, http:// downloads.frc.org/EF/EF15F47.pdf.

390. "Sec. 654. Policy Concerning Homosexuality in the Armed Forces," TITLE 10—ARMED FORCES, SUBTITLE A—GENERAL MILITARY LAW, PART II—PERSONNEL, CHAPTER 37—GENERAL SERVICE REQUIREMENTS, accessed March 12, 2016, http://uscode.regstoday. com/10USC_CHAPTER37.aspx.

391. Ryan Brown, "Top Intelligence Official: ISIS to Attempt U.S. Attacks This Year, CNN, February 9, 2016, http://www.cnn.com/2016/02/09/politics/ james-clapper-isis-syrian-refugees/.

392. Fred Gedrich, "World Threat Assessment Tells Obama: Time for 'Strategic Patience' Is Over," breitbart.com, February 18, 2016, http://www.breitbart.com/national-security/2016/02/18/ world-threat-assessment-tells-obama-time-for-strategic-patience-is-over/.

393. National Military Strategy, Department of Defense, 2015, p. 9, http:// www.jcs.mil/Portals/36/Documents/Publications/2015_National_ Military_Strategy.pdf.

394. "Homeland Sec: Terrorists Crossing U.S.-Mexico Border 'Not the Thing I Most Worry About'," breitbart.com, June 9, 2015, http://www.breitbart. com/big-government/2015/06/09/homeland-sec-terrorists-crossing-u-s-mexico-border-not-the-thing-i-most-worry-about/.

395. Raymond Ibrahim, "U.S. Mexican Border Porous to Jihadists," Clarion Project, May 15, 2012, http://www.clarionproject.org/analysis/us-mexican-border-porous-jihadists.

396. Stephen Dinan, "Agents nab Pakistanis with terrorist connections crossing U.S. border," *Washington Times*, December 30, 2015, http://www.washingtontimes.com/news/2015/dec/30/pakistanis-terrorist-connections-nabbed-us-border/.

397. Ibid.

398. Barack Obama, "Security and Privacy: In Search of a Balance," Delivered at the U.S. Department of Justice, Washington, D.C., January 17, 2014, VSOTD.COM, http://video.foxnews.com/v/3068164336001/obama-effort-to-strike-balance-between-privacy-security/#sp=show-clips .

399. As cited in Michael Chesbro, "A Brief History of U.S. Intelligence," American Board for Certification in Homeland Security, accessed March 17, 2016, http://www.abchs.com/ihs/SUMMER2014/ihs_articles_1.php.

400. Ibid.

401. Barack Obama, "Security and Privacy: In Search of a Balance," Delivered at the U.S. Department of Justice, Washington, D.C., January 17, 2014, VSOTD.COM, http://video.foxnews.com/v/3068164336001/obama-effort-to-strike-balance-between-privacy-security/#sp=show-clips.

402. James R. Clapper, "Worldwide Threat Assessment of the US Intelligence Community," Statement for the Record, Senate Select Committee on Intelligence, February 9, 2016, http://www.dni.gov/index.php/newsroom/testimonies/217-congressional-testimonies-2016/1313-statement-for-the-record-worldwide-threat-assessment-of-the-u-s-ic-before-the-senate-armed-services-committee-2016.

403. "Fentanyl," National Institute on Drug Abuse, accessed March 12, 2016, https://www.drugabuse.gov/drugs-abuse/fentanyl.

404. "ODNI FAQ," Office of the Director of National intelligence, accessed March 12, 2016, http://www.dni.gov/index.php/about/faq?start=3.

405. "Edward Snowden, Wikipedia, accessed March 12, 2016, https://en.wikipedia.org/wiki/Edward_Snowden.

406. George Gao, "What Americans Think about NSA Surveillance, National Security and Privacy," PewResearchCenter, May 29, 2015, http://www.pewresearch.org/fact-tank/2015/05/29/what-americans-think-about-nsa-surveillance-national-security-and-privacy/.

407. James Ball and Spenser Ackerman, "NSA Loophole Allows Warrantless Search for US Citizens' Emails and Phone Calls," *The Guardian*,

August 9, 2013, http://www.theguardian.com/world/2013/aug/09/nsa-loophole-warrantless-searches-email-calls.

408. Mitchell Shaw, "Why Should the Law-abiding Care About Electronic Surveillance?," *New American*, October 27, 2014, http://www.thenewamerican.com/tech/computers/item/19389-why-should-the-law-abiding-care-about-nsa-surveillance.

409. Jeremy Ravensky, "Snooping States: NSA Not Alone in Spying on Citizens," *Christian Science Monitor*, June 12, 2013, http://www.csmonitor.com/World/Global-Issues/2013/0612/Snooping-states-NSA-not-alone-in-spying-on-citizens.

410. Barack Obama, "Security and Privacy: In Search of a Balance," Delivered at the U.S. Department of Justice, Washington, D.C., January 17, 2014, VSOTD.COM, http://video.foxnews.com/v/3068164336001/obama-effort-to-strike-balance-between-privacy-security/#sp=show-clips.

411. "National Security Letter," Wikipedia, accessed March 12, 2016, https://en.wikipedia.org/wiki/National_security_letter.

412. Michael German, "The US Intelligence Community Is Bigger Than Ever, But Is It Worth the Cost?," DefenseOne, February 6, 2015, http://www.defenseone.com/ideas/2015/02/us-intelligence-community-bigger-ever-it-worth-it/104799/.

413. Ibid.

414. Ibid.

415. Ibid.

416. Samuel J. Rascoff, "Presidential Intelligence," *Harvard Law Review*, Vol. 129:622, p. 694.

417. Ibid, p. 694.

418. Ibid, p. 636.

419. Ibid, p. 640.

420. Shane Harris and Nancy Youssef, "Exclusive: 50 Spies Say ISIS Intelligence Was Cooked," Dailybeast.com, September 9, 2009, http://www.thedailybeast.com/articles/2015/09/09/exclusive-50-spies-say-isis-intelligence-was-cooked.html.

421. Samuel J. Rascoff, "Presidential Intelligence," *Harvard Law Review*, Vol. 129:622, p. 650.

422. Ibid, p. 663.

423. Ibid, p. 667–668.

424. Ibid, p. 662.

425. Ibid, p. 664.

426. Eric Lichtblau and Katie Benner, "Apple Fights Order to Unlock San Bernardino Gunman's iPhone," *New York Times*, February 17, 2016, http://www.nytimes.com/2016/02/18/technology/apple-timothy-cook-fbi-san-bernardino.html?_r=0 .

427. Ibid.

428. Glenn Greenwald and Murtaza Hussain, "Meet the Muslim American Leaders the FBI and NSA Have Been Spying on," theintercept.com, July 9, 2014, https://theintercept.com/2014/07/09/under-surveillance/?comments=1.

429. "Foreign Policy Quotes," BrainyQuote, accessed March 12, 2016, http://www.brainyquote.com/quotes/keywords/foreign_policy.html.

430. "Foreign Policy: What Now?," Policy Making American Interactions, American Government, accessed March 12, 2016, http://www.ushistory.org/gov/11a.asp.

431. Rebecca McCartney, "There Are 294 US Embassies and Consulates Around the World: We May Not Need All of Them," mic.com, September 14, 2012, http://mic.com/articles/14763/there-are-294-us-embassies-and-consulates-around-the-world-we-may-not-need-all-of-them#.t59XWEOA2.

432. "Zbigniew Brzezi ski," Goodreads, accessed March 12, 2016, http://www.goodreads.com/quotes/tag/foreign-policy.

433. John Hillen, "Foreign Policy by Map," *National Review*, February 23, 2015, https://www.nationalreview.com/nrd/articles/413255/foreign-policy-map.

434. Ibid.

435. Ibid.

436. Zeynep Taydas, Cigdem Kentmen and Larua Olson, "Faith Matters: Religious Affiliation and Public Opinion About Barack Obama's Foreign Policy in the "Greater" Middle East," *Social Science Quarterly*, Volume 93, Issue 5, December 2012, http://onlinelibrary.wiley.com/doi/10.1111/j.1540-6237.2012.00920.x/abstract.

437. Adam B. Lowther and Casey Lucius Lowther, "Identifying America's Vital Interests," International Affairs Forum, Volume 4, Issue 2, 2013, http://www.tandfonline.com/doi/abs/10.1080/23258020.2013.864884?journalCode=riaf20.

438. Ibid.

439. "Washington's Farewell Address 1796," the Avalon Project, Yale Law School, accessed March 12, 2016, http://avalon.law.yale.edu/18th_century/washing.asp.

440. Samuel Huntington, "The Erosion of American National Interests," *Foreign Affairs*, Vol. 76, No. 5(Sept/Oct 1997), 28–49.

441. Dennis M. Drew and Donald M. Snow, *Making Twenty-First-Century Strategy* (Maxwell AFB, AL: Air University Press, 2006), 34.

442. Mark P. Lagon, "Promoting Democracy: The Whys and Hows for the United States and the International Community," Council on Foreign Relations, February 2011, http://www.cfr.org/democratization/promoting-democracy-whys-hows-united-states-international-community/p24090.

443. Ibid.

444. Ibid.

445. Ibid.

446. Raymond Ibrahim, "U.S. Policy Made 2015 the Worst Persecution of Christians 'in Modern History,'" Gatestone Institute, March 15, 2016, http://www.gatestoneinstitute.org/7521/us-policy-christian-persecution.

447. Daniel Drezner, Rebooting Republican Foreign Policy, Council on Foreign Relations, January/February 2013, http://eds.b.ebscohost.com.pentagonlibrary.idm.oclc.org/eds/detail/detail?vid=4&sid=48611e5b-0dc1-4ccc-b1e3-4ebeeb30fdefpercent40sessionmgr110&hid=112&cbdata=JnNpdGU9ZWRzLWxpdmUpercent3d#AN=84474061&db=bth.

448. Ibid.

449. "George W. Bush: Foreign Affairs, Miller Center, accessed March 12, 2016, http://millercenter.org/president/biography/gwbush-foreign-affairs.

450. Ibid.

451. Fred Kaplain, "Obama's Way," *Foreign Affairs*, January/February 2016, https://www.foreignaffairs.com/articles/2015-12-07/obamas-way.

452. Katie Sanders, "Chris Wallace: Hillary Clinton Defended Syria's Assad as a 'Possible Reformer'," Punditfact, June 1, 2014, http://www.politifact.com/punditfact/statements/2014/jun/01/chris-wallace/chris-wallace-hillary-clinton-defended-syrias-assa/.

453. Don Fredrick, "A New Red Line," The Post & Email, September 17, 2014, http://www.thepostemail.com/2014/09/17/incompetent-or-treasonousperc entE2percent80percent8F/.

454. Karen DeYoung, "How the Obama White House runs foreign policy," *The Washington Post*, August 4, 2015, https://www.washingtonpost.com/world/national-security/how-the-obama-white-house-runs-foreign-policy/2015/08/04/2befb960-2fd7-11e5-8353-1215475949f4_story.html.

455. Ibid.

456. Ibid.

457. Ibid.

458. "What Is the Quote on the Statue of Liberty?," howtallisthestatueofliberty. org, accessed March 12, 2016, http://www.howtallisthestatueofliberty.org/ what-is-the-quote-on-the-statue-of-liberty/.

459. "Winston Churchill Quotes," BrainyQuote, accessed March 12, 2016, http://www.brainyquote.com/quotes/quotes/w/winstonchu135259. html#W5R3SLycS2SFwmEe.99.

460. Robert L. Maginnis, Never Submit: Will the Extermination of Christians Get Worse Before it Gets Better?, Defender, Crane, MO., 2015.

461. Interview with LTG William G. Boykin via telephone, April 14, 2015.

462. 22 U.S.C. § 8213. Investigations of violations of international humanitarian law

(A) IN GENERAL

The President, with the assistance of the Secretary, the Under Secretary of State for Democracy and Global Affairs, and the Ambassador-at-Large for War Crimes Issues, shall collect information regarding incidents that may constitute crimes against humanity, genocide, slavery, or other violations of international humanitarian law.

(B) ACCOUNTABILITY

The President shall consider what actions can be taken to ensure that any government of a country or the leaders or senior officials of such government who are responsible for crimes against humanity, genocide, slavery, or other violations of international humanitarian law identified under subsection (a) are brought to account for such crimes in an appropriately constituted tribunal.

(Pub. L. 110–53, title XXI, §2113, Aug. 3, 2007, 121 Stat. 531 .)

463. Terrence Jeffrey, "Will Obama and Kerry Declare ISIL Not Guilty of Anti-Christian Genocide?," CNS News, March 2, 2016, http://www.cnsnews.com/commentary/terence-p-jeffrey/ will-obama-and-kerry-declare-isil-not-guilty-anti-christian-genocide.

464. John Bacon, "Kerry: ISIS is committing genocide against religious sects," USA Today, March 17, 2016, http://www.militarytimes.com/story/ military/2016/03/17/isis-committing-genocide-state-kerry/81906218/.

465. The International Religious Freedom Act of 1998, 150th Congress, accessed March 12, 2016,http://www.state.gov/documents/ organization/2297.pdf.

466. Reihan Salam, "The Melting Pot," National Review, December 31, 2015, https://www.nationalreview.com/nrd/articles/428658/new-melting-pot.

467. Robert Tracci, "TRACCI: Six steps to end the border crisis," *The Washington Times*, July 11, 2014, http://www.washingtontimes.com/news/2014/jul/11/tracci-six-steps-to-end-the-border-crisis/.

468. Daniel Politi, "Iran's Khamenei: No Cure for Barbaric Israel but Annihilation," slate.com, November 9, 2014, http://www.slate.com/blogs/the_slatest/2014/11/09/iran_s_khamenei_israel_must_be_annihilated.html.

469. David Morgan, "Gates Criticizes NATO; How Much Does U.S. Pay?," *CBS News*, June 10, 2011, http://www.cbsnews.com/news/gates-criticizes-nato-how-much-does-us-pay/.

470. Raul Amoros, "The US Spends $35 Billion on Foreign Aid...But Where Does the Money Really Go?," mondoweiss.net, accessed March 12, 2016, http://mondoweiss.net/2015/11/spends-billion-foreign/#sthash.SpIVhnVg.eYe7bz3f.dpuf.

471. Robert Frank, "Americans Gave a Record $358 Billion to Charity in 2014," *CNBC*, June 16, 2015, http://www.cnbc.com/2015/06/16/americans-gave-a-record-358-billion-to-charity-in-2014.html.

472. Barbara Opall-Rome, "Israel Seeks Surge in US Security Support," *DefenseNews*, May 24, 2015, http://www.defensenews.com/story/defense/policy-budget/budget/2015/05/24/israel-assistance-funding-us-increase-fmf/27706775/.

473. Poncie Rusche, "Guess How Much of Uncle Sam's Money Goes to Foreign Aid. Guess Again!," NPR, February 10, 2015, http://www.npr.org/sections/goatsandsoda/2015/02/10/383875581/guess-how-much-of-uncle-sams-money-goes-to-foreign-aid-guess-again.

474. Cited from "Christian Economics," accessed March 12, 2016 http://www.allaboutworldview.org/christian-economics.htm and rendered with permission from the book, *Understanding the Times: The Collision of Today's Competing Worldviews* (Rev 2nd ed), David Noebel, Summit Press, 2006. Compliments of John Stonestreet, David Noebel, and the Christian Worldview Ministry at Summit Ministries.

475. "Christian Economics," allaboutworldview.org, accessed March 13, 2016, http://www.allaboutworldview.org/christian-economics.htm.

476. "What Is a 'Free Enterprise?," Investopedia.com, accessed March 13, 2016, http://www.investopedia.com/terms/f/free_enterprise.asp#ixzz40nqYnIzb.

477. "Unfair Competition," Wikipedia, accessed March 13, 2016, https://en.wikipedia.org/wiki/Unfair_competition.

478. "Living Wage," Soc 315—Social Welfare, Fall 2015, https://people.eou.

edu/socwelf/readings/week-6/living-wage/ and Jeanne Sahadi, "Where minimum wage is going up in 2016," *CNN*, December 23, 2016, http://money.cnn.com/2015/12/23/pf/minimum-wage-2016/.

479. David Ingram, "In What Ways Can the Government Encourage Business Activity?," smallbusiness.chron.com, accessed March 13, 2016, http://smallbusiness.chron.com/ways-can-government-encourage-business-activity-2282.html.

480. Stephen Dinan, "Obama Clean Energy Loans Leave Taxpayers in $2.2 Billion Hole," *The Washington Times*, April 27, 2015, http://www.washingtontimes.com/news/2015/apr/27/obama-backed-green-energy-failures-leave-taxpayers/?page=all.

481. "Characteristics of Minimum Wage Workers, 2014," Report 1054, BLS Reports, U.S. Bureau of Labor Statistics, April 2015, http://www.bls.gov/opub/reports/cps/characteristics-of-minimum-wage-workers-2014.pdf.

482. State Minimum Wages | 2016 Minimum Wage by State, National Conference of State Legislatures, January 1, 2016, http://www.ncsl.org/research/labor-and-employment/state-minimum-wage-chart.aspx#1.

483. "Characteristics of Minimum Wage Workers, 2014," Report 1054, BLS Reports, U.S. Bureau of Labor Statistics, April 2015, http://www.bls.gov/opub/reports/cps/characteristics-of-minimum-wage-workers-2014.pdf.

484. Simone Pathe, "Undisputed Facts About the Minimum Wage," *PBS News Hour*, January 1, 2015, http://www.pbs.org/newshour/making-sense/undisputed-facts-minimum-wage/.

485. Carey Nadeau and Amy Glasmeier, "Minimum Wage: Can an Individual or a Family Live on It?," Living Wage Calculator, Massachusetts Institute of Technology, January 1, 2016, http://livingwage.mit.edu/articles/15-minimum-wage-can-an-individual-or-a-family-live-on-it.

486. Simone Pathe, "Undisputed Facts about the Minimum Wage," *PBS News Hour*, January 1, 2015, http://www.pbs.org/newshour/making-sense/undisputed-facts-minimum-wage/.

487. "21.3 Percent of U.S. Population Participates in Government Assistance Programs Each Month," U.S. Census Bureau, May 28, 2015, https://www.census.gov/newsroom/press-releases/2015/cb15-97.html.

488. Ibid.

489. "Do You Trust the IRS?: The Beat (poll)," Cleveland.com, May 11, 2015, http://www.cleveland.com/business/index.ssf/2015/05/do_you_trust_the_irs_the_beat.html.

490. Holly Mangan, "What Is the Fair Tax Act Explained—Pros and Cons,"

Money Crashers, accessed March 13, 2016, http://www.moneycrashers. com/fair-tax-act-explained-pros-cons/.

491. "Flat Tax vs. Fair Tax," freedomworks.org, July 6, 2011, http://www. freedomworks.org/content/flat-tax-vs-fair-tax.

492. Richard Langworth, "Churchill on Taxes," richardlangworth.com, Hillsdale College, May 9, 2012, https://richardlangworth.com/taxes.

493. "The Truth about Corporate Tax Rates," forbes.com, March 25, 2015, http://www.forbes.com/sites/taxanalysts/2015/03/25/ the-truth-about-corporate-tax-rates/#7ab2d96420a5.

494. "Business Roundtable's Submission to Senate Committee on Finance Tax Reform Working Groups," Business Roundtable, April 15, 2015, http:// businessroundtable.org/sites/default/files/media-resources/ Businesspercent20Roundtablepercent20Submissionpercent20 topercent20SFCpercent20Workingpercent20Groupspercent202015 percent2004percent2015_finalpercent20forpercent20distribution.pdf.

495. Jay Ambrose, "Column: How High Business Taxes Hurt America," Tribune News Service, November 27, 2015,http://www.bendbulletin.com/ opinion/3736591-151/column-how-high-business-taxes-hurt-america#.

496. "Pulpit Commentary," as cited in http://biblehub.com/proverbs/16-3.htm.

497. Rachel Greszler, "Job Creation: Policies to Boost Employment and Economic Growth," Heritage Foundation, Issue Brief #4202, April 18, 2014, http://www.heritage.org/research/reports/2014/04/ job-creation-policies-to-boost-employment-and-economic-growth.

498. Ibid.

499. Ronald Reagan, "Reagan's First Inaugural: 'Government is not the solution to our problem; government is the problem,'" Heritage Foundation, accessed March 14, 2016, http://www.heritage.org/initiatives/first-principles/primary-sources/reagans-first-inaugural-government-is-not-the-solution-to-our-problem-government-is-the-problem.

500. Elena Holodny, "America's Path to Energy Independence in 2 Charts," Business Insider, September 1, 2015, http://www.businessinsider.com/ americas-path-to-energy-independence-in-charts-2015-9.

501. Chris Isidore, "U.S. Could Be Energy Independent Within Four Years," CNN Money, April 15, 2015, http://money.cnn.com/2015/04/15/ investing/us-energy-independence/.

502. Patti Domm, "US Is on Fast-Track to Energy Independence: Study," CNBC, February 11, 2013, http://www.cnbc.com/id/100450133.

503. Eric Rosenberg, "Can Fracking Survive at $60 a Barrel?," Investopedia, July

22, 2015, http://www.investopedia.com/articles/investing/072215/can-fracking-survive-60-barrel.asp.

504. Ivana Kottasova, "Europe Tries to Protect Steel Jobs with Tariffs on Chinese Imports," *CNN Money*, January 29, 2016, http://money.cnn.com/2016/01/29/news/economy/steel-china-europe-dumping-tariffs/.

505. Sonja Elmquist, "Chinese Steel Slapped by 236 percent U.S. Tariff Plan," BloombergBusiness, November 3, 2015, http://www.bloomberg.com/news/articles/2015-11-03/u-s-commerce-finds-china-steel-subsidies-of-as-much-as-236.

506. Miles Moore, "Titan, USW Pursue Duties on OTR Tires from Asia ," Rubber & Plastic News, January 27, 2016, http://www.rubbernews.com/article/20160127/NEWS/301259998/unfair-trade-practices-spark-move-titan-usw-seek-duties-on-off-road.

507. "Competition Policy and Enforcement in China," The U.S. China Business Council, 2014, https://www.uschina.org/reports/competition-policy-and-enforcement-china.

508. Colin Campbell, "TRUMP: Let's Slap a 45 percent Tariff on Chinese Imports," Business Insider, January 7, 2016,http://www.businessinsider.com/donald-trump-45-tariff-chinese-imports-china-2016-1.

509. Nelson Hultberg, "Free Trade vs. Fair Trade," Americans for a Free Republic, March 13, 2016, http://afr.org/free-trade-vs-fair-trade/.

510. Ibid.

511. Ibid.

512. Geoff Colvin, "Adm. Mike Mullen: Debt Is Still Biggest Threat to U.S. Security," *Fortune*, May 10, 2012, http://fortune.com/2012/05/10/adm-mike-mullen-debt-is-still-biggest-threat-to-u-s-security/.

513. Andrew Taylor, "Everything You Need to Know about the Federal Debt Limit," PBS Newshour, October 26, 2015, http://www.pbs.org/newshour/making-sense/guide-federal-government-debt-limit/.

514. "Stabilize the Debt," Committee for a Responsible Budget, accessed March 13, 2016, http://crfb.org/stabilizethedebt/.

515. Stephen Dinan, "Federal Deficit to Soar in 2016 After Ryan-Obama Tax Deal," *The Washington Times*, January 19, 2016, http://www.washingtontimes.com/news/2016/jan/19/federal-deficit-soar-2016-after-ryan-obama-tax-dea/?page=all.

516. "National Debt," Just Facts, accessed March 13, 2016, http://www.justfacts.com/nationaldebt.asp.

517. David Sherfinski, "John Kasich Pledges Balanced Federal Budget in Eight Years as President," *The Washington Times*, October 15,

2015, http://www.washingtontimes.com/news/2015/oct/15/
john-kasich-pledges-balance-federal-budget-eight-y/?page=all.

518. John Taylor, "From Thomas Jefferson to John Taylor, 26 November 1798," Founders Archives, accessed March 13, 2016, http://founders.archives.gov/documents/Jefferson/01-30-02-0398.

519. "Ronald Reagan," The American Presidency Project, accessed March 13, 2016, http://www.presidency.ucsb.edu/ws/?pid=42940.

520. Paul Gosar, "Balancing the Federal Budget: A (Not So) New Approach," The Hill, April 2, 2015, http://thehill.com/opinion/op-ed/237634-balancing-the-federal-budget-a-not-so-new-approach.

521. Eric Schnurer, "How to Cut the Deficit in Half Without Starting a Partisan Foodfight," The Atlantic, October 28, 2013, http://www.theatlantic.com/politics/archive/2013/10/how-to-cut-the-deficit-in-half-without-starting-a-partisan-foodfight/280883/.

522. A. Barton Hinkle, "How to Balance the Budget in One Easy Step," reason.com, September 7, 2015, http://reason.com/archives/2015/09/07/how-to-balance-the-budget-in-one-easy-st/print.

523. Frank Newport, "Americans Continue to Shift Left on Key Moral Issues," Gallup, May 26, 2015, http://www.gallup.com/poll/183413/americans-continue-shift-left-key-moral-issues.aspx?version=print.

524. Frank Newport, "Five Things We've Learned About Americans and Moral Values," Gallup, June 8, 2015, http://www.gallup.com/opinion/polling-matters/183518/five-things-learned-americans-moral-values.aspx.

525. Ann Carroll, "Poll: Failing to Read Bible Related to Declining Morals in US Society," Charisma News, March 27, 2013, http://www.charismanews.com/us/38848-poll-failing-to-read-Bible-related-to-declining-morals-in-us-society.

526. Pamela Rose Williams, "20 Important Bible Verses for Parents," What Christians Want to Know, accessed March 13, 2016, http://www.whatchristianswanttoknow.com/20-important-Bible-verses-for-parents/.

527. Mary Yerkes, "Conflict Resolution," Focus on the Family, accessed March 13, 2016, http://www.focusonthefamily.com/lifechallenges/relationship-challenges/conflict-resolution/a-Biblical-guide-to-resolving-conflict.

528. Scott Kroft, "Biblical Dating: How It's different from Modern Dating," Boundless, March 23, 2012, http://www.boundless.org/relationships/2012/Biblical-dating-how-its-different-from-modern-dating.

529. "Sex and Romance in the Bible," Answers from the Book, accessed

March 13, 2016, http://answersfromthebook.com/cgi-bin/aftb/articles?main_article=90.

530. Dave Miller, "America, Christianity and the Culture War (Part I), Apologetics Press, accessed March 13, 2016, http://www.apologeticspress.org/apcontent.aspx?category=7&article=1847#.

531. Ibid.

532. Ibid.

533. Ibid.

534. Ibid.

535. Barack Obama, "President Barack Obama's Inaugural Address," The White House, January 21, 2009, https://www.whitehouse.gov/blog/2009/01/21/president-barack-obamas-inaugural-address.

536. Dave Miller, "America, Christianity and the Culture War (Part II), Apologetics Press, accessed March 13, 2016, http://www.apologeticspress.org/apcontent.aspx?category=7&article=1853.

537. Ibid.

538. "Spiritual Heritage and Government Monuments," all about history, accessed March 13, 2016, http://www.allabouthistory.org/spiritual-heritage-and-government-monuments-faq.htm.

539. Dave Miller, "America, Christianity and the Culture War (Part III), Apologetics Press, accessed March 13, 2016, http://www.apologeticspress.org/apcontent.aspx?category=7&article=1901.

540. Charles Butts, "Obama 'a Judgment of God' on America, says Land," One News Now, August 5, 2015, http://www.onenewsnow.com/culture/2015/08/05/obama-a-judgment-of-god-on-america-says-land.

541. Ibid.

542. Ashleigh Axios, "Opening the Peoples House," The White House, September 15, 2014, https://www.whitehouse.gov/blog/2014/09/15/opening-peoples-house.

543. Michael Chapman, "Rev. Graham: Obama's Gay Activism 'Opened Wide' the 'Floodgates of Acceptance' of 'Sexual Immorality'," CNS News, September 11, 2015, http://www.cnsnews.com/blog/michael-w-chapman/rev-graham-obamas-gay-activism-opened-wide-floodgates-acceptance-sexual.

544. Ibid.

545. Sarah Zagorski, "Barack Obama Once Said "God Bless Planned Parenthood." How Can God Bless Selling Aborted Babies?," LifeNews.com, July 21, 2015, http://www.lifenews.com/2015/07/21/barack-obama-

once-said-god-bless-planned-parenthood-how-can-god-bless-selling-aborted-babies/.

546. "Black Genocide," accessed March 13, 2016, http://www.blackgenocide.org/black.html.

547. Sarah Zagorski, "Barack Obama Once Said "God Bless Planned Parenthood." How Can God Bless Selling Aborted Babies?," LifeNews.com, July 21, 2015, http://www.lifenews.com/2015/07/21/barack-obama-once-said-god-bless-planned-parenthood-how-can-god-bless-selling-aborted-babies/.

548. "The Muslim Brotherhood's "General Strategic Goal" for North America," discoverthenetworks.org, accessed March 13, 2016, http://www.discoverthenetworks.org/viewSubCategory.asp?id=1235.

549. Eileen F. Toplansky, "Obama's Moral Compass," *American Thinker*, May 20, 2012, http://www.americanthinker.com/articles/2012/05/obamas_moral_compass.html.

550. Ibid.

551. Ibid.

552. Adam Taylor, "The U.S. Keeps Killing Americans in Drone Strikes, Mostly by Accident," *The Washington Post*, April 23, 2015, https://www.washingtonpost.com/news/worldviews/wp/2015/04/23/the-u-s-keeps-killing-americans-in-drone-strikes-mostly-by-accident/.

553. Andrew Tilghman, "Here Are the New Rules for Transgender Troops," *Military Times*, June 30, 2016, http://www.militarytimes.com/story/opinion/2016/06/30/new-rules-transgender-troops/86555558/?utm_source=Sailthru&utm_medium=email&utm_campaign=DFN+EBB+7.1.16&utm_term=Editorial+-+Early+Bird+Brief.

554. Rowan Scarborough, "Army Soldiers Can 'Self-Identify,' Not Face Discrimination," *The Washington Times*, February 28, 2016, http://www.washingtontimes.com/news/2016/feb/28/army-soldiers-can-self-identify-not-face-bias/.

555. "A Clear and Present Danger: The Threat to Religious Liberty in the Military," Family Research Council, accessed March 12, 2016, http://downloads.frc.org/EF/EF15F47.pdf.

556. David French, "A President's Cultural Influence Is Profoundly Limited," *Christian Post*, May 6, 2015, http://www.christianpost.com/news/a-presidents-cultural-influence-is-profoundly-limited-138737/#hOkJEyC0lpID6EVj.99.

557. David Muhlhausen, "Why Are We Expanding the Federal Role

in Early-Childhood Education?," Heritage Foundation, April 24, 2014, http://www.heritage.org/research/commentary/2014/4/why-are-we-expanding-the-federal-role-in-early-childhood-education.

558. "The Consequences of Fatherlessness," National Center for Fathering, accessed July 3, 2016, http://www.fathers.com/statistics-and-research/the-consequences-of-fatherlessness/.

559. Ryan Anderson, "Marriage Matters: Consequences of Redefining Marriage," Heritage Foundation, Issue Brief #3879, March 18, 2013, http://www.heritage.org/research/reports/2013/03/why-marriage-matters-consequences-of-redefining-marriage.

560. Becky Sweat, "10 Practical Ways to Teach Your Children Right Values," Beyond Today, March 28, 2008, http://www.ucg.org/the-good-news/10-practical-ways-to-teach-your-children-right-values.

561. Ibid.

562. Chris Stewart, "Who's Responsible for the Education of Your Child?," Education Post, June 3, 2015, http://educationpost.org/whos-responsible-for-the-education-of-your-child/.

563. "We Can't Keep Holding Schools Responsible for the Education of Our Children—Parents Matter Too," The Conversation, June 16, 2015, http://theconversation.com/we-cant-keep-holding-schools-responsible-for-the-education-of-our-children-parents-matter-too-43159.

564. "Pope to Parents: You Are Responsible for Educating Your Children," National Catholic Register, May 20, 2015, http://www.ncregister.com/daily-news/pope-to-parents-you-are-responsible-for-educating-your-children/#ixzz41bfmj2gZ.

565. Penny Starr, "Intact Families—Not Government Social Programs—Most Beneficial to Children and Society, Group Finds," CNS News, February 12, 2013, http://cnsnews.com/news/article/intact-families-not-government-social-programs-most-beneficial-children-and-society.

566. Edward Wells and Joseph Rankin, "Families and Delinquency: A Meta-Analysis of the Impact of Broken Homes," Social Problems 38 (1991): 71–89.

567. Patrick Darby, Wesley Allan, Javad Kashani, Kenneth Hartke and John Reid, "Analysis of 112 Juveniles Who Committed Homicide: Characteristics and a Closer Look at Family Abuse," Journal of Family Violence 13 (1998): 365–374.

568. Cynthia C. Harper and Sara S. McLanahan, "Father Absence and Youth Incarceration," Journal of Research on Adolescence 14 (2004): 369-397.

569. D. Wayne Osgood and Jeff Chambers, "Social Disorganization Outside the Metropolis: An Analysis of Rural Youth Violence," *Criminology* 38 (2000): 81-115.

570. Wendy Manning and Kathleen Lamb, "Adolescent Well-Being in Cohabiting, Married, and Single-Parent Families," *Journal of Marriage and Family* 65(2003): 876-893.

571. Jeannie A. Fry, "Change in Family Structure and Rates of Violent Juvenile Delinquency," thesis submitted to the faculty of Virginia Polytechnic Institute and State University in partial fulfillment of the requirements for the degree of: Master of Science in Sociology James E. Hawdon, Chair Donald J. Shoemaker Jeanne Mekolichick, May 17, 2010 Blacksburg, Virginia, https://theses.lib.vt.edu/theses/available/etd-05192010-095240/unrestricted/Fry_JF_T_2010.pdf.

572. Chuck Donovan, "A Marshall Plan for Marriage: Rebuilding Our Shattered Homes, Heritage Foundation, June 7, 2011, http://www.heritage.org/research/reports/2011/06/a-marshall-plan-for-marriage-rebuilding-our-shattered-homes.

573. Ibid.

574. "Community Policing," Bureau of Justice Statistics, accessed March 13, 2016, http://www.bjs.gov/index.cfm?ty=tp&tid=81.

575. "Juveniles, Delinquency Prevention," crimesolutions.gov, National Institute of Justice, accessed March 13, 2016, https://www.crimesolutions.gov/TopicDetails.aspx?ID=62#Overview.

576. "Chuck Colson's Final Speech," Colson Center Library, March 30, 2012, http://www.colsoncenter.org/search-library/search?view=searchdetail&id=21209.

577. Judges 17:6, Matthew Henry's Concise Commentary, as cited in http://biblehub.com/commentaries/mhc/judges/17.htm.

578. Judges 17:6, Adam Clark's Commentary, as cited in http://www.studylight.org/commentary/judges/17-6.html.

579. Brian Graden, "The Coarsening of Culture," FrontLine, *PBS*, accessed March 13, 2016, http://www.pbs.org/wgbh/pages/frontline/shows/cool/themes/coarse.html.

580. Ibid.

581. Proverbs 22:6, Matthew Henry's Concise Commentary, as cited in http://biblehub.com/commentaries/proverbs/22-6.htm.

582. Brian Graden, "The Coarsening of Culture," FrontLine, *PBS*, accessed March 13, 2016, http://www.pbs.org/wgbh/pages/frontline/shows/cool/themes/coarse.html.

583. "12 Ways to Protect Your Child From Stress," ahaparenting.com, accessed March 13, 2016, http://www.ahaparenting.com/parenting-tools/family-life/protective-parenting.

584. Felicia Alvarez, "9 Most Dangerous Apps for Kids," crosswalk.com, June 13, 2014, http://www.crosswalk.com/family/parenting/kids/9-most-dangerous-apps-for-kids.html.

585. "The Constitution of the United States," The U.S. National Archives, accessed March 13, 2016, http://www.archives.gov/exhibits/charters/print_friendly.html?page=constitution_transcript_content.html&title=Thepercent20Constitutionpercent20ofpercent20thepercent20United percent20Statespercent3Apercent20Apercent20Transcription.

586. "The Declaration of Independence," The U.S. National Archives, accessed March 13, 2016, http://www.archives.gov/exhibits/charters/print_friendly.html?page=declaration_transcript_content.html&title=NARA percent20percent7Cpercent20Thepercent20Declaration percent20ofpercent20Independencepercent3Apercent20Apercent 20Transcription.

587. David Barton, "The Founding on Jesus, Christianity and the Bible," Wallbuilders, May 2008, http://www.wallbuilders.com/LIBprinterfriendly.asp?id=8755.

588. Thomas Jefferson, *The Writings of Thomas Jefferson* (Washington D. C.: The Thomas Jefferson Memorial Association, 1904), Vol. XIII, p. 292-294. In a letter from John Adams to Thomas Jefferson on June 28, 1813.

589. John Adams, *The Works of John Adams, Second President of the United States*, Charles Francis Adams, editor (Boston: Little, Brown and Company, 1856), Vol. X, p. 254, to Thomas Jefferson on April 19, 1817.

590. John Adams, *Works*, Vol. III, p. 421, diary entry for July 26, 1796.

591. John Adams, *Works*, Vol. II, pp. 6–7, diary entry for February 22, 1756.

592. John Adams, *Works*, Vol. X, p. 85, to Thomas Jefferson on December 25, 1813.

593. John Adams and John Quincy Adams, *The Selected Writings of John and John Quincy Adams*, Adrienne Koch and William Peden, editors (New York: Alfred A. Knopf, 1946), p. 292, John Quincy Adams to John Adams, January 3, 1817.

594. *Life of John Quincy Adams*, W. H. Seward, editor (Auburn, NY: Derby, Miller & Company, 1849), p. 248.

595. John Quincy Adams, An Oration Delivered Before the Inhabitants of the Town of Newburyport at Their Request on the Sixty-First Anniversary of

the Declaration of Independence, July 4, 1837 (Newburyport: Charles Whipple, 1837), pp. 5–6.

596. From the Last Will & Testament of Samuel Adams, attested December 29, 1790; see also Samuel Adams, *Life & Public Services of Samuel Adams*, William V. Wells, editor (Boston: Little, Brown & Co, 1865), Vol. III, p. 379, Last Will and Testament of Samuel Adams.

597. *Letters of Delegates to Congress: August 16, 1776–December 31, 1776*, Paul H. Smith, editor (Washington DC: Library of Congress, 1979), Vol. 5, pp. 669-670, Samuel Adams to Elizabeth Adams on December 26, 1776.

598. From a Fast Day Proclamation issued by Governor Samuel Adams, Massachusetts, March 20, 1797, in our possession; see also Samuel Adams, *The Writings of Samuel Adams*, Harry Alonzo Cushing, editor (New York: G. P. Putnam's Sons, 1908), Vol. IV, p. 407, from his proclamation of March 20, 1797.

599. Samuel Adams, *A Proclamation for a Day of Public Fasting, Humiliation and Prayer*, given as the Governor of the Commonwealth of Massachusetts, from an original broadside in our possession; see also, Samuel Adams, *The Writings of Samuel Adams*, Harry Alonzo Cushing, editor (New York: G. P. Putnam's Sons, 1908), Vol. IV, p. 385, October 14, 1795.

600. Samuel Adams, Proclamation for a Day of Fasting and Prayer, March 10, 1793.

601. Samuel Adams, Proclamation for a Day of Fasting and Prayer, March 15, 1796.

602. Josiah Bartlett, Proclamation for a Day of Fasting and Prayer, March 17, 1792

603. Gunning Bedford, *Funeral Oration Upon the Death of General George Washington* (Wilmington: James Wilson, 1800), p. 18, Evans #36922.

604. Elias Boudinot, *The Life, Public Services, Addresses, and Letters of Elias Boudinot*, J. J. Boudinot, editor (Boston: Houghton, Mifflin & Co., 1896), Vol. I, pp. 19, 21, speech in the First Provincial Congress of New Jersey.

605. Elias Boudinot, *The Age of Revelation* (Philadelphia: Asbury Dickins, 1801), pp. xii–xiv, from the prefatory remarks to his daughter, Susan, on October 30, 1782; see also *Letters of the Delegates to Congress: 1774-1789*, Paul H. Smith, editor (Washington, D. C.: Library of Congress, 1992), Vol. XIX, p. 325, from a letter of Elias Boudinot to his daughter, Susan Boudinot, on October 30, 1782; see also, *Elias Boudinot, The Life Public Services, Addresses, and Letters of Elias Boudinot* (Boston and New York: Houghton, Mifflin, and Company, 1896), Vol. I, p. 260–262.

606. Elias Boudinot, *The Age of Revelation, or the Age of Reason Shewn to be An Age of Infidelity* (Philadelphia: Asbury Dickins, 1801), p. xv, from his "Dedication: Letter to his daughter Susan Bradford."

607. Jacob Broom to his son, James, on February 24, 1794, written from Wilmington, Delaware, from an original letter in our possession.

608. From an autograph letter in our possession written by Charles Carroll to Charles W. Wharton, Esq., September 27, 1825.

609. Lewis A. Leonard, *Life of Charles Carroll of Carrollton* (New York: Moffit, Yard & Co, 1918), pp. 256-257.

610. Kate Mason Rowland, *Life of Charles Carroll of Carrollton* (New York: G.P. Putnam's Sons, 1890), Vol. II, pp. 373-374, will of Charles Carroll, Dec. 1, 1718 (later replaced by a subsequent will not containing this phrase, although he reexpressed this sentiment on several subsequent occasions, including repeatedly in the latter years of his life).

611. Journal of the House of the Representatives of the United States of America(Washington, DC: Cornelius Wendell, 1855), 34th Cong., 1st Sess., p. 354, January 23, 1856; see also: Lorenzo D. Johnson, Chaplains of the General Government With Objections to their Employment Considered (New York: Sheldon, Blakeman & Co., 1856), p. 35, quoting from the House Journal, Wednesday, January 23, 1856, and B. F. Morris, The Christian Life and Character of the Civil Institutions of the United States(Philadelphia: George W. Childs, 1864), p. 328.

612. Reports of Committees of the House of Representatives Made During the First Session of the Thirty-Third Congress (Washington: A. O. P. Nicholson, 1854), pp. 6–9.

613. From the Last Will & Testament of John Dickinson, attested March 25, 1808.

614. John Dickinson, *The Political Writings of John Dickinson* (Wilmington: Bonsal and Niles, 1801), Vol. I, pp. 111–112.

615. From his last will and testament, attested on September 21, 1840.

616. Benjamin Franklin, *Works of Benjamin Franklin*, John Bigelow, editor (New York: G.P. Putnam's Sons, 1904), p. 185, to Ezra Stiles, March 9, 1790.

617. Benjamin Franklin, *Works of the Late Doctor Benjamin Franklin* (Dublin: P. Wogan, P. Byrne, J. More, and W. Janes, 1793), p. 149.

618. Elbridge Gerry, *Proclamation for a Day of Thanksgiving and Praise*, October 24, 1810, from a proclamation in our possession, EAI #20675.

619. Elbridge Gerry, *Proclamation for a Day of Fasting and Prayer*, March 13, 1811, from a proclamation in our possession, Shaw #23317.

620. Elbridge Gerry, *Proclamation for a Day of Fasting and Prayer*, March 6, 1812, from a proclamation in our possession, Shaw #26003.

621. John M. Mason, A Collection of the Facts and Documents Relative to the Death of Major General Alexander Hamilton (New York: Hopkins and Seymour, 1804), p. 53.

622. John M. Mason, A Collection of the Facts and Documents Relative to the Death of Major General Alexander Hamilton (New York: Hopkins and Seymour, 1804), pp. 48–50.

623. Alexander Hamilton, *The Works of Alexander Hamilton*, John C. Hamilton, editor (New York: John F. Trow, 1851), Vol. VI, p. 542, to James A. Bayard, April, 1802; see also, Alexander Hamilton, *The Papers of Alexander Hamilton*, Harold C. Syrett, editor (New York: Columbia University Press, 1977), Vol. XXV, p. 606, to James A. Bayard, April 16, 1802.

624. *Independent Chronicle* (Boston), November 2, 1780, last page; see also Abram English Brown, *John Hancock, His Book* (Boston: Lee and Shepard, 1898), p. 269.

625. John Hancock, *A Proclamation For a Day of Public Thanksgiving 1791*, given as Governor of the Commonwealth of Massachusetts, from an original broadside in our possession.

626. John Hancock, *Proclamation for a Day of Public Thanksgiving*, October 28, 1784, from a proclamation in our possession, Evans #18593.

627. John Hancock, *Proclamation for a Day of Public Thanksgiving*, October 29, 1788, from a proclamation in our possession, Evans #21237.

628. John Hancock, *Proclamation for a Day of Fasting and Prayer*, March 16, 1789, from a proclamation in our possession, Evans #21946.

629. John Hancock, *Proclamation for a Day of Thanksgiving and Praise*, September 16, 1790, from an original broadside in our possession.

630. John Hancock, *Proclamation for a Day of Fasting and Prayer*, February 11, 1791, from a proclamation in our possession, Evans #23549.

631. John Hancock, *Proclamation for a Day of Fasting, Prayer and Humiliation*, February 24, 1792, from a proclamation in our possession, Evans #24519.

632. John Hancock, *Proclamation for a Day of Public Thanksgiving*, October 25, 1792, from an original broadside in our possession.

633. John Hancock, *Proclamation for Day of Public Fasting, Humiliation and Prayer*, March 4, 1793, from a broadside in our possession.

634. From his last will and testament, attested April 16, 1779.

635. A. G. Arnold, *The Life of Patrick Henry of Virginia* (Auburn and Buffalo: Miller, Orton and Mulligan, 1854), p. 250.

636. William Wirt, *Sketches of the Life and Character of Patrick Henry* (Philadelphia: James Webster, 1818), p. 402; see also George Morgan, *Patrick Henry* (Philadelphia & London: J. B. Lippincott Company, 1929), p. 403.

637. Patrick Henry, *Patrick Henry: Life, Correspondence and Speeches*, William Wirt Henry, editor (New York: Charles Scribner's Sons, 1891), Vol. II, p. 632, addendum to his resolutions against the Stamp Act, May 29, 1765.

638. Patrick Henry, *Patrick Henry: Life, Correspondence and Speeches*, William Wirt Henry, editor (New York: Charles Scribner's Sons, 1891), Vol. II, p. 592, to Archibald Blair on January 8, 1799.

639. Will of Patrick Henry, attested November 20, 1798.

640. Samuel Huntington, *A Proclamation for a Day of Fasting, Prayer and Humiliation*, March 9, 1791, from a proclamation in our possession, Evans #23284.

641. James Iredell, *The Papers of James Iredell*, Don Higginbotham, editor (Raleigh: North Carolina Division of Archives and History, 1976), Vol. I, p. 11 from his 1768 essay on religion.

642. William Jay, *The Life of John Jay* (New York: J & J Harper, 1833), Vol. I p. 518, Appendix V, from a prayer found among Mr. Jay's papers and in his handwriting.

643. William Jay, *The Life of John Jay* (New York: J. & J. Harper, 1833), Vol. I, pp. 519-520, from his Last Will & Testament.

644. William Jay, *The Life of John Jay* (New York: J & J Harper, 1833), Vol. II, p. 386, to John Murray, April 15, 1818.

645. John Jay, *The Correspondence and Public Papers of John Jay, 1794-1826*, Henry P. Johnston, editor (New York: Burt Franklin, 1890), Vol. IV, pp. 494, 498, from his "Address at the Annual Meeting of the American Bible Society," May 13, 1824.

646. William Jay, *The Life of John Jay* (New York: J. & J. Harper, 1833), Vol. I, pp. 457-458, to the Committee of the Corporation of the City of New York on June 29, 1826.

647. John Jay, *John Jay: The Winning of the Peace. Unpublished Papers 1780-1784*, Richard B. Morris, editor (New York: Harper & Row Publishers, 1980), Vol. II, p. 709, to Peter Augustus Jay on April 8, 1784.

648. William Jay, *The Life of John Jay* (New York: J. & J. Harper, 1833), Vol. II, p. 266, to the Rev. Uzal Ogden on February 14, 1796.

649. William Jay, *The Life of John Jay* (New York: J. & J. Harper, 1833), Vol. II, p. 376, to John Murray Jr. on October 12, 1816.

650. Thomas Jefferson, *The Writings of Thomas Jefferson*, Albert Bergh, editor (Washington, D. C.: Thomas Jefferson Memorial Assoc., 1904), Vol. XV, p. 383, to Dr. Benjamin Waterhouse on June 26, 1822.

651. Thomas Jefferson, *The Writings of Thomas Jefferson*, Alberty Ellery Bergh, editor (Washington D.C.: The Thomas Jefferson Memorial Association, 1904), Vol. XII, p. 315, to James Fishback, September 27, 1809.

652. Thomas Jefferson, *Memoir, Correspondence, and Miscellanies from the Papers of Thomas Jefferson*, Thomas Jefferson Randolph, editor (Boston: Grey & Bowen, 1830), Vol. III, p. 506, to Benjamin Rush, April 21, 1803.

653. Thomas Jefferson, *The Writings of Thomas Jefferson*, Albert Ellery Bergh, editor (Washington, D.C.: The Thomas Jefferson Memorial Association, 1904), Vol. XIV, p. 385, to Charles Thomson on January 9, 1816.

654. Edwards Beardsley, *Life and Times of William Samuel Johnson* (Boston: Houghton, Mifflin and Company, 1886), p. 184.

655. E. Edwards Beardsley, *Life and Times of William Samuel Johnson* (Boston: Houghton, Mifflin and Company, 1886), pp. 141–145.

656. William Kent, *Memoirs and Letters of James Kent*, (Boston: Little, Brown, and Company, 1898), pp. 276–277.

657. Hugh A. Garland, *The Life of John Randolph of Roanoke* (New York: D. Appleton & Company, 1853), Vol. II, p. 104, from Francis Scott Key to John Randolph.

658. James Madison, *Letters and Other Writings of James Madison* (New York: R. Worthington, 1884), Vol. I, pp. 5-6, to William Bradford on November 9, 1772.

659. James Madison, *The Papers of James Madison*, William T. Hutchinson, editor (Illinois: University of Chicago Press, 1962), Vol. I, p. 96, to William Bradford on September 25, 1773.

660. *Letters of Delegates to Congress: November 7, 1785-November 5, 1786*, Paul H. Smith, editor (Washington DC: Library of Congress, 1995), Vol. 23, p. 337, James Manning to Robert Carter on June 7, 1786.

661. *Letters of Delegates to Congress: May 1, 1777 - September 18, 1777*, Paul H. Smith, editor (Washington DC: Library of Congress, 1981), Vol. 7, pp. 645–646, Henry Marchant to Sarah Marchant on September 9, 1777.

662. Kate Mason Rowland, *Life of George Mason* (New York: G. P. Putnam's Sons, 1892), Vol. I, p. 373, Will of Colonel George Mason, June 29, 1715 (this will was later replaced by the will below.)

663. Will of George Mason, attested March 20, 1773.

664. Bernard C. Steiner, One Hundred and Ten Years of Bible Society Work in Maryland, 1810-1920 (Maryland Bible Society, 1921), p. 14.

665. Bernard C. Steiner, One Hundred and Ten Years of Bible Society Work in Maryland, 1810-1920 (Maryland Bible Society, 1921), p. 14.

666. A. J. Dallas, *Reports of Cases Ruled and Adjudged in the Courts of Pennsylvania* (Philadelphia: P. Byrne, 1806), p. 39, Respublica v. John Roberts, Pa. Sup. Ct. 1778.

667. William B. Reed, *Life and Correspondence of Joseph Reed* (Philadelphia: Lindsay and Blakiston, 1847), Vol. II, pp. 36–37.

668. *Collections of the New York Historical Society for the Year 1821* (New York: E. Bliss and E. White, 1821), pp. 32, 34, from "An Inaugural Discourse Delivered Before the New York Historical Society by the Honorable Gouverneur Morris, (President,) 4th September, 1816."

669. *Letters of Delegates to Congress: February 1, 1778-May 31, 1778*, Paul H. Smith, editor (Washington DC: Library of Congress, 1982), Vol. 9, pp. 729-730, Gouverneur Morris to General Anthony Wayne on May 21, 1778.

670. Jedidiah Morse, A Sermon, Exhibiting the Present Dangers and Consequent Duties of the Citizens of the United States of America, Delivered at Charlestown, April 25, 1799, The Day of the National Fast (MA: Printed by Samuel Etheridge, 1799), p. 9.

671. From his last will and testament, attested January 28, 1777.

672. James Otis, *The Rights of the British Colonies Asserted and Proved* (London: J. Williams and J. Almon, 1766), pp. 11, 98.

673. Robert Treat Paine, *The Papers of Robert Treat Paine*, Stephen T. Riley and Edward W. Hanson, editors (Boston: Massachusetts Historical Society, 1992), Vol. I, p. 48, Robert Treat Paine's Confession of Faith, 1749.

674. From the Last Will & Testament of Robert Treat Paine, attested May 11, 1814.

675. Robert Treat Paine, *The Papers of Robert Treat Paine*, Stephen T. Riley and Edward W. Hanson, editors (Boston: Massachusetts Historical Society, 1992), Vol. I, p. 49, Robert Treat Paine's Confession of Faith, 1749.

676. *United States Oracle* (Portsmouth, NH), May 24, 1800.

677. Charles W. Upham, *The Life of Timothy Pickering* (Boston: Little, Brown, and Company, 1873), Vol. IV, p. 390, from his prayer of November 30, 1828.

678. Mary Orne Pickering, *Life of John Pickering* (Boston: 1887), p. 79, letter from Thomas Pickering to his son John Pickering, May 12, 1796.

679. From his last will and testament, attested October 8, 1807.

680. *Collected Letters of John Randolph of Roanoke to Dr. John Brockenbrough*, Kenneth Shorey, editor (New Brunswick: Transaction Books, 1988), p. 17, to John Brockenbrough, August 25, 1818.

681. Hugh A. Garland, *The Life of John Randolph of Roanoke* (New York: D. Appleton & Company, 1853), Vol. II, p. 99, to Francis Scott Key on September 7, 1818.

682. Hugh A. Garland, *The Life of John Randolph of Roanoke* (New York: D. Appleton & Company, 1853), Vol. II, p. 374.

683. Hugh A. Garland, *The Life of John Randolph of Roanoke* (New York: D. Appleton & Company, 1853), Vol. II, p. 106, to Francis Scott Key, May 3, 1819.

684. Benjamin Rush, *The Autobiography of Benjamin Rush*, George W. Corner, editor (Princeton: Princeton University Press, 1948), pp. 165–166.

685. Benjamin Rush, *Letters of Benjamin Rush*, L. H. Butterfield, editor (Princeton, New Jersey: American Philosophical Society, 1951), Vol. I, p. 475, to Elias Boudinot on July 9, 1788.

686. Benjamin Rush, *Letters of Benjamin Rush*, L. H. Butterfield, editor (Princeton, NJ: Princeton University Press, 1951), Vol. II, p. 936, to John Adams, January 23, 1807.

687. Benjamin Rush, *Essays, Literary, Moral and Philosophical* (Philadelphia: Thomas and William Bradford, 1806), p. 84, Thoughts upon Female Education."

688. Benjamin Rush, *Essays, Literary, Moral & Philosophical* (Philadelphia: Thomas & Samuel F. Bradford, 1798), p. 112, "A Defence of the Use of the Bible as a School Book."

689. Benjamin Rush, *Letters of Benjamin Rush*, L. H. Butterfield, editor (Princeton, NJ: Princeton University Press, 1951), Vol. I, p. 521, to Jeremy Belknap on July 13, 1789.

690. Benjamin Rush, *Essays, Literary, Moral & Philosophical* (Philadelphia: Thomas & Samuel F. Bradford, 1798), p. 93, "A Defence of the Use of the Bible as a School Book." See also Rush, Letters, Vol. I, p. 578, to Jeremy Belknap on March 2, 1791.

691. Benjamin Rush, *Essays, Literary, Moral & Philosophical* (Philadelphia: Thomas & Samuel F. Bradford, 1798), p. 93, "A Defence of the Use of the Bible as a School Book;" see also Rush, *Letters*, Vol. I, p. 578, to Jeremy Belknap on March 2, 1791.

692. Benjamin Rush, *Essays, Literary, Moral & Philosophical* (Philadelphia: Thomas & Samuel F. Bradford, 1798), pp. 94, 100, "A Defence of the Use of the Bible as a School Book."

693. Lewis Henry Boutell, *The Life of Roger Sherman* (Chicago: A. C. McClurg and Company, 1896), pp. 271–273.

694. *Correspondence Between Roger Sherman and Samuel Hopkins* (Worcester, MA: Charles Hamilton, 1889), p. 9, from Roger Sherman to Samuel Hopkins, June 28, 1790.

695. *Correspondence Between Roger Sherman and Samuel Hopkins* (Worcester, MA: Charles Hamilton, 1889), p. 10, from Roger Sherman to Samuel Hopkins, June 28, 1790.

696. *Correspondence Between Roger Sherman and Samuel Hopkins* (Worcester, MA: Charles Hamilton, 1889), p. 26, from Roger Sherman to Samuel Hopkins, October, 1790.

697. *The Globe* (Washington DC newspaper), August 15, 1837, p. 1.

698. Will of Richard Stockton, dated May 20, 1780.

699. John Sanderson, *Biography of the Signers to the Declaration of Independence* (Philadelphia: R. W. Pomeroy, 1824), Vol. IX, p. 333, Thomas Stone to his son, October 1787.

700. Joseph Story, *Life and Letters of Joseph Story*, William W. Story, editor (Boston: Charles C. Little and James Brown, 1851), Vol. II, p. 8.

701. Joseph Story, *Life and Letters of Joseph Story*, William W. Story, editor (Boston: Charles C. Little and James Brown, 1851), Vol. I, p. 92, March 24, 1801.

702. Caleb Strong, Governor of Massachusetts, *Proclamation for a Day of Fasting, Prayer and Humiliation*, February 13, 1813, from a proclamation in our possession, Shaw #29090.

703. Zephaniah Swift, *The Correspondent* (Windham: John Byrne, 1793), p. 135.

704. The Autobiography of Benjamin Rush; His "Travels Through Life" together with his Commonplace Book for 1789-1813, George W. Carter, editor (New Jersey: Princeton University Press, 1948), p. 294, October 2, 1810.

705. Jonathan Trumbull, *Proclamation for a Day of Fasting and Prayer*, March 9, 1774, from a proclamation in our possession, Evans #13210.

706. Last will and testament of Jonathan Trumbull, Sr., attested on January 29, 1785.

707. Jonathan Trumbull, Governor of Connecticut, *A Proclamation for a Day of Public Thanksgiving*, October 12, 1770, from a proclamation in our possession.

708. George Washington, *The Writings of Washington*, John C. Fitzpatrick, editor (Washington: Government Printing Office, 1932), Vol. XV, p. 55, from his speech to the Delaware Indian Chiefs on May 12, 1779.

709. George Washington, *The Writings of Washington*, John C. Fitzpatrick, editor (Washington: Government Printing Office, 1932), Vol. XI, pp. 342-343, General Orders of May 2, 1778.

710. George Washington, *The Writings of George Washington*, John C. Fitzpatrick, editor (Washington: Government Printing Office, 1932), Vol. 5, p. 245, July 9, 1776 Order.

711. George Washington, The Last Official Address of His Excellency George Washington to the Legislature of the United States (Hartford: Hudson and Goodwin, 1783), p. 12; see also The New Annual Register or General Repository of History, Politics, and Literature, for the Year 1783 (London: G. Robinson, 1784), p. 150.

712. Daniel Webster, Mr. Webster's Speech in Defence of the Christian Ministry and in Favor of the Religious Instruction of the Young. Delivered in the Supreme Court of the United States, February 10, 1844, in the Case of Stephen Girard's Will (Washington: Printed by Gales and Seaton, 1844), p. 41.

713. Daniel Webster, *The Works of Daniel Webster* (Boston: Little, Brown and Company, 1853), Vol. I, p. 44, A Discourse Delivered at Plymouth, on December 22, 1820.

714. Daniel Webster, Address Delivered at Bunker Hill, June 17, 1843, on the Completion of the Monument (Boston: T. R. Marvin, 1843), p. 31.

715. Daniel Webster, Address Delivered at Bunker Hill, June 17, 1843, on the Completion of the Monument (Boston: T. R. Marvin, 1843), p. 31.

716. Noah Webster, *History of the United States* (New Haven: Durrie and Peck, 1832), p. 300, ¶ 578.

717. Noah Webster, *History of the United States* (New Haven: Durrie & Peck, 1832), p. 339, "Advice to the Young," ¶ 53.

718. Noah Webster, *History of the United States* (New Haven: Durrie & Peck, 1832), p. 339, "Advice to the Young," 53.

719. Noah Webster, *History of the United States* (New Haven: Durrie and Peck, 1832), p. 6.

720. Noah Webster, *A Collection of Papers on Political, Literary, and Moral Subjects* (New York: Webster and Clark, 1843), p. 291, from his "Reply to a Letter of David McClure on the Subject of the Proper Course of Study in the Girard College, Philadelphia. New Haven, October 25, 1836."

721. Noah Webster, *The Holy Bible… With Amendments of the Language* (New Haven: Durrie & Peck, 1833), p. v.

722. K. Alan Snyder, *Defining Noah Webster: Mind and Morals in the Early*

Republic(New York: University Press of America, 1990), p. 253, to James
Madison on October 16, 1829.

723. John Witherspoon, *The Works of John Witherspoon* (Edinburgh: J. Ogle,
1815), Vol. V, p. 255, Sermon 15, "The Absolute Necessity of Salvation
Through Christ," January 2, 1758.

724. John Witherspoon, *The Works of John Witherspoon* (Edinburgh: J. Ogle,
1815), Vol. V, p. 245, Sermon 15, "The Absolute Necessity of Salvation
Through Christ," January 2, 1758.

725. John Witherspoon, *The Works of John Witherspoon* (Edinburgh: J. Ogle,
1815), Vol. V, p. 248, Sermon 15, "The Absolute Necessity of Salvation
Through Christ," January 2, 1758.

726. John Witherspoon, *The Works of John Witherspoon* (Edinburgh: J. Ogle,
1815), Vol. V, p. 276, Sermon 15, "The Absolute Necessity of Salvation
Through Christ' January 2, 1758.

727. John Witherspoon, *The Works of John Witherspoon* (Edinburgh: J. Ogle,
1815), Vol. V, p. 267, Sermon 15, "The Absolute Necessity of Salvation
Through Christ," January 2, 1758.

728. John Witherspoon, *The Works of John Witherspoon* (Edinburgh: J. Ogle,
1815), Vol. V, p. 278, Sermon 15, "The Absolute Necessity of Salvation
Through Christ," January 2, 1758.

729. John Witherspoon, *The Works of the Reverend John
Witherspoon* (Philadelphia: William W. Woodward, 1802), Vol. III, p. 42.

730. *Letters of Delegates to Congress: January 1, 1776-May 15, 1776*, Paul H.
Smith, editor (Washington DC: Library of Congress, 1978), Vol. 3, pp.
502-503, Oliver Wolcott to Laura Wolcott on April 10, 1776.